Robots in Space

NEW SERIES IN NASA HISTORY

Steven J. Dick, *Series Editor*

Related Books in the Series

Single Stage to Orbit: Politics, Space Technology, and the Quest for Reusable Rocketry
Andrew J. Butrica

NASA and the Space Industry
Joan Lisa Bromberg

Space Policy in the Twenty-First Century
edited by W. Henry Lambright

The Space Station Decision: Incremental Politics and Technological Choice
Howard E. McCurdy

Faster, Better, Cheaper: Low-Cost Innovation in the U.S. Space Program
Howard E. McCurdy

High-Speed Dreams: NASA and the Technopolitics of Supersonic Transportation, 1945–1999
Erik M. Conway

Robots in Space

Technology, Evolution, and Interplanetary Travel

ROGER D. LAUNIUS
and
HOWARD E. MCCURDY

The Johns Hopkins University Press
Baltimore

Printed in the United States of America on acid-free paper
2 4 6 8 9 7 5 3 1

The Johns Hopkins University Press
2715 North Charles Street
Baltimore, Maryland 21218-4363
www.press.jhu.edu

Library of Congress Cataloging in Publication Data
Launius, Roger D.
Robots in space : technology, evolution, and interplanetary travel /
Roger D. Launius and Howard E. McCurdy
p. cm.
Includes bibliographical references and index.
ISBN-13: 978-0-8018-8708-6 (hardcover : alk. paper)
ISBN-10: 0-8018-8708-9 (hardcover : alk. paper)
1. Robots. 2. Space robotics. I. McCurdy, Howard E. II. Title.
TJ211.L38 2007
629.4—dc22 2007019374

A catalog record for this book is available from the British Library.

For our nieces

Contents

Acknowledgments

Whenever two people attempt to complete a history project such as this, they by necessity draw on the work of earlier investigators and incur a good many intellectual debts. The authors would like to acknowledge the assistance of these individuals. The contributions of several key people allowed us to conduct our research and write this book. In addition to publishing a seminal 2003 article on the subject of the postbiological universe and organizing the 2005 Critical Issues in the History of Spaceflight symposium at which we first presented our work, NASA Chief Historian Steven J. Dick made available the extensive resources of the NASA History Division. For their many contributions in helping us complete this project we wish especially to thank Dick (who edits the series at Johns Hopkins University Press in which this book appears); archivists Jane Odom, Colin Fries, and John Hargenrader, who helped track down information and correct inconsistencies; Stephen J. Garber and Glen Asner, who offered valuable advice; and Nadine Andreassen, who served as a contact point for the whole of NASA.

We wish to thank the staff of the Smithsonian Institution's National Air and Space Museum: the director, Gen. John R. Dailey, and deputy director, Donald A. Lopez; Director of Collections and Research Ted A. Maxwell; and the staff of the Division of Space History, Paul E. Ceruzzi, Martin Collins, James David, David H. DeVorkin, Jennifer Levasseur, Cathy Lewis, Jo Ann Morgan, Valerie Neal, Allan A. Needell, Michael J. Neufeld, Toni Thomas, Margaret Weitekamp, Frank Winter, and Amanda Young. The museum's archival and library staff also assisted in numerous ways. We wish to thank the staffs of the NASA Headquarters Library and the Center for Aerospace Information who provided assistance in locating materials; and archivists at various presidential libraries and the National Archives and Records Administration who aided with research efforts. At American University, William LeoGrande, Ivy Broder, and Neil Kerwin made available a full-year sabbatical that allowed one of us to concentrate on writing portions of the book, while the faculty and staff of the Department of Public Administration and Policy, including David Rosenbloom, Robert Boynton, Robert Durant, Kimberly Martin, Janet Nagler, Renee Howatt, and Alycia Ebbinghaus, absorbed work that would have otherwise diverted time from this task.

We also thank the staff of the Johns Hopkins University Press. As always, Senior Acquisition Editor Robert J. Brugger was a decided help in bringing this book to fruition. So too was Martin Schneider, our copy editor, and Andre Barnett, our production editor. Our research assistants, Maeve Monvalvo, Natalia Moustafina, Nicholas Limparis, Suzanne Roosen, Katy Smith, and Jennifer Troxell, collected materials from many sources and helped us interpret them.

In addition to these individuals, we wish to acknowledge the following scholars who provided intellectual inspiration and aided in a variety of ways: Buzz Aldrin, Joel Achenbach, William Sims Bainbridge, Richard Berendzen, Haym Benaroya, Gregory Benford, Rodney Brooks, Frances Brown, Richard H. Buenneke, Glenn E. Bugos, William E. Burrows, Andrew J. Butrica, W. Bernard Carlson, Robert A. Casanova, Erik M. Conway, Tom D. Crouch, Walt Cunningham, Richard DalBello, Leonard David, Anthony Duignan-Cabrera, Peter H. Diamandis, Clay Durr, Mohammad S. El-Genk, Robert W. Farquhar, Richard Faust, James Rodger Fleming, Jack Fox, Slava Gerovitch, Michael H. Gorn, Chris Hables Gray, G. Michael Green, Barton C. Hacker, Roger Handberg, James R. Hansen, Albert A. Harrison, Peter L. Hays, Noel W. Hinners, David A. Hounshell, Scott Hubbard, Dennis R. Jenkins, Dana J. Johnson, Stephen B. Johnson, Thomas D. Jones, Kathy Keltner, Yoji Kondo, Sylvia K. Kraemer, Martin P. Kress, John Krige, W. Henry Lambright, W. David Lewis, Byran Lilley, John M. Logsdon, Laura E. Lovett, Paul D. Lowman, Valerie J. Lyons, W. Patrick McCray, Neil M. Maher, Hans Mark, Greg Maryniak, Wendell Mendell, David A. Mindell, Nicholas de Monchaux, James Oberg, David Ost, Scott Pace, Sidney Perkowitz, Ian Pryke, Stephen J. Pyne, Alex Roland, Eligar Sadeh, John B. Sheldon, Asif A. Siddiqi, Marcia S. Smith, Robert W. Smith, Ted Swanson, Harley Thronson, Jannelle Warren-Findley, and Edward J. Weiler.

None of these individuals will agree with everything we have written in this book. Such is the nature of scholarly discourse and a healthy marketplace of ideas. We hope that at least some of our thoughts will advance the discussion of robotic and human spaceflight and the future of these contesting points of view.

INTRODUCTION

A False Dichotomy

In the fall of 2000 we traveled to Boston to tape a television program on space exploration, discussing a book we had just completed, *Imagining Space.* The book contained fantastic images anticipating the wonders of space exploration juxtaposed with photographs of actual accomplishments. A favorite set of images opened the chapter on the exploration of Mars. On the left side, the book displayed a 1949 painting by the renowned space artist Chesley Bonestell, depicting water flowing toward the setting sun as might be seen by a person standing on the polar ice cap of Mars. On the right side, a full-page photograph of the Ares Vallis flood plain appeared, taken in the summer of 1997 by the *Mars Pathfinder* lander on actual Martian soil.[1] We closed a subsequent chapter with a painting by Pat Rawlings depicting an astronaut in a space suit bending down to retrieve the *Sojourner* rover that the *Pathfinder* lander had delivered to Mars—the first human on that planet greeting the first robot to arrive on that distant world. Those images and similar ones depict a central issue in the course of space exploration—the relative emphasis given to human spaceflight as opposed to expeditions conducted by robotic or automated craft. On the airplane flight back to Washington, we resolved to examine this issue in more detail and to do so in a wider time frame than the hundred-year period our book had allowed.

When asked to comment on the virtues of "manned" and "unmanned" spaceflight, leaders of the National Aeronautics and Space Administration issue a consistent reply. The venture, they insist, will be a cooperative one. Space exploration will be accomplished by "robots and humans together."[2] NASA's position is well represented by the Rawlings painting in which an astronaut retrieves the robot that helped open the way for humans to explore Mars. Over the course of space exploration, however, cooperation has progressively given way to competition. People advancing proposals for activities in space increasingly view humans and robots as competitors in the celestial realm.

As we examined the issue of humans versus robots, what we found startled us. The debate over humans and robots in space does not well represent the full range of possible alternatives, especially when one anticipates developments over long periods of time. Pitting humans against robots, we found, produces a false dichotomy. The issue is multi-sided, with approaches like "manned" and "unmanned" giving way to less conventional concepts as exploration activities mature.

PHASES OF SPACEFLIGHT

As children of the mid–twentieth century, we were raised on visions that placed astronauts and space cadets squarely at the center of exploration. Humans were clearly in charge. We met robots, to be sure, such as Gort in the 1951 science fiction classic *The Day the Earth Stood Still* or the silly robot in the 1960s television series *Lost in Space.* As youngsters encountering space science for the first time, however, we knew that human beings would be needed to manage the machines that people sent into space. The technology of our youth required it. Humans would change the vacuum tubes in communication satellites, we assumed, a thought encouraged by writers no less perceptive than Arthur C. Clarke, who invented the communication satellite concept and insisted that such switching stations would need to be manned. We marveled at the diorama painted by Chesley Bonestell, appearing in the March 22, 1952, issue of *Collier's* magazine, that helped launch public interest in space travel. Astronauts were everywhere—piloting a winged space shuttle, tending a large rotating space station, and driving space tugs. Between the shuttle and the space station, astronauts serviced an automated space observatory, a precursor of the Hubble Space Telescope. Why were astronauts crawling over the automated observatory? They were needed, Wernher von Braun assured us in the accompanying article, to change the film.[3]

At the beginning of the space age, both popular culture and the state of technology demanded a strong human presence in space. Given the primitive state of space technology, machines did not operate well when so far removed from human control. Disseminators of popular science and science fiction encouraged people to believe that humans would pilot spaceships to exotic destinations. Rocket technology and large-scale project management facilitated this goal. "Man will conquer space soon," editors of a national magazine assured us, slighting the role that women would eventually play as well.[4] Humans stood at the center of the vision of spaceflight presented to the public at large. Popular culture, that is, images presented through print and visual media, typically gave us familiar themes reassembled in new settings. In this respect, the popular culture of spaceflight was clearly one with humans in control. The resulting classical approach to space exploration reached its zenith during the early stages of space exploration, when NASA engineers working on Project Apollo sent the first humans to the Moon.

As we matured, so did space technology. Cold War inventions overcame many of the obstacles restricting robotic flight. Scientists learned how to beam images back from orbiting observatories and reconnaissance satellites, creating a science of remote sensing that absolved the need for humans on board. Engineers developed solid-state transistors and placed them in telecommunication satellites, allowing automated messengers to relay signals for long periods of time without human repair. Advances in microelectronics and miniaturization significantly reduced the cost of automated flight relative to human endeavors. Using deep space networks, scientists and engineers learned how to maintain contact with spacecraft on vast journeys, allowing humans to remain behind. Computer technology improved. An increasingly large number of scientists, journalists, and historians of technology began to question the need for humans in space.

A popular culture of robotics also arose. Isaac Asimov published his three laws of robotics. Philip Dick explored the propensity of robots to rebel against their creators. The murderous HAL 9000 computer received a starring role in the 1968 movie *2001: A Space Odyssey.*[5] Popular culture transmits values, assumptions, and practices through the most broadly disseminated forms of entertainment and communication. It is a subset of the culture at large—the behaviors, traits, and social practices that characterize a society or group of people. Humans create the cultures in which they live through social repetition, invention, and the practices they embrace. As the twentieth century progressed, images of robots and other automatic devices became more familiar to the general public and as much a part of the general culture as the slightly older visions of humans piloting spacecraft.

Together, culture and technology favored a second phase of space exploration, one dominated by machines under the control of human beings working at flight centers on Earth. Such missions are commonly termed *unmanned,* a grievous slight to the women who have worked to build spacecraft and risked their lives on missions in space. We prefer the term *robotic,* although that does not quite describe the nature of the mechanics involved. In an excellent survey of robotic spacecraft, journalist and space historian Jim Oberg characterizes the nature of these creations. In general, a robot is a machine under human control that can perform tasks similar to those performed by human beings. To fully qualify within this definition, a space robot needs the ability to manipulate or touch other objects, capture images, and move around. In brief, it needs arms, eyes, and legs—although wheels or maneuvering jets make fine substitutes for legs.[6]

Many of the automated spacecraft sent on cosmic missions during the first fifty years of spaceflight possessed some but not all of these characteristics. Oberg calls them "proto-robots": the first stages in the process of creating machines with extensive human capabilities. The *Surveyor* spacecraft that preceded Americans to the Moon possessed motorized arms that dug into the lunar surface, obtaining lunar samples for study and testing the surface to assure engineers that it was solid enough for humans to land and stand. The Soviet *Lunokhod* vehicles, the first robotic spacecraft to rove across an extraterrestrial body under remote control, had metal spoked wheels. Soviet engineers also built a series of landers with robotic arms and drilling mechanisms. The robotic arms deposited lunar samples in capsules, which, despite a few failures, brought the precious material home without the direct intervention of human beings. The two *Viking* landers that NASA placed on the Martian surface in 1976 used arms to collect and deposit soil in automated biological laboratories. Without any direct human help, the robotic landers searched for evidence of living organisms. Mobility arrived with the rollout of NASA's *Sojourner* rover, part of the 1997 Mars Pathfinder mission, followed by the more sophisticated rovers *Spirit* and *Opportunity* in 2004. Soviet engineers built rovers designed to traverse Martian surfaces like mechanical bugs on cross-country skis during the 1990s. They would have produced interesting results had the rigors of interplanetary flight not prevented their arrival.[7]

We refer to such devices as *robotic,* meaning that they possess at least some of the material characteristics of robots, complete tasks too tedious or dangerous for humans to perform, and receive instructions from human beings overseeing their work from distant sites. We extend the term to a wide range of mechanical devices operating in space that contain sensory mechanisms and the ability to carry out scientific investigations without humans on site, including satellites. We also use the term *automated,* which refers to devices capable of carrying out work with a minimum amount of human intervention. Both terms are offered as substitutes for the term *unmanned* and hence apply to a great variety of satellites, spacecraft, and instruments working in space. The terminology is imperfect but, we hope, adequate for explaining the distinctions involved.

The growing capabilities of robotic spacecraft during the first half-century of spaceflight clearly surprised the advocates of human travel. Human flight advocates wanted astronauts to return to the Moon and venture to Mars, yet technology and culture encouraged an alternative history in which humans stayed close to Earth while robotic spacecraft ventured beyond.

Perhaps humans from Earth will someday return to the Moon and walk across the surface of Mars. We cannot say with certainty that government support for the classical vision of human spaceflight is done. Trips as far as Mars are technically achievable, the motivation is powerful, and the cost would not exceed the sums spent on national defense. Nations other than the United States, for reasons of national prestige, may send humans to the Moon and beyond. Even if government declines to go, private entrepreneurs may discover methods of human spaceflight unconsidered by engineers employed by large tax-supported bureaucracies. Nonetheless, the technical, financial, and cultural forces that shape support for space activities do not favor such adventures. They favor a limited role for humans in space and an expanding presence of robotic machines.

Looking beyond Mars, the prospects for human flight of the classical sort are dim. Humans are simply not well suited for long-duration space travel, especially to objects whose climactic conditions and radiation levels differ markedly from those found on Earth. Technology and culture favor a role for human spaceflight that under the most favorable circumstances appears limited to short excursions to the Moon and perhaps a remote outpost on Mars, with remotely operated craft surveying the remainder of the solar system.

The grand vision of spaceflight, moreover, reaches beyond the local solar system. In science fiction and popular science, it extends to the whole galaxy. Additionally, it takes place over time periods that are geological in length. Visionaries such as Robert Goddard and Carl Sagan expected space travel to last for millennia, over vast distances, even to a time when the Earth might become uninhabitable. In our book *Imagining Space,* we anticipated space activities a mere fifty years into the future, barely touching on potential activities beyond. When we examine space exploration in longer time spans over galactic distances, the underlying conflict between human and robotic approaches breaks down, and new approaches emerge.

In fact, a third approach, spurred by interest in extra-solar planets, has already begun. It is characterized by scientific investigations through the electromagnetic spectrum using telescopes that can see a much wider array of features than are visible by the human eye, conducted as frequently through Earth-based instruments as from objects located in space. Discovery of the first extra-solar planet around a sun-like star, 51 Pegasi b, occurred as a result of observations undertaken at the Observatoire de Haute-Provence.[8] The much-publicized search for extraterrestrial intelligence (SETI) began in 1960 at the National Radio Astronomy Observatory in West Virginia, when astronomer Frank Drake pointed the radio telescope at Epsilon Eridani and listened for signals that might have been dispatched by a technological civilization residing thereby. Observations from the Hubble Space Telescope and the infrared Spitzer Space Telescope have been used to detect extra-solar planets, while other work continues from the ground.

Examination of extra-solar objects through the electromagnetic spectrum eliminates the necessity of dispatching either robots or humans on long and tedious voyages; observers merely wait for electromagnetic signals traveling at the speed of light to reach Earth. Automated telescopes are utilized in such pursuits. Significantly, many of those instruments, programmed to conduct their operations without the continual presence of human operators, are located not in space but on Earth. In such cases, automation is utilized not to alleviate the need for humans in space but to eliminate the purely human inconveniences of long nights spent behind cold instruments.

Astronomers have already captured images of planetary objects around nearby stars.[9] Using advanced observation techniques, they will someday produce an image of a faraway blue-and-white planet with liquid water and a breathable atmosphere. It seems inevitable. Such a discovery will certainly spur interest in closer observation, revitalizing the dream of galactic space travel. Yet neither humans nor robots in their conventional forms are adequately suited for journeys of such magnitude. Robots by definition require human supervision, an impossible requirement for spacecraft light years removed from terrestrial control centers. Humans as well are unlikely to go. Barring the fanciful vision of multigenerational spaceships, humans simply do not live long enough to complete such voyages.

A robot so dispatched would need the cognitive capabilities of a well-educated human being. A human crew sent to another solar system would need the perseverance of machines. In the face of such requirements, the characteristics of humans and robots begin to fuse in strange and fascinating ways. A fourth alternative arises, one that crosses the barriers between conventional human and robotic flight.

People like Goddard and Sagan, who contemplated the ultimate purpose of space exploration, saw it leading to a state of near-immortality. Earthly life forms will spread themselves throughout the galaxy; Sagan and Goddard foresaw a human diaspora, the species surviving for as long as the stars might burn. Such an achievement would encompass an incredibly long period of time, certainly more than the 2 million years that human beings and their closest relatives have occupied the Earth. Contemplation of space travel over such periods is called Stapledonian thinking, crediting the British philosopher Olaf Stapledon who wrote novels and essays that examined the consequences of very long-term activities in space.[10] The vast time periods involved required Stapledon to account for biological and cultural change. Simply stated, people change. The physical form and cultural interests of any creatures who initiate space travel will not remain the same if they persist long enough in the enterprise. In this sense, the life forms that complete the task of exploring and colonizing the galaxy may not resemble the creatures who initiate the task. Given the life spans of sun-like stars, the time periods available for such change are extraordinary. Stapledon dealt in billions of years.

Such speculation might seem, given the time periods involved, almost irrelevant. Within a human civilization that has lasted only a few thousand years—only a few hundred in its technological form—who would want to contemplate changes over hundreds of millions of years? The natural process of biological change proceeds so slowly that it is hardly discernable. The most ardent supporters of "transhumanist" or "postbiological" thinking, however, do not think that such adaptations will take that long. Given the pace of advances in technology, they believe that alterations could appear within the next century. The alterations might take the form of artificial intelligence computers that are smarter than human beings or biologically reengineered human beings with exceptionally long life spans.

Prospective alterations such as these radically alter the robotic-versus-human spaceflight debate. Joined with a growing interest in galactic exploration, it prompts a reconsideration of the relationship between robots and human beings. The clear lines of distinction between people and machines disappear in transhumanist and postbiological thought. Discussion of this subject, much of which takes place over the Internet, has been preoccupied with the implications for earthly existence. As part of our plan for a book on robotic and human spaceflight, we resolved to apply the concepts to space.

To summarize, we view the "human/machine" debate as a false dichotomy. It may be useful for understanding the early history of space travel, perhaps for the first one hundred years, but it is a weak framework for time periods much beyond. In its place, we present a scenario of space travel consisting of at least four phases: a period in which humans venture into space in a classical exploration mode, a period during which robotic technologies provide increasing advantages to machines supervised by humans, a transitory phase during which investigations of relatively nearby areas of the universe are made through different regions of the electromagnetic spectrum, and a period during which the characteristics of human and robotic exploration begin to merge. The phases overlap and compete for public attention. The varying emphases at any one time are largely determined, we believe, by culture and technology.

The dominant vision of space exploration, in which humans with the assistance of machine servants complete heroic journeys into the cosmos, is already outmoded. It may persist for a few more years, but it is technologically and culturally archaic. The robotic point of view, with its emphasis on machine subservience, is likewise doomed. It may endure for longer than the human flight vision, given its technological advantages. Yet it too is equally undercut by technological and cultural forces. Most particularly, the robotic alternative suffers from the continuing desire of intelligent beings to extend their presence into the heavens as well as emerging trends in biotechnology and artificial intelligence. Note that by robotic technology we mean mechanical servants under human supervision, not independent machines.

In a strange and unexpected way, the original NASA vision of humans and robots exploring space together may be prophetic. The original vision links two conventional paradigms—astronauts in space suits and robotic servants spreading their presence throughout the solar system. Its architects view humans and robots as separate entities. The transhumanist or postbiological approach abolishes that separation. Humans and robots explore space together, but in forms that merge the most useful characteristics of each. This may seem like science fiction, but it is certainly interesting to contemplate. Do not forget that space travel itself was wholly fictional for at least one hundred years before it began.

THE PLAN OF THE BOOK

Chapter 1 lays out the history of the human-versus-robotic spaceflight debate in its various parameters. Despite characterizations of the issue as a debate, direct exchanges between advocates of the two points of view are not plentiful. As a result, much of the history concerns events, technological developments, and social movements particular to each.

In spite of its antiquated nature, the human spaceflight perspective continues to dominate civil space policy. Chapter 2 analyzes the persistence of this vision, tracing the desire for human space travel to its utopian appeal, the desire to start life anew in fresh and different places. This motivation is not likely to disappear even as new paradigms emerge. In chapter 3, we analyze the manner in which advocates of the human spaceflight vision managed to incorporate this perspective in the organizational mandate of the publicly funded U.S. National Aeronautics and Space Administration.[11] At the beginning of the space age, a large government agency with access to significant funds became the principal force for accomplishing the classic human spaceflight dream.

Like the dominant human vision, the quest for robotic flight draws on an equally strong but alternative set of beliefs. Popular presentations of the machine perspective are explored in chapter 4. Set largely in works of science fiction, the popular culture of robotics posits a set of relationships more relevant to the industrial age than to the postindustrial one, weakening its further appeal. Actual experiences from fifty years of spaceflight are analyzed in chapter 5, a period during which various developments permitted substantial advances in autonomous technology relative to the human flight alternative.

The remaining chapters examine future challenges. Chapter 6 assesses the prospects for interstellar flight, an essential part of the overall vision and an undertaking likely to weaken both the traditional human and robotic alternatives. Chapter 7 enters the strange new world of artificial intelligence and biotechnology and offers the suggestion, drawing upon discussions of transhumanism and a postbiological universe, that humans may reengineer themselves or their machines in ways that make space travel much easier. Chapter 8 summarizes the main observations of the book, presents some options that would advance the cause of spaceflight in its various forms, and discusses a new flight paradigm based on computer intelligence or biotechnology that might emerge soon.

For much of the twentieth century, space travel existed wholly in the imaginations of the people who envisioned it. Not until the late 1950s and early 1960s did the first satellites and humans fly in space. For most people living at that time, space travel remained "that Buck Rogers stuff," a reference to a popular comic strip and movie serial character providing a form of fantasy no more believable than ghostly apparitions or atom-powered trains.

The classical vision cast humans in the central role, a perspective that few early commentators on space travel effectively challenged. The potential for robotic flight seemed especially unbelievable, given familiar limitations on machine technology and the general social distrust of machines present in the early twentieth century. Yet the dominant vision of spaceflight, especially the image of humans "conquering" space, sat on a thin reality base.[12] It owed more to the importation of often inappropriate analogies than to a sober assessment of possible technologies.

Fifty years into the age of space travel, a few features are clear. Most of the revelation lies ahead, especially if space travel occurs on the time scales required for cosmic investigation. Still, some observations can be made. To that end, our book examines the dominant visions of spaceflight and offers some speculative thoughts on the principal alternatives that may emerge.

CHAPTER ONE

The Human / Robot Debate

For many years, people advancing the dominant approach to space travel envisioned humans leaving the Earth and traveling to distant spheres. The steps in this approach were first advanced by purveyors of science fiction and then fully articulated in scientific publications designed for popular consumption. Appearing in the mid–twentieth century, the vision became quite familiar. Humans would construct large, winged rocketships capable of flying into space. The rocketships would propel humans to a large, usually rotating space station. From the space station, humans would board spacecraft bound for the Moon, where they would construct a lunar base. With the experience gained from long stays on the Moon, earthlings would venture to Mars. On Mars they would establish research bases and eventually colonies where large numbers of *Homo sapiens* would live and work. With the technology gained, humans would explore the outer planets of the solar system and eventually venture to livable spheres around other stars. Humans would become a spacefaring species, leaving the cradle in which they were born for worlds far beyond.

Over time the vision came to dominate both popular culture and public policy.[1] It provided the technological basis for the classic science fiction film *2001: A Space Odyssey,* released in 1968 as Americans were preparing to land on the Moon. It inspired a series of government commission reports, including the 1969 statement of the Space Task Group and the 1986 report produced by the National Commission on Space. The vision encouraged Presidents George H. W. Bush and George W. Bush to propose space exploration programs to take humans back to the Moon and beyond. "We will build new ships to carry man forward into the universe," the second President Bush announced in January 2004, "to gain a foothold on the moon and to prepare for new journeys to the worlds beyond our own."[2]

For nearly as many years, proponents of spaceflight have contemplated a second scenario. This approach leaves little room for human flight, depending instead upon a collection of robotic spacecraft to conduct the work of scientific investigation and exploration. Proponents of this approach envision a multitude of space telescopes probing the universe in many ways, joined with robotic spacecraft observing the planets and their moons and returning samples to Earth. Eventually, according to this scenario, robots built by humans will journey to nearby solar systems. Humans would remain behind, analyzing the scientific data collected by these exploring machines.

James Van Allen, one of the principal proponents of this approach, argues that "almost all of the space program's important advances in scientific knowledge have been accomplished by hundreds of robotic spacecraft." Van Allen designed the instrument carried on the first U.S. satellite, *Explorer I,* that discovered the radiation belts bearing his name. Proponents of robotic flight disparage the vision motivating a human presence in space. "In a dispassionate comparison of the relative values of human and robotic spaceflight," Van Allen observes, "the only surviving motivation for continuing human spaceflight is the ideology of adventure." After all, few of the Earth's six billion inhabitants will actually fly into space. "For the rest of us, the adventure is vicarious and akin to that of watching a science fiction movie."[3]

Between the romantic vision of human settlement and the investigative preference for machines, a third position appears. Humans and robots will explore space together, some believe. This position is closely associated with the human spaceflight movement—a compromise of sorts—and underlies the work of the U.S. civil space agency, NASA. Advocates of collaboration advance a vision frequently encountered in works of science fiction—spaceship captains working with robotic companions who serve as machine servants to the people in charge. Robotic technology may advance to very high levels, the advocates of collaboration agree, but humans will always be needed to repair and supervise the machines. Even if machines could wholly supplant human beings, this group would still want to send people. To them, migration of the species off of planet Earth provides one of the primary justifications for spaceflight. Humans are a migratory species with the technical ability to adapt to a wide variety of circumstances in which other creatures would not be able to survive. Machines assist in this process, but they are not its ultimate justification.

In the governmental realm, especially in the United States, public officials set up government-financed space programs in which engineers constructing piloted spacecraft and scientists designing robotic spacecraft could cooperate. That state of cooperation quickly gave way to competition. People advocating human and robotic missions came to view themselves as rivals pursuing limited funds; technical factors favored increased separation between the two approaches.

Advocates of robotic missions believe that technological advances will permit the construction of machines of ever-increasing autonomy, thereby weakening one of the principal justifications for a human presence in space. The prospects are certainly intriguing. Advances in robotics have occurred more rapidly than first imagined. Concurrently, the achievement of human spaceflight capabilities proved to be fantastically expensive and more time-consuming than anticipated. These developments, in conjunction with the receding social traditions supporting human flight, have favored the robotic point of view, at least through the first half-century of space travel. Fifty years into the enterprise, human capabilities still exceed machine skills, but machines are both less expensive and advancing in capability, and we care less if they are lost in their endeavors. This situation may change in the future, as scientists and engineers create machines with more human-like qualities, even to the point of installing biological parts on them.

THE HUMAN SPACEFLIGHT VISION

As a cultural phenomenon, human spaceflight has its roots in the European efforts to explore the Earth, proceeding more or less continuously since the expansion of European civilization in the fifteenth century. That period, marked by European efforts to find a short route to China, incorporates Ferdinand Magellan's three-year circumnavigation of the globe and the voyages of Christopher Columbus. In a manner analogous to the frustrations encountered by early spaceflight advocates searching for easy pathways into the extraterrestrial realm, Magellan (who died during his expedition) discovered that vast barriers prevented easy transit westward across the seas.[4]

The European experience was not the only one that took place on Earth. Chinese explorers under the Ming emperor Yongle launched seven great voyages of discovery at the beginning of the fifteenth century. They reached the west coast of India, the east coast of Africa, and parts of the Middle East. As partisans of spaceflight like to point out, conservative Chinese leaders forced the explorers to abandon their efforts some sixty years before Columbus sailed, ending what proved to be a period of remarkable innovation and discovery. Modern exploration advocates warn that a certain and similar decline would follow any decision by a modern leader to abandon human space travel.[5]

Terrestrial exploration in the classical mode continued into the early twentieth century, through the "heroic age" of polar exploration. Under the classical model, explorers in small ships typically crossed vast bodies of water to explore unfamiliar lands. During their expeditionary periods, often lasting as much as three years, explorers were cut off from direct communication with their home ports. Full reports of seafaring discoveries thus depended upon the safe return of at least some of the vessels and the crew. This classical method of exploration ended with the advent of mechanized expeditions in the third decade of the twentieth century, notably Richard E. Byrd's airplane expedition to Antarctica in 1929.[6]

Significantly, the era of modern rocketry began just as the traditional approach to terrestrial exploration ended. In 1926, Robert Goddard launched the first liquid-fueled rocket from a farmer's field in central Massachusetts. The rocket provided a means by which classical methods of exploration, with their various social and scientific advantages, might continue in a new realm. Not by accident, President John F. Kennedy referred to the cosmos as "this new sea" in explaining why humans needed to move across the realm.[7]

The rise of rocketry, coupled with the decline of traditional expeditions on the home planet, helped to motivate an explosion in works of fiction describing similar travel in the extraterrestrial realm. Hugo Gernsback launched *Amazing Stories,* the first pulp magazine devoted wholly to science fiction, in 1926. E. E. Smith's *Skylark of Space,* the first work of fiction to employ the formula of classical sagas like *The Iliad* and *The Odyssey* in a galactic setting, appeared in 1928. Fritz Lang produced the first film to deal realistically with space travel, *Frau im Mond,* in 1930. The comic strip character Flash Gordon appeared in 1934. Enthusiasts launched rocket societies in Russia (1924), Germany (1927), the United States (1930), Great Britain (1933), and France (1938). The American Interplanetary Society (later renamed the American Rocket Society) was founded by a group of science fiction writers hoping to make their fantasies real.[8]

Works of fiction dealing with human space travel are a modern phenomenon, closely linked to the disappearance of traditional terrestrial exploration tales and the advent of rocketry. Only a small number of stories dealing with extraterrestrial travel, such as those penned by Jules Verne and H. G. Wells, appeared prior to the twentieth century, while fewer and odder still were those produced before the nineteenth.[9] Especially in the United States, science fiction helped to familiarize the general public with the idea of space travel and to provide a cultural foundation for its pursuit before practical efforts began. Without the contributions of science fiction, advocates of real spaceflight would have encountered a much less informed audience, both within the general public and in the corridors of governmental power.

Cultural expectations regarding human space travel were further amplified by the advent of aviation, marked by the first successful flight of a powered heavier-than-air machine in 1903. Aviation was well enough developed by the mid-1920s that government leaders around the world felt confident in establishing mechanisms to encourage its commercial growth. In the United States, Congress passed legislation that paid private companies to fly the mail. The action offered entrepreneurs a secure financial base, from which they began to develop the modern airline industry. Airplane enthusiasts proclaimed what has been characterized as a "winged gospel," beliefs anticipating the vast positive effects that atmospheric flight would visit upon the lives of ordinary people.[10]

Principal among the beliefs was the notion that anyone could fly. At a time when airplanes were small and flying seemed confined to the brave and foolhardy, the anticipation of air travel as a form of mass transportation seemed absurd. Yet aviation advocates continued to promote the idea. Atmospheric flight would not be confined to a small elite, advocates predicted, but would extend to ordinary people. The experience, moreover, would transform society, its advocates foretold, ushering in a new age of mobility and commercial opportunity, even promoting equality between the sexes.[11]

Aviation enthusiasts anticipated a wide range of social improvements arising from what began as a technological change. Using the analogy of another new technology, the automobile, aviation advocates predicted the widespread ownership of personal airplanes. Forecasts of "an airplane in every garage" failed to materialize; most airborne travelers were flown rather than flew.[12] Yet the general vision of mass aviation continued. Scarcely fifty years after the successful test of the Wright brothers' original *Flyer* at Kitty Hawk, North Carolina, the appearance of the modern jetliner fulfilled the prophesy, as ordinary people sped across oceans and continents in a new age of mobility with its associated opportunities.

Supporters of human spaceflight believe that a similar future waits for humans in the extraterrestrial realm. In their minds, rocketry will permit a new and essentially endless era of extraterrestrial exploration, leading to technologies that will allow anyone to fly in space. Humans in large numbers, according to this vision, will travel to new places, just as in previous periods technological innovations allowed people to traverse the Earth.

The romanticized memories of terrestrial exploration, joined with the advent of rocketry and the promise of aviation, created a human spaceflight vision of remarkable durability. The vision was strongly utopian, an appealing feature to people disturbed by earthly imperfections. From the founding of religious and ideological communes to the impulses that supported continental migration, utopian thought has repeatedly influenced the societies that embrace spaceflight. Utopianism encompasses the desire to leave corrupt places behind and form more perfect societies in distant lands. As will be seen in the words of its advocates, the human spaceflight vision embraces utopian doctrines in many important ways.

The rocketry that permitted human spaceflight also allowed the nations pursuing it to develop rockets powerful enough to dispatch weapons of mass destruction. Space travel by astronauts and machines on peaceful missions became a means of demonstrating rocket capability without resorting to more warlike missile launches and deployment. Rocket technology also provided a methodology for inspecting the activities of potential adversaries. Both factors led military leaders in the United States and Soviet Union, nominally charged with the protection of national security, to advance proposals for expeditions of discovery. Prior to the formation of the National Aeronautics and Space Administration, officials in the U.S. Air Force presented plans designed to "achieve an early capability to land a man on the moon and return him safely to earth."[13] Wernher von Braun, still a civilian employee in the Army Ballistic Missile Agency, aggressively promoted the advantages of human flights to the Moon and Mars. While acknowledging the scientific benefits flowing from such endeavors, military officers and their civilian employees pointed out the security advantages spacecraft offered as bomb carriers, troop transports, and reconnaissance posts in space.[14]

Spaceflight advocates argued that all of these benefits required human crews. Given the state of robotic technology at that time, nearly every military use of space—from reconnaissance platforms to missile stations on the Moon—was thought to need human operators at the site.[15] Within both the United States and the Soviet Union, people using space to demonstrate national capability placed a premium on human missions. A robotic mission to the Moon, even one that returned samples to Earth, was not viewed as having the same social impact as humans walking on the Moon. The evidence supporting this doctrine may have been thin, but advocates of human flight pervasively believed it and convinced many political leaders as well. The cost advantages of robotic flight, which existed at the beginning and grew more pronounced as flights progressed, essentially disappeared under the force of this message.

As a consequence of these developments, human spaceflight came to occupy a central position in the civil space efforts of the spacefaring powers. The ultimate objective of the U.S. national space effort, in the words of one early planning document, was "exploration of the moon and the nearby planets," a task to be done by human beings.[16] Announcing the orbiting of the first artificial satellite, Soviet leaders predicted that it would "pave the way to interplanetary travel."[17] A special advisory group repeated a commonly transmitted message when they concluded in 1958 that "although it is believed that a manned satellite is not necessary for the collection of environmental data in the vicinity of the earth, exploration of the solar system in a sophisticated way will require a human crew."[18] Space advocates further encouraged the belief that public support for spending vast sums of taxpayer money on space required a human presence. An early participant in the U.S. space effort repeated a widely held belief when he offered the observation that "if the mission was manned, people cared deeply, and if only instruments flew, interest was lessened and somewhat remote."[19]

In the Soviet Union, control of human space exploration remained in military hands.[20] In the United States, President Dwight D. Eisenhower assigned it to NASA. The civil agency was also charged with the conduct of space expeditions using automated spacecraft.[21] The merging of space science and human flight engineering in a single agency created an uneasy alliance, since advocates of civil space exploration did not agree on the reasons for its pursuit. Eisenhower approved Project Mercury in 1958 for the purpose of testing the effects of spaceflight on human subjects, an inquiry in which the military retained substantial interest. Could humans function in the space environment, or would they be immobilized by weightlessness or disorientation? Project Apollo, approved by Kennedy in 1961, was largely an engineering challenge. While scientific investigation of the Moon did occur, it did not dominate the flight protocols as much as the engineering challenges of developing the rockets and spacecraft necessary to land and return. NASA officials reaffirmed the uneasy alliance between science and human flight three years after the formation of the civil space agency when they officially renamed the division devoted to robotic flight. They called it the Office of Space Science, clearly delineating it from the office devoted to "manned space" efforts.[22]

Shortly after the 1969 landing on the Moon, advocates of human spaceflight reaffirmed the traditional vision. Their plans appeared in the report of the Space Task Group, a special advisory group appointed by President Richard M. Nixon. Members of the group announced that "NASA has the demonstrated organizational competence and technology base, by virtue of the Apollo success and other achievements, to carry out a successful program to land man on Mars within 15 years."[23] With a human Martian expedition as its centerpiece, the Space Task Group proposed an ambitious program of winged spaceships, large space stations, a lunar base, and a liftoff date of 1986 for the first human mission to Mars. Simultaneously, the group proposed a wide-ranging program of automated activities including a large space telescope and a robotic Grand Tour of the outer solar planets.

Presidential advisers in the Nixon White House viewed these proposals as excessively grandiose and contemplated actions that would have effectively closed down the human flight effort after the last lunar landing. Nixon saw no need to spend the vast sums of money required to complete the agenda contained in the recommendations of the Space Task Group. Yet he was reluctant to discontinue the human flight program altogether. Such a move, budget adviser Caspar Weinberger warned in a memorandum, would confirm the belief "that our best years are behind us. . . . America should be able to afford something besides increased welfare, programs to repair our cities, or Appalachian relief and the like." On the memo, Nixon scribbled, "I agree with Cap."[24]

Members of the Space Task Group understood that opposition to the human flight effort in the United States was growing. "There has been increasing public reaction over the large investments required to conduct the manned flight program," they wrote. "Scientists have been particularly vocal about these high costs and problems encountered in performing science experiments as part of Apollo, a highly engineering oriented program in its early phases."[25] In a Solomonesque decision, Nixon refused to approve the grand vision of human exploration proposed by the Space Task Group but allowed NASA to continue human spaceflight with one new initiative. Members of the Space Task Group suggested that "much of the negative reaction to manned space flight, therefore, will diminish if costs for placing and maintaining man in space are reduced." Nixon concurred and approved a single new human flight initiative designed to "take the astronomical costs out of astronautics."[26] NASA engineers struggled unsuccessfully to achieve that mandate through the development of the reusable space shuttle. Soviet engineers pursued a similar goal with their *Buran* space shuttle in the 1980s, abandoning the effort after a single robotically piloted orbital flight. Neither the U.S. Space Shuttle nor the Soviet *Buran* reduced the cost of space transportation.[27]

Advocates of space travel nonetheless continued to issue appeals lauding the inspirational effects of human flight. "We have a continuing urge to chart new paths and explore the unknown," NASA Administrator James M. Beggs observed during the early 1980s while advancing his agency's case for an orbiting space station that might provide a staging point for lunar and planetary expeditions. "It is clear that if we ever lose this urge to know the unknown, we would no longer be a great nation."[28] Proposing that humans return to the Moon and organize an expedition to Mars, President George H. W. Bush in 1989 observed that "space is the inescapable challenge to all the advanced nations of the Earth."[29] Fifteen years later, George W. Bush repeated his father's proposal, announcing that "mankind is drawn to the heavens for the same reason we were once drawn into unknown lands and across the open sea. We choose to explore space because doing so improves our lives, and lifts our national spirit."[30]

Advocates of human space travel would like to return to the Moon and organize expeditions to Mars. The more enthusiastic of them believe that humans could transform that planet's atmosphere so as to create moderate temperatures, a condition conducive to human settlement and self-sufficient colonies.[31] By settling Mars, humans would become a multiplanetary species. Given sufficient time, *Homo sapiens* would learn how to move to hospitable planets around nearby stars, continuing the migratory patterns that have spread the species to destinations far from the site of its original continental home. Interplanetary dispersion of the sort that characterizes the expansion of the species across the face of the Earth provides a fundamental justification for human space travel. As an earthly species, humans expanded and thrived because they learned how to live under terrestrial conditions less hospitable than those in the locales where they first arose. Similar opportunities, advocates believe, await them in space. "It is important for the human race to spread out into space for the survival of the species," proclaimed British astrophysicist Stephen Hawking. In the same way that continental dispersion helped early humans survive local catastrophes, planetary dispersion would help the human race survive global ones. "Life on Earth is at the ever-increasing risk of being wiped out by a disaster," Hawking warned, "such as sudden global warming, nuclear war, a genetically engineered virus or other dangers we have not yet thought of." Hawking recommended that humans begin by settling Mars, then look for planets around other star systems "as nice as Earth."[32]

Among its most devoted advocates, the urge to fly in space draws upon cultural forces far different than those motivating robotic flight. Historically, government leaders have supported expeditions of discovery for reasons that go well beyond science or curiosity about the unknown. They have sought national prestige, external colonies, expanded trade, the spread of religion, and security.

To this list, spaceflight advocates add survival of the species. At its best, scientific discovery is one of many motivating factors. In its most limited form, science is an incidental byproduct, a passenger on a voyage undertaken for broader reasons.[33] In this sense, human spaceflight in the twentieth century became a means for reinspiring a romanticized pursuit of terrestrial exploration, with all of the benefits such efforts seemed to confer.

HUMANS AND ROBOTS TOGETHER

Within their vision, advocates of human spaceflight needed to account for the presence of machines. They surrounded themselves with machines, from the rockets that propelled their spacecraft to the traveling capsules that carried the crews. Machines like the airplane transformed modern terrestrial exploration. Machines like the automobile expanded human freedom. As much as advocates of human flight might have wished to reestablish a heroic age of exploration, they needed a place for machines.

To this end, many human flight advocates adopted a unique perspective: they embraced togetherness. "NASA will send human and robotic explorers as partners," wrote the authors of the 2004 Vision for Space Exploration guidelines. "Robotic explorers will visit new worlds first." Once the machines have surveyed local conditions, "human explorers will follow." The policy reiterates the position expressed by an early U.S. advisory committee whose members reported that "a great part of the unmanned program for the scientific exploration of space is a necessary prerequisite to manned flight."[34]

This particular view of robotic flight is a product of industrial age thinking and the accompanying emphasis upon machines that serve humankind. In appearance, it seems reasonable, but in practice the viewpoint contains a substantial contradiction. The machine age ended the classical era of terrestrial exploration that advocates of human flight hope to resurrect in space with their quaint vision of humans and machines working together.

The heroic age of terrestrial exploration relied upon human capabilities to a much greater extent than the mechanistic methods that supplanted it. When the Norwegian explorer Roald Amundsen reached the South Pole in 1911, he did so with the assistance of four team members, fifty-two dogs, and four sleds. He reached his base camp along the Ross Sea in a three-masted schooner made of oak and tropical greenheart. When Amundsen left Norway in 1910, only one person knew whether he was heading for the North Pole or the South Pole.[35]

The 1929 Byrd expedition, by contrast, reached its base at Little America with an armada of mechanization. Byrd's was the best-equipped expedition ever to arrive at Antarctica up to that point. Nonetheless, in Byrd's expedition, humans were still firmly in charge. This appropriately conformed to the optimistic vision of mechanization prevalent at that time. Machines existed to improve the human condition, relieving people of the meaner aspects of manual labor so that they could lead better lives. In essence, Byrd substituted airplane engines for dogs.

Many of the early robot stories offered for popular consumption adopted this humanistic point of view. In such works of fiction, robots serve as human companions, carrying out work too tedious or delicate for humans to do. Fictional robots perform housework, sit with children, operate power-generating stations, and assist with the piloting of spacecraft. Robots are mechanized servants, a product of Edwardian-era thinking and its emphasis upon the proper relationship of household servants and masters. As servants, robots do not exist to supplant humans, they exist to help them perform human responsibilities. From the fictional robots created by Isaac Asimov to the amusing compatriots R2D2 and C3PO in the *Star Wars* sagas, humans and robots work together with humans in the dominant roles.

Asimov in particular emphasized the robot-as-servant. In practically all of his stories, robots and humans cooperate in situations similar to the relationship between Amundsen and his dogs. Asimov railed against what he called "the Frankenstein complex," the belief that robots would escape from personal supervision and revolt against their masters. They were just machines, he insisted, dangerous if improperly used but always under human control.[36]

Engineers instituted this relationship in a number of twentieth-century spaceflight activities, most notably the Surveyor missions to the Moon and the Hubble Space Telescope. NASA officials undertook the Surveyor project in 1960, shortly before the official decision to land humans on the Moon. Surveyor was a three-legged robot designed to land on the Moon, test the lunar soil, and undertake a series of scientific experiments. Scientists had warned the engineers organizing the human flights to the Moon that the lunar surface might be covered with a thick mantle of dust into which any approaching spacecraft would hopelessly sink. The five Surveyor spacecraft that landed on the Moon disproved this proposition and collected significant data on surface conditions in preparation for human visitation. The last Surveyor completed its mission in 1968, the year preceding the one in which humans reached the lunar surface. The second human crew to reach the Moon landed within walking distance of *Surveyor 3*, from which the humans retrieved parts for further inspection back on Earth. The supposed symbiotic relationship between the Surveyor and Apollo missions provided a conceptual model for people anticipating a close relationship between humans and machines.[37]

NASA officials followed a similar path in the development of the Hubble Space Telescope. They could have built a fully automated instrument, launching it with an irretrievable rocket into an orbit inaccessible by human beings. Instead, engineers configured the telescope so that it could be placed in the payload bay of the NASA Space Shuttle and delivered to an orbit where subsequent astronauts could perform repairs. This decision proved fortuitous when, following launch, scientists discovered a spherical aberration in the telescope's primary mirror. On a subsequent flight astronauts installed a large instrument containing ten mirrors that corrected the problem, a costly but essential repair. Between 1993 and 2002, astronauts visited the Hubble four times, installing new instruments and replacing gyroscopes, recorders, the power control unit, and solar arrays, thereby extending the life and capability of the orbiting telescope. The result was a remarkable harvest of scientific data that contributed significantly to astronomical knowledge.[38]

Space advocates envisioned similar relationships continuing on the Moon and planets. One of the more creative proposals for future lunar cooperation involves a collection of six-legged robots that walk away from their individual landing zones. The machines, called Habots, would be programmed to rendezvous after some designated period of exploration. Meeting at the predetermined site, they would link themselves together in such a manner as to produce a hollow shell, release oxygen into the newly formed interior, and receive human visitors. Once humans completed their work, the Habots would separate and resume their autonomous investigations.[39] In a similar fashion, human flight advocates have advanced a variety of proposals for teleoperated robots on Mars. From secure positions on an orbiting station or a Martian surface base, humans would drive robotic explorers across the local terrain. The drivers would be at Mars, not on Earth. The rovers would take risks not permitted human explorers, while the humans could drive the rovers without the time delays that beset earlier attempts to supervise rovers from control stations on Earth.[40]

The image of humans and robots working together in space is an enticing one. It contains substantial deficiencies, however. Like the vision of human exploration to which it is joined, it draws upon social traditions that are rapidly receding in the postindustrial age. Technologically, it is challenged by space robots that work in an increasingly autonomous fashion. Machines working alone threaten to replace humans and machines working together. Cooperation between humans and robots in space, moreover, requires a relationship between engineers and scientists on Earth that has been hard to maintain. Historically, the relationship between people working on robotic space activities and human flight has been marked by increasing degrees of separation.

FROM COOPERATION TO DISTRUST

At the dawn of the space age, robotic spaceflight efforts served the same objectives as the human flight effort. The U.S. policy document advancing the case for America's first artificial satellite justified the undertaking with language commonly applied to human endeavors. Such an accomplishment, the White House document read, would confer "considerable prestige and psychological benefits" to the nation achieving it. It would have substantial value for national security in areas such as "defense communication and missile research" as well as "the political determination of free world countries to resist Communist threats." Not incidentally, the document noted, such a satellite would produce important information on conditions in the ionosphere that could be used for scientific, communication, and military purposes.[41]

Broad justifications such as these encouraged early advocates of a balanced space effort to draw few distinctions between human and robotic flight. Concerning the occasional disagreements between advocates of human and robotic flight, Oran Nicks, a NASA engineer who served as director of lunar and planetary programs for the first thirty robotic missions to the Moon, Mars, and Venus, offered the following observation. As a pilot and aeronautical engineer who joined NASA shortly after its founding and became the overall director for many of NASA's earliest lunar and planetary missions, he represented both camps in the human versus robots debate. Nicks wrote: "The truth is that there were no such things as unmanned missions; it was merely a question of where man stood to conduct them. In some cases he sent his instruments and equipment into space while controlling them remotely, and in other cases he accompanied them in spacecraft equipped with suitable life support systems. It should be noted that the spurious arguments over the merits of manned versus unmanned missions were never between machines and men, but between men and men."[42]

NASA supporters labored to convince scientists that a mix of human and robotic flights enhanced both. Quintessential science entrepreneur Lloyd V. Berkner urged his colleagues in the science community to recognize that the Apollo program would be completed with or without them, so scientists might as well take advantage of the substantial funds that were being made available. In the spring of 1961 Berkner oversaw the preparation of a Space Studies Board report sponsored by the revered National Academy of Sciences lauding "Man's Role in the National Space Program."[43] It was followed some months later by a press release rammed through the Space Studies Board giving scientific blessing to the proposition that "from a scientific standpoint there seems little room for dissent that man's participation in the exploration of the Moon and planets will be essential."[44]

John E. Naugle, who succeeded Homer Newell as NASA associate administrator for space science, likewise insisted that everyone gained from a large human spaceflight program. In his view, human flight activities provided political cover for the frequently neglected space science program. If not for human spaceflight, Naugle asserted, money otherwise available for space science would disappear into a fiscal black hole.[45]

Space scientists such as Newell, Naugle, and Berkner helped to establish a substantial presence for robotic space activities within the NASA budget. During the 1960s, Congress directed about 25 to 30 percent of NASA spending to space science annually. From this foundation, funds for science projects of a robotic sort grew to 43 percent by the 1990s. A community of scientists dedicated both to NASA missions and robotics matured.[46] Indeed, some NASA supporters viewed the persistent criticism from scientists like James Van Allen as an ungrateful response, given the governmental largess lavished upon him and his fellow scientists. As one official told a group of agency public affairs officers in 1998, "NASA made Van Allen, and now all he does is condemn us."[47]

Guided by the philosophy of cooperation, NASA executives prepared to merge the human and robotic flight programs at the new Goddard Space Flight Center under construction in Greenbelt, Maryland. The human flight group would have moved from its original home at the Langley Research Center in Hampton, Virginia, which housed Project Mercury, while satellite research would be transferred from facilities such as the Naval Research Laboratory in Washington, D.C. Geographically, NASA executives would have permitted no distinction between human and robotic flight.

This arrangement was abrogated in 1961, when human flight advocates decided to construct an entirely separate center near Houston, Texas. "It became clear," NASA's director of space flight, Abe Silverstein, observed, "that Goddard should direct the unmanned satellite program and a wholly new center be created for the manned spaceflight program." Engineers and astronauts began moving to the Manned Spacecraft Center, as it was originally called, in 1962, giving it a characteristic style quite different from that found at the more scientifically oriented Goddard Space Flight Center, which managed only robotic flights.[48]

Although official policy required continued cooperation between the human and robotic flight efforts, divisions repeatedly appeared. NASA officials assigned the Surveyor program, the effort to soft-land robots on the Moon, to the Jet Propulsion Laboratory (JPL) in Pasadena, California. Like the Goddard Center, JPL was dominated by scientists with little experience in human flight. Officially, NASA officials set up the Surveyor project to aid Apollo managers and engineers in their work of safely landing astronauts on the surface of the Moon. Expanding this mandate, JPL scientists succeeded in installing instruments on the Surveyor spacecraft that could carry out substantial scientific investigations. It was one of many instances in which space science efforts rode "piggyback" on projects undertaken for other purposes.[49]

The forces promoting separation revealed themselves during an early congressional visit to the Manned Spacecraft Center in Houston. A revealing exchange took place between Congressman Joseph Karth, a Minnesota Democrat, and Max Faget, one of the lead engineers on Project Apollo. Oran Nicks relates the story: "The Congressman, seated next to Max Faget at lunch, asked Max about Surveyor's importance to Apollo, in a context that implied that Surveyor had been funded largely on the basis of its probable importance to Apollo Lunar Module design. Faget, never known for pulling his punches, flatly told Karth they really were not depending on Surveyor. In fact, Max told him they had plans of their own for obtaining the necessary data by orbital reconnaissance with manned vehicles before committing to landing."[50] NASA executives sternly reprimanded Faget for his admission. Officials at the Manned Spacecraft Center were more circumspect in their future statements. Nonetheless, according to Nicks, human flight engineers continued to maintain that their flight objectives could not depend upon robotic projects "over which they had no control."[51]

Further divisions appeared during the controversy over the Manned Orbiting Laboratory, or MOL. The project was a U.S. Air Force undertaking designed to establish a small orbiting laboratory from which a two-person crew could monitor foreign military threats and verify conformance with nuclear arms agreements. Air Force officers viewed the project as the principal means by which the military would introduce soldiers into space. They designed a small pressurized canister and appointed their own astronaut corps. Partisans of robotic flight, housed largely in the intelligence community and possessing little interest in piloted observation, insisted that the functions proposed for MOL could be carried out more reliably with automated reconnaissance satellites.

The issue was decided largely on technical grounds. Officials in the office of the Secretary of Defense weighed the advantages of automated reconnaissance satellites, flown from the ground, against observation platforms occupied by soldiers. Advances in remote sensing—the ability to collect information from automated satellites—favored the robotic alternative. Consequently, Secretary of Defense Melvin Laird cancelled the MOL program and disbanded the Air Force astronaut corps in 1969, inaugurating a lengthy period during which the U.S. military relied exclusively upon robotic technology for national security needs.[52]

In both the United States and the Soviet Union, scientists pressed ahead with an ambitious series of robotic flybys, orbiters, and landers. NASA's *Mariner 4* became the first spacecraft to fly by Mars in 1965, returning the initial closeup pictures of that sphere. *Mariner 9* became the first spacecraft to circle another planet when it entered an orbit around Mars in 1971. The Pioneer twins left Earth for the outer planets in 1972 and 1973, followed by *Voyagers 1* and *2* in 1977. In 1975, the Soviet Union's *Venera 9* became the first spacecraft to transmit images from the surface of Venus. This was followed by the landings of the U.S. *Vikings 1* and *2* on Mars in 1976. By 1979, automated probes had inspected Jupiter and Saturn. As Apollo flight activities wound down with the 1975 U.S./Soviet Apollo-Soyuz rendezvous, momentum in the robotic field created what one commenter characterized as a "golden age of planetary exploration."[53]

Human spaceflight activity in the United States should have resumed in 1978 with the first orbital test of NASA's Space Shuttle. Technical obstacles delayed the first flight until 1981, however, and NASA executives had to seek three supplemental appropriations from Congress to pay for unexpected difficulties. Cost growth on the Space Shuttle, along with the prospect of future spending on a NASA space station, precipitated a severe reaction from space scientists. Writing in both *Scientific American* and the *New York Times* in 1986, physicist James Van Allen complained that excessive spending on human spaceflight had crowded out robotic missions of great merit. In the *Scientific American* article, a lengthy attack on human spaceflight, Van Allen provided a specific list of science missions that had been terminated, postponed, or cut back in order to help pay for the shuttle overruns. While some of the projects indeed vanished, such as the U.S. mission to Comet Halley, most reappeared once money became available, as with the Galileo mission to Jupiter and the Compton Gamma Ray Observatory. The restorations did not stop Van Allen from attacking leaders of the human flight program for engaging in what he called a "slaughter of the innocents," a phrase that captured the growing distrust many scientists felt toward human flight priorities. Such cuts, Van Allen insisted, had produced "a severe chill on scientific and other civilian activities in space." In the *New York Times* article, he reiterated his judgment that "well-founded scientific and utilitarian uses of space technology were suffering severely because of NASA's overriding emphasis on manned flight." Within the scientific community, physicists and astronomers tend to be the most persistent critics of human spaceflight.

What began as a vision of mutual cooperation had morphed, after twenty-five years of spaceflight, into a climate of division and distrust. "I believe," Van Allen announced, that "the progressive loss of U.S. leadership in space science can be attributed . . . largely to our excessive emphasis on manned space flight."[54] The spirit of mutual cooperation that had characterized projects such as the Hubble Space Telescope in the United States dwindled. The Hubble was proposed during a time when partisans of human flight had convinced many scientists that the shuttle would provide cheap and frequent access to space. In fact, difficulties with the shuttle program delayed the launch of the Hubble for more than four years.[55] Officials configured the next two telescopes in the sequence of Great Observatories—the Compton Gamma Ray Observatory and the Chandra X-ray Observatory—for launch on the Space Shuttle but did not plan for their servicing or repair by astronauts.[56] Critics characterized proposals for a shuttle-launched fourth observatory (the Spitzer Space Telescope, which had started out as the Space Infrared Telescope Facility) as "a disastrously bad idea."[57] Scientists designing the Spitzer Space Telescope foreswore both shuttle launches and astronaut repair.

For technical reasons, scientists placed the Spitzer Space Telescope in an Earth-trailing heliocentric orbit, beyond the reach of shuttle astronauts. Even if it had been set in a near-Earth orbit, the telescopes would not have been shuttle-launched. Loss of the *Columbia* in February 2003 resulted in the removal of all scientific payloads from the shuttle manifest (with the possible exception of one last Hubble serving mission), just as commercial and military payloads had been removed following the 1986 *Challenger* accident. Prior to 1986, shuttle astronauts deployed reconnaissance satellites, retrieved scientific payloads, and repaired communication satellites, actions that allowed a closer integration of humans and machines. Under the more restrictive shuttle launch policy, all of those machines had to be made more autonomous.[58]

Confronted with such events, some of the strongest advocates of human spaceflight began to express concerns about the direction of that program. Faget, once the chief engineer for human spaceflight vehicles, suggested that the Space Shuttle "ought to be flown unmanned." This would require a wholly automated system for what was initially designed as a human flight vehicle. (Soviet engineers attained such a capability with their *Buran* space shuttle and made one orbital flight without humans on board.) According to Faget, non-piloted flights for scientific and commercial payloads would save weight and, "if nothing else, would give the engineers the opportunity to experiment in improving the systems without risking human life." The necessity of protecting the human crew, Faget said, was "a big inhibitor" on innovation. He also criticized NASA executives who insisted that the International Space Station (ISS) be permanently occupied by human beings. Their approach, he observed, was "getting nowhere" and did not serve "any vital interests." In 1982 Faget and a group of ex-NASA employees formed Space Industries, Inc., and proposed the construction of an industrial space facility (ISF), a pressurized spacecraft that could be used to manufacture new materials in space. The ISF would be visited by astronauts, but most of the time it would carry on its work in an automated fashion without people on board. To Faget, a "man-tended" facility made more sense; that the ISS required human occupancy meant that its first task would always be oriented toward maintaining the safety of its crew.

NASA executives advanced commercial manufacturing as one of the principal functions of their orbiting space station. They envisioned factories in space with astronauts producing microgravity products. Struggling to find a sufficient financial base for the ISF, Faget asked for government help. The projected cost of the NASA space station, however, was growing rapidly. Faget's funding request, in his own words, represented "a major threat to the continuation of the space station program. It had to be killed, and they killed it."[59] The ISF proposal died in 1989 when NASA executives officially refused to sign a statement agreeing to lease space on the largely automated facility. To many observers, the incident provided another example of the manner in which human flight commitments pushed out innovative proposals for automated facilities.

The space age began with advocates of human flight advancing a vision of cooperation in which humans and machines explored space together. By the late twentieth century, that vision had devolved into competition, separation, and distrust. Many space scientists, especially those interested in physics and astronomy, came to view human flight initiatives as intrusions into monies otherwise reserved for space science projects. Military officers concluded that reconnaissance and other national security satellites could be operated through remote control. Advocates of human exploration expressed frustration with what seemed to be a stalled space program restrained by a combination of accidents, escalating costs, and aversion to risk. More than a few human flight advocates came to question NASA's ability to sustain the effort and complete the vision of large stations, lunar bases, and expeditions to Mars.

Increasingly, as the space program matured, advocates of the two points of view divided into separate camps. In a 2004 survey of space leaders, Jon D. Miller noted a "crystallization of policy attitudes" along human and robotic lines. "What do you think should be NASA's number one priority?" he asked. Thirty-five percent of the space policy leaders he interviewed expressed a strong preference for robotic programs; 27 percent wanted to revitalize and expand the human flight program. As Miller wrote, "It is clear that there is a strong division between leaders who value major unmanned programs such as the space telescopes and the Earth monitoring programs and leaders who see the primary mission of NASA to be manned space exploration."[60]

This dichotomization of spaceflight began shortly after the first missions into space and has been a perennial issue ever since. The debate grew more heated as robotic capabilities advanced. Homer E. Newell, NASA's director of space science programs between 1958 and 1973, commented on the manner in which space science and human exploration seemed to diverge:

> For space science one of the most difficult problems of leadership, both inside and outside NASA, concerned the manned spaceflight program. Underlying the prevailing discontent in the scientific community regarding this program was a rather general conviction that virtually everything that men could do in the investigation of space, including the moon and planets, automated spacecraft could also do and at much lower cost. This conviction was reinforced by the Apollo program's being primarily engineering in character. Indeed, until after the success of *Apollo 11*, science was the least of Apollo engineers' concerns. Further, the manned project appeared to devour huge sums, only small fractions of which could have greatly enhanced the unmanned space science program.[61]

Lee A. DuBridge, a renowned scientist and presidential adviser, likewise disparaged human space exploration. DuBridge possessed a conservative, almost libertarian belief that science would be tainted by government contracts and influence. NASA administrator James E. Webb tried to convince DuBridge that Project Apollo would yield important scientific data and that DuBridge should support it.[62] He never did.

COMPARATIVE ADVANTAGES

Protectors of the original vision insist that humans play a useful role in space. In the competition between humans and robots, they say, humans work smarter and faster than machines. An early paper emphasizing the importance of Apollo astronauts to the exploration of the Moon summarized this position: "Man becomes useful in complex, multi-use vehicles and operations; or where a high degree of judgment, discrimination and selectivity are required; or his manual dexterity and analytical capabilities are required."[63] Apollo astronauts controlled the landings on the Moon, fine-tuned orbital maneuvers, and reacted to emergencies (such as the lightning strike that caused a temporary loss of *Apollo 12* spacecraft control shortly after launch). They deployed instruments, repaired equipment, and investigated surface conditions on the Moon. On the last three missions, they drove electrically powered lunar roving vehicles. In the judgment of the people assessing Project Apollo, human beings accomplished in hours what automated landers and rovers took much longer to do.

The history of lunar exploration supports this judgment. On the last three expeditions to the Moon, during surface operations totaling slightly less than nine days, astronauts working from lunar roving vehicles conducted traverses totaling 91 kilometers (56 miles). They ventured as far from their lunar lander as 7.6 km (4.7 miles)—about the distance they could safely walk if their buggy broke down.[64] During the same period, the Soviet Union dispatched two robotic lunar rovers and three sample return missions. Flight controllers in the Soviet Union drove the rovers. They experienced far more difficulties driving their rovers than the Apollo astronauts encountered with their moon buggies, in part a consequence of the six-second delay between electronic transmissions and execution. The cumulative traverses of the remotely driven Soviet rovers totaled 48 km (25 miles) and took fourteen months.[65]

Robotic technology improved significantly in the following decades, with substantial progress and impressive achievements. In the summer of 1997, flight controllers at NASA's Jet Propulsion Laboratory drove the automated *Sojourner* rover off its landing platform onto the Ares Vallis flood plain on Mars. This was a substantial achievement. Moving at the sluggish pace of 0.4 meters per minute, however, the *Sojourner* rover was not designed to rove more than 500 meters (0.3 miles) from its landing site. The mission nonetheless excited further improvements. Technological advances in hazard-avoidance procedures allowed the 2004 Martian rovers *Spirit* and *Opportunity* far greater mobility. After three years of surface operations, *Spirit* had traversed 7.1 km (4.4 miles), while *Opportunity* had logged 10.3 km (6.4 miles).[66] In comparative terms, however, the two rovers took years to cover the territory typically traversed by Apollo astronauts in a single day, encouraging spaceflight advocates to press their case for even more expeditions to follow—ones with humans as well as machines.

In the eyes of human flight advocates, rovers were wonderful machines, but no match for human drivers. A famous 2004 experiment sponsored by the Defense Advanced Research Projects Agency (DARPA) pitted thirteen robots in a fast-paced race to complete a 142-mile course laid out on the Mojave Desert. The resulting contest resembled a scenario from early rocketry. The robotic vehicles flipped, got stuck, collided with obstacles, and caught fire. The vehicle with the best performance completed less than 6 percent of the course. Notwithstanding their advanced electronics, science writer Joel Achenbach observed that the autonomous vehicles "had trouble figuring out fast enough the significance of obstacles that a two-year-old human recognizes immediately." Defense Department officials, nonetheless, continued to encourage the development of smart vehicles that could travel without the necessity of a human driver or remote control. Members of Congress agreed and declared that one-third of all combat aircraft and ground vehicles be robotic by 2015.[67]

Even if scientists construct rovers that move as fast as human beings, people will still need to explore, advocates of the classic spaceflight vision maintain. Machines can search for life on the Martian surface, Mars Society president Robert Zubrin observes, but the most favorable repository of life is likely to be found underground, in aquifers a kilometer or more below the surface. The search for life, he insists, will require drilling in many places and building surface laboratories where water samples can be cultured and analyzed. Says Zubrin, "this is a job for humans."[68]

In any assessment of performance, overall effectiveness is influenced by cost. In that respect, human flight does not do well. The U.S. Congress provided $820 million for the mission that placed the rovers *Spirit* and *Opportunity* on the surface of Mars in 2004. The cost of the expeditions that landed humans on the Moon, expressed in 2004 dollars, totaled approximately $150 billion. The Apollo astronauts may have gone five times as far in just nine days, but they required 180 times as much money to do so. Advocates of human flight note that in spite of disadvantages imposed by their high cost, human missions still remain cost-effective because they can accomplish more work. To a certain extent, this is true. Although the human lunar and Mars rover missions lack perfect equivalency, a statistical comparison that equalizes the relative costs of the two missions suggests that the Apollo astronauts still covered more ground in less time than the rovers.[69]

Based on the achievements of Apollo astronauts, advocates of human flight continued to argue that a civil space program uniting people and machines would outperform one relying upon robots alone. The high cost of moving humans through space sobered everyone up. No factor inhibits the future of human space travel more than the contemplation of its cost. The persistent inability of spaceflight engineers to reduce the high cost of building spacecraft capable of housing human beings and launching them through space significantly retarded that enterprise in the late twentieth century and continues to impose the principal challenge to human endeavors in the twenty-first.

In 1972, spaceflight engineers set out to construct a winged Space Shuttle that—among its many objectives—was supposed to reduce the cost of space access significantly. In 1984, President Ronald W. Reagan directed NASA to assemble a permanently occupied space station within one decade at a predicted cost of $8.8 billion. In 1989, President George H. W. Bush recommended a "Space Exploration Initiative" that would have returned humans to the Moon and sent them on to Mars. All three of these mega-projects floundered on issues of cost, confining astronaut travel to the region immediately above the Earth's surface.

Regarding the space shuttle program, NASA officials promised a fleet of five spacecraft that could carry out 580 missions over a twelve-year period at a total cost of about $16 billion. This would have reduced the expense of reaching space substantially in comparison to existing launch vehicles, thereby advancing the cause of human spaceflight. In fact, NASA flight engineers and their contractors spent $85 billion through 2002 building, upgrading, and flying the Space Shuttle. This sum had purchased, as of that year, 110 flights. The average cost per flight exceeded $770 million. Even adjusting the figures to account for inflation, the gap between expectation and achievement was enormous. The space station program suffered similar difficulties. Afflicted by frequent redesign, schedule delays, and technical difficulties, cost estimates for the space station ballooned from $8.8 billion to more than $61 billion.[70]

The International Space Station and the U.S. Space Shuttle were conceived as practical, economical investments in infrastructure that would enhance the ability of humans to move into and through space. The Space Shuttle was conceived as a device providing easy, inexpensive transport of humans and cargo into orbit, while the original plans for the ISS provided a staging platform for deep space exploration. NASA officials promised to complete the two programs in less than twenty years and save money in the process. In reality, the two programs took twice that time to complete and accumulated expenses well in excess of the amounts spent to take humans to the Moon. Rather than expedite the vision of human space travel, the programs substantially retarded it.[71]

Concern with the high cost of human spaceflight sank the 1989 Space Exploration Initiative. This proposal, which envisioned a lunar base and human expedition to Mars, suffered a premature demise when public officials leaked an internal NASA cost estimate suggesting that the initiative would cost upwards of $400 billion to achieve. Most of that was for the sustenance of NASA as an institution, critics asserted. This amount was viewed as excessive by U.S. lawmakers, who at the time had constrained the civil space and aeronautics budget to less than $15 billion annually. Normally a strong supporter of NASA efforts, Senator Barbara A. Mikulski (D-Md.) bluntly declared, "We're essentially not doing Moon-Mars."[72]

Throughout this period, the cost of various robotic missions fell. Taking advantage of advances in computer capacity and electronic miniaturization, space scientists designed automated spacecraft that were substantially smaller and more capable than their robotic predecessors. Smaller spacecraft could be launched on smaller rockets, reducing the associated transportation costs.[73]

Concurrently, the cost of human spaceflight remained stubbornly high. Following the announcement of the Vision for Space Exploration program in 2004, NASA officials decided to incur Apollo-style expenses to return humans to the Moon. As a whole, the first return would not cost as much as Project Apollo. The United States possessed field centers and much technological know-how whose expense already had been charged to the Apollo undertaking, reducing the cost of the undertaking by about 45 percent compared to Project Apollo. Still, the burden would be high. NASA Administrator Michael Griffin estimated that a lunar return using Apollo technology would take thirteen years and roughly $104 billion.[74]

Buoyed by the cost reductions sweeping the robotic flight program, planners at the Johnson Space Center during the mid-1990s suggested that NASA could complete a human expedition to Mars for less than $50 billion. The key element in this low-cost plan lay in the potential ability of engineers to place robotic processing facilities on the Martian surface that could manufacture fuel from local resources for the return voyage.[75] Yet when asked ten years later to prepare an actual cost estimate for a return expedition to the Moon, a much more accessible destination, NASA planners produced a plan that cost more than twice as much as the low-cost Mars scenario.

Repeatedly, advocates of human space travel issued cost and schedule estimates that were excessively large or impossible to achieve. Had they been able to reduce the financial burden imposed by their space activities, human flight advocates might have placed their movement in a commanding position. Instead, they conceded advantage after advantage to proponents of robotic flight. While astronauts continued to perform useful functions in space, the overall expense of doing so relative to robotic activities undercut the movement favoring human spaceflight considerably.

ROBOTIC AUTONOMY

In the same manner that high cost inhibits the future prospects for human spaceflight, the inability to operate autonomously hinders robotic flight. High cost and lack of autonomy provide the principal counterpoints in the human versus robot debate.

Robotic spacecraft, by definition, require human supervision. As noted in the earlier statement by Oran Nicks, the term *unmanned* is a misnomer. Humans have been involved in every robotic spaceflight since the enterprise began, the only difference being that the humans control the spacecraft from remote locations. Conversely, spacecraft with humans on board operate with a great deal of autonomy or "pilotless control." For much of its reentry into the Earth's atmosphere, the Space Shuttle with humans on board flies itself according to imbedded programs even though pilots are at the controls.[76]

The essential distinction in the debate between "manned" and "unmanned" spaceflight concerns the degree of autonomy afforded the spacecraft and its potential human crew. In their ideal scenario, advocates of robotic flight would like machines to achieve levels of autonomy such that robots could carry out space activities with much less ground control than imposed on human expeditions. If scientists and engineers could achieve such levels of machine autonomy, the prospects for robotic flight would advance considerably.

Such levels of autonomy would permit robotic expeditions into the solar system as sophisticated as those anticipated by advocates of human flight. With machines doing the work, the missions could go to places more adverse than humans could endure. Nearly autonomous robots would rove and drill and return samples from distant sites. A NASA proposed mission to the icy moons of Jupiter suggests the challenges involved. Scientists believe that the Jovian satellite Europa harbors a subterranean sea. Though possessing a frozen surface, Europa undergoes a gravitational tug-of-war with its parent planet that may create sufficient energy to melt the ice below. The same forces may have provided Europa with a molten core, creating a source of energy around which life forms might have appeared. Europan creatures could be similar to those inhabiting the areas surrounding thermal vents on the Earth's ocean floors, where some scientists suspect terrestrial life began. Such organisms would not need surface sunlight to thrive, only a source of subterranean energy. Some scientists believe that the probabilities of discovering life on Europa are as good as, if not better than, the chances of locating life on Mars.

To look for life on Europa, scientists would like to land a robotic spacecraft on the moon's icy crust. A cryobot, named for the extremely cold temperatures at which it would work, would melt a path through the ice, descending perhaps three kilometers while trailing a surface communications line. Encountering liquid water, the cryobot would cease its downward journey and dispatch a second robot, a free-traveling hydrobot, that would swim off in search of thermal vents. The hydrobot would scan the ocean floor while communicating its findings back to Earth through the cryobot and its surface station.[77]

Such a mission would require degrees of robotic autonomy not yet attained. It would pose substantial technical challenges. Yet it is difficult to envision the mission being accomplished in an alternative way by a human crew, given the inhospitable conditions involved. If astronauts in spaceships managed to reach Europa and did not succumb to the frigid temperatures there (estimated at minus 260 degrees Fahrenheit on the surface ice), their bodily tissues would be irreparably damaged by the intense radiation belts surrounding Jupiter.

How much autonomy would be necessary to push the advantage to machines? The answer, generally speaking, is as much autonomy as would be attained in spacecraft with humans on board. Consider as an illustration one of the principal challenges of space exploration—the landing of a spacecraft on another sphere. Human landings on the Moon required the combined efforts of pilots on board, flight controllers in Houston, and lines of computer code written in advance. During the descent of *Apollo 11*, a fault in the radar altimeter tripped an automatic alarm in the onboard computer. Commander Neil Armstrong, with the approval of flight controllers in Houston, overrode the alarm and landed safely on the Moon.[78]

As with the Apollo expeditions to the Moon, the landing of robots on distant bodies is controlled by lines of code written in advance of touchdown. In the case of robotic landings, humans cannot intervene to correct last-minute errors. The case of the Mars Polar Lander is instructive. The spacecraft was designed to land on the Martian south pole using a succession of pyrotechnic devices, a braking shield, a parachute, and pulse-modulated descent engines. Magnetic sensors on the landing pads fixed to the landing legs recorded the exact moment of touchdown, requiring an immediate main engine shutdown. A main engine that continues to fire for more than a few microseconds after touchdown can cause a lander to tip over. The landing sequence for the Mars Polar Lander took place autonomously, without instantaneous Earth control.

Unfortunately, the magnetic sensors tended to sense a touchdown signal at the moment when the spacecraft deployed its landing legs. This occurred about five thousand feet above the ground—not a good point at which to disable the main engines. Engineers recognized this idiosyncrasy. Before the flight, they programmed the spacecraft computer to ignore the spurious signal. The complexity of the computer code, however, created an unanticipated contingency. An erroneous line of code caused the spacecraft to recognize the premature shutdown signal as a true shutdown signal at an altitude of 130 feet, when the spacecraft's radar altimeter instructed the spacecraft's computer to prepare to land.

An astronaut sensing a main engine shutdown during the final stages of descent would recognize the severity of the situation and override the command. The Mars Polar Lander did not. When such anomalies occur, robotic spacecraft are programmed to flip themselves into "safe mode" and request help from ground controllers. With signals taking fourteen minutes to reach Earth, the Polar Lander did not have time to pause and receive new commands. So it crashed.[79]

Humans work to prevent such errors by carefully testing spacecraft components before launch, by running spacecraft commands through ground simulators once the craft is in space, and by assembling investigation teams to review anomalies after accidents occur. A series of commands, such as those that might be sent to a robotic spacecraft to begin a landing sequence, must be carefully tested in ground simulators to ensure that the spacecraft responds in the desired way. A special investigation team set up to investigate the crash of Mars Polar Lander, working with records and equipment on Earth, found the errant code and the faulty ground test that failed to disclose it.

The amount of hazard-sensing capability and operational autonomy necessary to operate a robotic spacecraft increases with its distance from Earth. The 2005 landing of the *Huygens* probe on Saturn's moon Titan was carried out with a communication gap of more than one hour. Eventually, scientists would like to dispatch robotic spacecraft to explore planets around stars that are light years away. Such spacecraft would need to be completely autonomous. News of an arrival at any destination not would reach Earth until years after it occurred.

Producers of *Alien Planet,* a 2005 television documentary envisioning a robotic mission to a life-supporting planet six and a half light years from Earth, analyzed the requirements for such an expedition. The arriving spacecraft, as large as a nuclear attack submarine, contains three robotic explorers designed to use the planet's atmosphere to brake and land. Possessing a lifting body shape similar to that found on the U.S. Space Shuttle, the first explorer detaches itself from the mother ship and descends. In a piercing insight, the producers address the consequences of an accident on the first try. The first explorer disintegrates during its landing approach. Rather than call home for new instructions, a process that would take thirteen years for a single round-trip transmission, the spacecraft and its probes "think on their own." The two remaining explorers descend and successfully land. "You're sitting there hoping that they're going to make the right decision and come forth with the right answer," notes one of the commentators, but "you can't do a thing to help them at this moment."[80]

Advocates of robotic flight envision a future in which machines achieve such high levels of autonomy that they leave humans behind. Humans design the machines and receive the benefits of discovery but do not accompany the machines on their cosmic voyages. This is an intriguing possibility. Should the cost of transporting humans through space remain persistently high while advances in technology permit increasing levels of spacecraft autonomy, the need for human space travel would diminish considerably relative to the prospects for robotic flight.

INTRIGUING PROSPECTS

Support for autonomous robotic flight exists within NASA's civil space program, especially at the Goddard Space Flight Center and the Jet Propulsion Laboratory. The official vision emanating from NASA headquarters, however, remains one of humans and robots working together. NASA's principal task, to quote the opening lines from a certain famous television series, is "to boldly go where no man has gone before."[81] The implication, clearly, is that humans do the going—with the help of their machines.

Military space efforts proceed from a very different perspective and, as a consequence, favor robotics to a much higher degree. Some of the most innovative work on robotic autonomy is being done by people addressing national security needs. Conscious of the dangers of combat, modern security doctrine proceeds from the premise that machines should go where humans dare not tread. Whereas the civil space effort promotes the advance of humans into the cosmos, military technology seeks to send no human beings where they do not absolutely need to go. The work resulting from this doctrine is strikingly unique.

As a first step, military officials are attempting to harden the automated space satellites that serve national needs. Military planners worry that the technological foundation for U.S. national security—as well as for much of the world economy—could be compromised by anti-satellite weaponry. At the present time, robotic satellites possess practically no means of repelling hostile attacks. A special government commission headed by Donald Rumsfeld shortly before he became secretary of defense examined military space capabilities and explained how such attacks could occur. A foreign power might launch a small, relatively low-cost microsatellite, setting it into an orbit that shadowed a defense or commercial machine. A microsatellite with a mass of a few kilograms would be very hard to detect until it attacked its prey. Alternatively, a foreign power or terrorist group might launch and explode a thermonuclear device a few hundred kilometers above the Earth's atmosphere. While this would produce little direct damage on the Earth's surface, the ambient radiation would be sufficient to disable nearby satellites. As an example of potential vulnerability, commission members recounted the 1998 failure of the *Galaxy IV* communications satellite. While not the result of a hostile act, this malfunction shut down "80 percent of U.S. pagers, as well as video feeds for cable and broadcast transmission, credit card authorization networks and corporate communication systems." The satellite-based global positioning system, upon which both military commanders and commercial users depend, is similarly vulnerable. Proposed countermeasures include satellites with the capability to undertake evasive maneuvers, microsatellites that can repair big satellites, and satellites that can fire on approaching missiles or killer satellites. Nearly all of these countermeasures are robotic in nature.[82]

Advances in automation such as those required for satellite defense draw on robotic technologies being developed for military forces closer to the ground. The U.S. military already uses drones—pilotless aircraft that resemble model airplanes—to locate enemy combatants and guide weapons toward them. Military planners have developed special operations robots with tractor treads and eyes that have preceded troops into caves and will be used in advance of troops engaged in urban warfare, one of the most dangerous situations that recruits can face. One year after the disastrous 2004 Mojave robot race, research teams returned with much more advanced vehicles that surprised onlookers by finishing the challenging course in impressive times.[83]

Plans for the future use of military robots are surprisingly exotic. The U.S. Air Force has already equipped its Predator RQ-1 robotic reconnaissance drone with Hellfire missiles. Viewing the battlefield through a camera pod on the nose of the twenty-eight-foot-long drone, operators thousands of miles away can aim at and destroy targets on the ground. The U.S. military is very interested in swarm technology, which can be used to sweep minefields or conduct search-and-rescue operations. Swarms consist of mobile robots, often very small, working in a fashion that mimics the behavior of insects. In one version, an armada of eight-legged "robo-lobsters" would precede troops onto beaches, roads, and fields. Each of the ten-pound robots would be programmed to locate one mine and sit next to it. A single signal would cause all robots and their adjacent mines to explode. Military swarm technology formed the basis for Michael Crichton's science fiction novel *Prey,* which describes the miscalculated effort of scientists at a government-funded research laboratory to produce a cloud of nanotechnology sensors that, like mosquitoes flying in formation, are able to use individual electronic eyes in a collective fashion to hunt. A bullet fired at the swarm has the same effect as a rifle discharged into a cloud of bugs—essentially none. Although it sounds like science fiction, military officials have funded research intended to produce insect-sized robots that can work in swarms.[84]

Such technologies would advance robotic capabilities considerably. Some view these developments as advancing the vision of humans and robots working together, in the sense of robots helping people in the field by preceding them into dangerous realms. Yet many of these technologies have the opposite effect. They widen the separation between robots and human operators—in some cases from a few feet to thousands of miles. Imagine a battlefield robot equipped with a rifle searching a cave under the guidance of an operator in another country. Applied to space, the enabling technologies might substantially reduce the need for direct human control.

Stranger still are developments arising from the juncture of robotics and biotechnology. No natural barrier in the development of robots restricts their construction solely to machine parts. As part of their sponsored research, officers from the U.S. Defense Advanced Research Projects Agency (DARPA) have encouraged researchers to link machines to biological entities such as insect parts. In one experiment, researchers used tissue from moth antennae to locate explosive devices. Such tissue may provide a more sensitive medium for identifying explosive substances than mechanical sensors. In a battlefield setting, living tissue could be attached to tiny flying robots, although other researchers have attached machine parts to whole insects. At the opposite end of that spectrum, humans are being equipped with mechanical parts that permit them to operate in robotic fashion. One proposal involves a crustacean-like exoskeleton that at once provides combatants with body armor, enhanced mobility, and sensory extension devices.[85]

Advances in biotechnology, propelled by military pressures to create more advanced robots, are accelerating the development of entities that are part organism and part machine. The implications for space travel are profound. Technological developments might permit the use of living creatures equipped with robotic parts. Alternatively, machines might become so sophisticated that they approximate human beings in their behavior. Is it alive or is it a machine? An entity so constructed would advance the vision of humans and robots exploring space together—really together. The social metaphor for future space exploration would not be Luke Skywalker and his pleasant robotic companions R2D2 and C3PO, but the Terminator. Portrayed by Austrian actor Arnold Schwarzenegger (who years later would become the governor of California), the Terminator is a powerful machine with human skin that is dispatched back through time to deter a rebellion against the race of superintelligent computers that created it.[86]

For robotic technology, military research is a powerful driver. Defense departments have far more money to spend than civil space agencies, and the necessities imposed by security missions are more immediate. Significantly, both the Russian and Chinese space programs are run by the military, permitting an easier transfer of technologies and ideas than in the United States.

For many years, combatants in the "human/machine" debate have envisioned humans and robots as separate entities. Advocates of cooperation foresee the two entities exploring space together, yet still as master and servant, owner and slave. Advocates of autonomy anticipate machines exploring space alone. Challenging these two points of view, another now appears. Technological developments may reveal a new paradigm that collapses the distinctions between biological entities and their machines.

Closely aligned with this vision is the concept of humans and robots exploring space together. From this perspective, humans and machines remain separate entities. Robots are mechanical devices; humans are made of flesh. Advocates of this point of view expect robots to precede and eventually accompany humans into the cosmos. This particular vision of robotics is socially rooted in the experience of the industrial revolution and the concepts of servitude surrounding it. It may be expensive, but advocates of human and robotic cooperation insist that this is the most effective method for exploring space.

As space exploration matured, an alternative vision of robotic flight emerged, gradually separating itself from the earlier image of humans and robots exploring together. Initially, the vision was associated with the desire to use machines (that is, robots) to do work in space too dangerous or expensive for humans to perform. Socially, it is most closely associated with the spirit of scientific discovery and the desire to understand the unknown. In this regard, it does not foresee an extensive role for humans in exploring the extraterrestrial realm but rather confines them in perpetuity to their home planet, a disheartening prospect to human flight advocates. Looking forward, the robotic perspective is associated with advances in microelectronics and computer technology. It envisions the development of machines of ever-increasing technological autonomy that possess the ability to perform activities (especially in the areas of sight and problem-solving) that increasingly approximate human skills.

The first vision (human flight) motivates the civil space effort of nations like the United States, Russia, and China that aspire to major spacefaring status; the latter vision (robots) dominates the scientific community, especially among physicists and astronomers, and motivates most military activities in space.

Within this framework, the debate between advocates of human and robotic spaceflight resolves itself into three forms: humans versus robots, and whether it is the destiny of human beings to continue the patterns of exploration and migration that have characterized the expansion of the species across the surface of the earth; humans and robots, and whether such a joint effort (the official NASA position) is more cost-effective than a program dominated by robots alone (the military experience); and finally, robots alone, and whether advances in machine intelligence and biotechnology will allow the development of autonomous entities (such as cyborgs) that do not fit the conventional definition of robots as machines under human control. In its simplest forms, the debate can be seen as one that pits humans against robots; humans and robots together against robots alone; and robots against something more advanced. We shall have much more to say about the third form—the transformational aspects of the debate—as the book progresses.

CHAPTER TWO

Human Spaceflight as Utopia

In the 1830s an astute French interpreter of United States society, Alexis de Tocqueville, observed that Americans had a "lively faith in human perfectibility" and that as a society they believed they were "a body progressing" rather than one declining or stable.[1] If anything de Tocqueville understated this belief, for the concept of America as a Utopia in the making has permeated the national ideology since before the birth of the republic. From Thomas Jefferson's stirring statement in the Declaration of Independence requiring governments to promote the unalienable rights of "Life, Liberty and the pursuit of Happiness,"[2] to Crosby, Stills, Nash, and Young's lyrical refrain "We can change the world, rearrange the world. It's dying—to get better,"[3] the quest for Utopia has been a major subtext of American life. America is a land characterized by faith in progress, belief that the future will be better than the past, and optimism about humanity's ability to better its condition.[4]

The strong utopian impulse so prevalent in American intellectual thought and political life finds its counterpart in the nation's space community. Tapping into the deep wellspring of American utopian ideals, those who argue for an aggressive human flight program base their case at a fundamental level on the positive social changes thought to result from that effort. By linking their advocacy to fundamental utopian values, spaceflight enthusiasts have been able to justify considerable funding for their grand designs despite their relatively small numbers.[5]

The ideology of utopia has had a powerful effect on the contest between human and robotic spaceflight. Utopian thought implies a process of starting over. In the American experience, it has been closely associated with the process of founding new communities beyond the corrupting influence of old societies. From the religious settlements of New England to the utopian communities of the American West, utopianism has implied migration. Extended to space, the utopian ideal likely requires the creation of new communities well removed from Earth in the same way that the New World was separated from the Old World. This necessitates human migration as the core activity. Limitation of spaceflight to robotic explorers cannot fulfill this impulse. Robots might be useful servants—even the modern equivalent of slaves making human lives luxurious—but the central goal for space travel, as utopians see it, remains humans in space. Scientific understanding gained from automated spacecraft is decidedly less important from this point of view. Advocates of robotic flight, with their frequent appeals to its scientific virtues, consistently fail to address this fundamental impulse.

The particular circumstances affecting spaceflight in the United States, notably the separation of civil from military space activities, created a forum in which advocates of space migration could press their vision without the extensive encumbrance posed by the necessities of national security. To the extent that the United States leads the world in space exploration, it also leads in defining that enterprise in utopian terms.

The few U.S. scientists who have tried to criticize the value of human spaceflight have encountered a utopian wall of opposition. In the summer of 2004, esteemed space scientist James A. Van Allen asked the poignant question, "Is human space flight obsolete?" He commented:

> My position is that it is high time for a calm debate on more fundamental questions. Does human space flight continue to serve a compelling cultural purpose and/or our national interest? Or does human space flight simply have a life of its own, without a realistic objective that is remotely commensurate with its costs? Or, indeed, is human space flight now obsolete? . . . Risk is high, cost is enormous, science is insignificant. Does anyone have a good rationale for sending humans into space?[6]

Space scientists like Van Allen place scientific discovery at the center of their spaceflight agenda. Their primary (often almost sole) purpose in exploring space is to advance science—to know the unknown, to search for extraterrestrial life, and to unlock the secrets of the cosmos. From a strictly rational perspective, in their view, machines provide a better means for completing that endeavor. When astronauts climb on board, most scientists believe, the safe return of human beings displaces science as the prime mission objective. With the exception of medical research on human subjects, the scientific community prefers robotic spacecraft for scientific purposes. They believe that human crews interfere with the process of scientific discovery while significantly increasing the cost of the individual missions.

Advocates of human flight emphasize a wholly different agenda, one in which utopian colonization displaces science as the primary purpose of spaceflight. As one observer using the pseudonym "Hans L. D. G. Starlife" noted on an Internet message board where Van Allen's arguments arose:

> Sure, if it's all about science, you can always raise these questions. But it's not, and it never has been—whatever the scientists themselves try to make us believe. The human expansion into space is about totally different things—although like many times before, it isn't fully apparent until we can see it in the light of history. . . .
>
> In a very long-range perspective, it's easy to see that these ventures simply make up the path of evolution for Human civilization, not much different from how biological evolution works. Indeed, Human space flight is precisely what Van Allen argues it's not: it does and should have a life of its own. Now is the time to once and for all separate the case for Human space flight with the case for science. These are two different agendas—both worthwhile—and sometimes crossing their paths, but having their own sets of motives and rationales![7]

To its most faithful believers, human spaceflight is about making human civilization anew, placing it in the mold of the best ideas about founding settlements beyond Earth. It is—in reality always has been—about creating a technological Utopia.

EARLY SPACE ADVOCATES AND THE UTOPIAN IDEAL

From the very beginning of public discussion about the possibilities of human space exploration, those advocating aggressive efforts have tapped a rich wellspring of utopian ideology in human history. Perhaps the most consistently eloquent argument in favor of spaceflight as a utopian ideal has been expressed in the context of human destiny. One of the earliest serious thinkers about the prospect of exploring space was the Russian Konstantin Tsiolkovsky. He repeatedly raised the utopian argument. An obscure schoolteacher in a remote part of Tsarist Russia, Tsiolkovsky stated in 1911 that "the Earth is the cradle of the mind, but we cannot remain forever in the cradle."[8] American space enthusiasts latched onto this catchy phrase to support their efforts to move humans into space in search of a better, ultimately perfect, future for humanity.[9] The powerful human destiny argument has been repeated over and over since the time of Tsiolkovsky.

The American rocket pioneer Robert H. Goddard also wrote effectively about breaking the bonds of Earth to achieve the full potential of the human spirit. A native of Worcester, Massachusetts, Goddard had a surprisingly metaphysical approach to the cause of human spaceflight. As a boy, while his family was staying at the suburban home of friends in Worcester on October 19, 1899, he climbed into an old cherry tree to prune its dead branches. Instead of pruning, he began daydreaming. As he wrote later: "It was one of the quiet, colorful afternoons of sheer beauty which we have in October in New England, and as I looked toward the fields at the east, I imagined how wonderful it would be to make some device which had even the possibility of ascending to Mars, and how it would look on a small scale, if sent up from the meadow at my feet." From that point on, Goddard enthusiastically pursued the idea of spaceflight as a necessary part of human destiny. He wrote in his diary, "existence at last seemed very purposive." In addition, October 19 became "Anniversary Day," noted in his diary as his personal holiday. He went on to tie space exploration to a utopian vision of the future. At his high school oration in 1904, he summarized his life's perspective: "It is difficult to say what is impossible, for the dream of yesterday is the hope of today and the reality of tomorrow." Later he added, "Every vision is a joke, until the first man accomplishes it."[10]

At one point Goddard envisioned intergalactic arks taking biological matter, along with human knowledge, to new homes throughout the vastness of the Milky Way. The occupants, perhaps no more than protoplasmic seeds of earthly life, would be held in some form of suspended animation, and then awakened eons later upon reaching their destination. "It has long been known," he wrote, "that protoplasm can remain inanimate for great periods of time, and can also withstand great cold, if in the granular state." There, amid the stars, earthly life forms would reestablish themselves, replicating the best that human society had to offer.[11]

Thoughts of human destiny on new and perfect worlds have been extended far beyond Goddard's basic vision in numerous ways. Clearly, one of the most influential has been the community of science fiction aficionados and futurists. While many early science fiction writers were hacks writing for a specialized but lucrative market, a few broke the boundaries of the genre and made significant contributions to public perceptions of space travel. Among the most influential were Robert A. Heinlein, Isaac Asimov, Arthur C. Clarke, and Ray Bradbury, all taking pains to make their science fiction believable as reality, exciting as works of literature, and reflective of possibilities that might take place in space. They found a ready audience in the United States, tapping into the utopian beliefs permeating the nation, as Americans wrestled with the prospects of life in a modern technological world. All told their stories within the context of a galactic politics easily recognizable to persons familiar with the superpower conflicts of their time, in which their central characters triumphed over the dangers of technology, often with the help of benevolent alien beings. Often, the quest for Utopia served as a centerpiece for the story.[12]

One particularly powerful explanation of humanity's utopian destiny is expressed in Arthur C. Clarke's *Childhood's End,* a book supposedly written for young people, although few adults have found the contents to be in any way immature. Clarke posits the possibility that small changes in the genetic code of certain primates thousands of years ago transformed them from primitive beasts into beings capable of producing great art, wonderful music, and timeless literature. *Childhood's End* raises the question: Is there yet another key residing inside human beings that will transform the species once again? In the novel, a group of galactic Overlords, alien visitors to Earth, facilitate this transformation. While some might question Clarke's theme of species perfection given the violence and hatred in the world, his vision of a perfect world helped to energize the community of space exploration advocates for more than fifty years.[13]

No one has been more consistent in clamoring for spaceflight as the means to achieving Utopia than Ray Bradbury. In a literary career spanning more than a half-century, he has quipped his way to stardom as a shaper of the future. He told students at Brown University in 1995, in a public statement of his utopian beliefs, that they must make the world a better place. "Go to the edge of the cliff and jump off," he said. "Build your wings on the way down."[14] He added that they had a responsibility to "recreate the world in your own image and make it better for your having been here." His belief in the destiny of humanity to become a better species by exploring space is best summarized in his comment to the National School Board Association in 1995: "We are beholden to give back to the universe. . . . If we make landfall on another star system, we become immortal."[15]

Bradbury eloquently agitated for Utopia in his first major book, *The Martian Chronicles,* published in 1950. This book describes the first attempts of earthlings to explore and colonize Mars; the constant thwarting of their efforts by gentle, telepathic Martians; the eventual colonization; and finally the extraterrestrial effect of a massive nuclear war on Earth. As much a work of social criticism as science fiction, *The Martian Chronicles* is an allegory for the tragic moral blindness of humankind and reflects some of the prevailing anxieties of America in the early atomic age of the 1950s. There is a special charm to Bradbury's style in this work that exudes both sentimental nostalgia and powerful idealism.

When a small group of humans escape war-torn Earth toward the end of the book, Bradbury describes how they succeed in cutting themselves off from their former home and creating a new and perfect society on Mars. Appropriately, they call themselves *Martians*. While hope for a better life on Mars proves too optimistic for early pioneers, it bears fruit among these sadder, wiser humans—no longer colonists but true immigrants. While the native Martians are gone, these newcomers willingly embrace Mars and respect its heritage. Bradbury believed the same would become true of humanity as a whole when the species heeded the clarion call of Mars and worked to create a new and perfect society.[16]

Bradbury's urge to flee from perceived destruction toward a better life is longstanding in human history. How far to flee, however, has changed over time, as the technological reach of civilization changes. For centuries, limited by the ability to travel great distances, those who sought the perfect society journeyed into the "wilderness" to seek their "promised land." With the development of transoceanic transportation in the fifteenth century, utopians could place an entire ocean between their settlements and the status quo ante. The Puritans who journeyed to New England, the Quakers who founded Pennsylvania, the Moravians who came from Germany to western Pennsylvania and the mountains of North Carolina, the Hutterites, the Huguenots, and a host of others founded societies that encompassed their fundamental ideals. Later on, the Mormons, in their quest for Zion, separated themselves from mainstream society and established their own state in the American West. All of these groups sought to put thousands of miles between the established order and their utopian experiments. Extrapolating this into the era of space travel, it makes sense to extend this distance to planetary proportions. Bradbury's characters in *The Martian Chronicles* do just that, and wisely so, since in Bradbury's novel the earthlings destroy themselves in a nuclear war.[17]

Equally powerful were the several science fiction films that addressed the creation of utopias with the aid of alien and (in some cases) robotic intelligence. In 1951, film producer-director Robert Wise released *The Day the Earth Stood Still*. In this influential film, the benevolent alien Klaatu, supported by a powerful robot, warns the leaders of Earth's warring nations to control their aggressive arms race or suffer destruction for the good of the universe. Wise stressed the importance of this film as a commentary about humanity and the need to make a better world. "The whole purpose of it," said Wise, "was for Klaatu to deliver that warning at the end. I feel very strongly in favor of what the movie says. It's very much of a forerunner in its warning about atomic warfare, and it shows that we must all learn to get along together."[18] This and other films excited the public with the possibility that spaceflight, exploration, alien contact, and robotic intelligence could offer positive developments for humankind. It is often easy to forget that these sophisticated visions of space travel and its transforming effects occurred before the first artificial satellite had even pierced the skies.

WERNHER VON BRAUN, ARYAN UTOPIA, AND SPACEFLIGHT

Wernher von Braun, the single most important promoter of America's space effort in the 1950s and 1960s, captured the essence of American utopianism and used it to justify an aggressive space exploration agenda. Although he was a German immigrant to the United States after World War II—or perhaps precisely for that reason—von Braun had a remarkable understanding of American culture. He spoke often of "the challenge of the century," which he envisioned as the struggle to continue American exploration and settlement and the creation of a perfect society in a new land. "For more than 400 years the history of this nation has been crammed with adventure and excitement and marked by expansion," he said of his adopted home. "Compared with Europe, Africa, and Asia, America was the New World. Its pioneer settlers were daring, energetic, and self-reliant. They were challenged by the promise of unexplored and unsettled territory, and stimulated by the urge to conquer these vast new frontiers." Americans needed the space frontier, both physically and spiritually, von Braun insisted, and greater efforts in moving beyond the Earth would lead to a society in which "right relationships" prevailed.[19] He believed this was "as inevitable as the rising of the sun; man has already poked his nose into space and he is not likely to pull it back. . . . There can be no thought of finishing, for aiming at the stars—both literally and figuratively—is the work of generations, and no matter how much progress one makes, there is always the thrill of just beginning."[20]

Von Braun never wavered in his commitment to creating a better and ultimately perfect society in space. In a 1976 speech to the National Space Institute one year before his death, von Braun predicted a bright future for humanity if it embarked on the high frontier of space. He said that space would "offer new places to live—a chance to organize a new interplanetary society, and make fresh beginnings."[21]

There was a dark side to this agenda as well. A critical component of his message suggests dystopia as much as Utopia. Dystopia is Utopia's evil twin, often characterized by tyranny and oppression. Unlike Willy Ley, another influential space advocate of that time, von Braun did not flee the German Nazi regime when it came to power. He remained in Germany through World War II, joined the Nazi Party, and oversaw the design of intermediate-range ballistic missiles constructed with the use of forced labor from concentration camps. The less savory aspects of the German rocketeers are often overlooked in the telling of the space exploration tale.[22]

Most people are familiar at a cursory level with the Aryan policies adopted by the Nazi regime, notably the forced removal and extermination of Jews and other minorities. Less well known are the utopian roots of those policies and their relationship to space travel. The general Aryan myth advanced the belief that the Nordic peoples of Europe descended from an ancient and highly creative civilization. In one version of the myth, the civilization occupied the fabled nation of Atlantis, located at the northern reaches of the world, near present-day Greenland and Iceland. Astonishingly, the more mystical proponents of this tale suggested that the blond-haired, blue-eyed Aryans were not related to *Homo sapiens* at all but were instead extraterrestrials from another star system who had founded a colony on Earth. Their settlement, Atlantis, was destroyed in a galactic war, and the surviving Aryans fled to the Middle East, whence the Nordic races of Europe were thought to originate. Nazism represented an attempt to separate the Aryan line from the other peoples of the Earth and restore it to its previous levels of creativity. In the mystical version, the Aryan movement sought the recreation of a racially pure community of extraterrestrial beings.[23]

No evidence suggests that von Braun accepted this extraterrestrial tale. Though he longed for extraterrestrial travel, he was a technical rationalist more interested in engineering principles and rocketry than in Aryan mythology.[24] Whatever the case, von Braun's message of new beginnings resonated with a significant number of Americans after he arrived in the United States. Yet it found little support among the senior officials of the U.S. government who possessed the funds to accomplish it. Arguments about human destiny, the creation of new societies on other planets, and the full achievement of human potential rarely evoke political support from governmental leaders. While officials in the Eisenhower administration appreciated the engineering skills of von Braun and his German rocket team, they found his enthusiasm for spaceflight misguided and naive. For his part, President Eisenhower had no interest in the ethereal objectives of human exploration. At a February 1958 legislative leadership meeting, he expressed his basic lack of interest in space exploration. He would rather have a good intermediate-range ballistic missile, he confessed, than put humans on the lunar surface. "We didn't have any enemies on the moon," he noted.[25]

In October 1952, von Braun, then technical director of the Army Ordnance Guided Missile Development Group, spoke at a Hayden Planetarium symposium on the future of space travel. He opened his speech by stating that "the conquest of space represents the outstanding challenge to science and technology of the age in which we live." Space was there to be explored and settled, he insisted, and the United States should lead the way. At the time, von Braun was working on a detailed plan to send humans to Mars. Such a journey was an integral part of human destiny, he felt, and would transform humanity's place in the universe. He envisioned a time when an open, boundless condition would exist for all humanity, in which justice would rule and all would have enough of the necessities of life. Space exploration could bring that, von Braun believed, helping people to live together in greater harmony than ever before by ending the need to compete for resources. Von Braun stopped just short of criticizing the U.S. government, for which he was then working, for failing to provide the visionary leadership necessary to realize these dreams.

Also speaking out was Milton W. Rosen, director of the Naval Research Laboratory's Viking rocket project. Rosen characterized spaceflight as difficult and expensive work, unlikely to progress rapidly. He attached no special significance to space exploration beyond the need to develop a launch capability to orbit a few satellites for scientific purposes.[26]

Most of those attending the Hayden Planetarium conference probably agreed with Rosen. *Time* magazine even ran a lengthy feature story on the alternative visions of space exploration offered by von Braun and Rosen. The story characterized von Braun as "the major prophet and hero (or wild propagandist, some scientists suspect) of space travel" whose visionary goals were laudable but ultimately unachievable, quixotic.[27] While von Braun's utopian vision of spaceflight gained few adherents among scientists, it played well among the public at large. Those who embraced it undertook an almost religious pilgrimage to convince Americans that it could come true. In articles for major periodicals, on television and radio, at the Disneyland theme park, and at countless public appearances, von Braun enthusiastically espoused his dreams for a spacefaring people who would transform the human race.[28]

Among the scientists and engineers who worked for NASA in its formative years, only von Braun's name is well known to the American public. While some scientists and engineers criticized von Braun for his blatant promotion of spaceflight (and himself), his efforts changed public attitudes toward spaceflight. Media observers noted the favorable public response to three Disney television programs on spaceflight and stated that "the thinking of the best scientific minds working on space projects today" went into them—meaning von Braun's—"making the picture[s] more fact than fantasy."[29]

Despite all attempts to do so, including appeals to national security, von Braun and the growing number of spaceflight enthusiasts failed to gain the favor of the Eisenhower administration.[30] One critic complained that von Braun "is trying to sell the U.S. a spaceflight project disguised as a means of dominating the world."[31] Eisenhower told senior aides that he was "getting a little weary of Von Braun's publicity seeking."[32]

As often occurs in public affairs, a precipitating event energized the creation of a larger coalition that forced policy change. The Soviet launches of *Sputniks 1* and *2* and the spectacular failure of the U.S. rocket *Vanguard TV3* launch in late 1957 created a pro-space majority of sufficient strength to overcome the reluctance of space conservatives. The majority consisted of pro-space true believers, scientists interested in space research, members of the armed services who wanted to use space for military ends, corporate personalities and industrial engineers hoping to expand the aerospace industry and obtain government contracts, and politicians such as Lyndon Johnson and John F. Kennedy hoping to benefit from the symbolic resonances of the space race.[33]

Eisenhower never intended to create a civilian agency devoted to such far-reaching utopian ends. His purpose for a separate civilian agency was to ensure that space would be "used only for peaceful purposes," a somewhat disingenuous statement, since military space spending in the United States has historically equaled or outpaced civil space spending.[34] To Eisenhower, "peaceful purposes" meant that the United States and other nations could use space for military reconnaissance satellites without the threat of hostile attack. The existence of a civil space agency with scientific activities of international interest reinforced Eisenhower's position that the United States be allowed unencumbered access to space. Eisenhower had no interest in flights to the Moon or in Martian colonies, but the creation of NASA inadvertently provided the most powerful forum in the world for people with those dreams.[35]

The Sputnik crises of 1957 and 1958 yielded one of the most sweeping governmental reorganizations undertaken at the federal level between the New Deal and the security overhaul after the September 11, 2001, terrorist attacks in New York and Washington. As a direct result of this crisis atmosphere, space advocates gained a civil agency into which they could carry their vision. The agency's

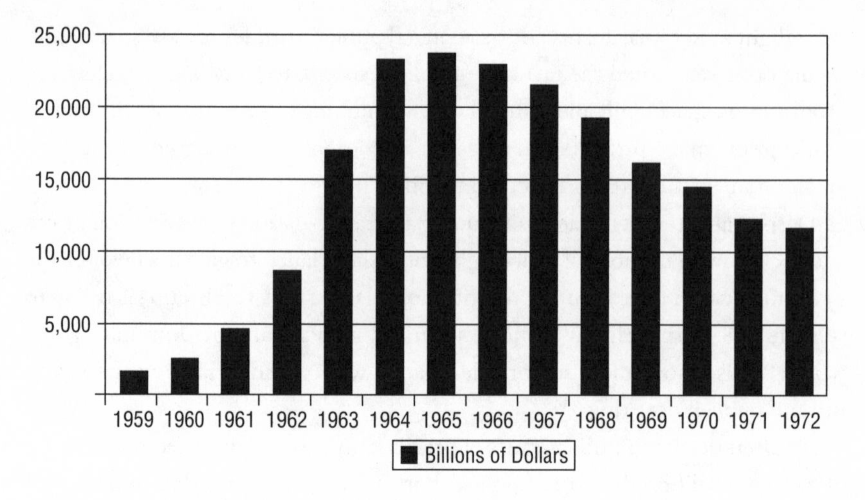

NASA Budget Adjusted for Inflation (2004 billions of dollars)
SOURCE: *Aeronautics and Space Report of the President, Fiscal Year 1998* (Washington, D.C.: NASA Annual Report, 1999), appendixes E-1a and E-1b.

congressional mandate declared that NASA's undertakings "should be devoted to peaceful purposes for the benefit of all mankind."[36] Utopian visions of space exploration emerged in numerous ways during the agency's early history. In early 1958, von Braun and his staff cobbled together a strategic plan that foresaw a space station, a human expedition to the Moon, a permanent lunar base, and a human expedition to Mars—all within twenty years. T. Keith Glennan, the first NASA administrator, contended with the space utopians from his first days in office, complaining on several occasions that "space cadets" were everywhere and that he had to fight a rearguard action to keep NASA in sync with the president's directives. On July 11, 1960, for instance, Glennan confided in his diary that a set of briefings oriented toward program planning "give every evidence of making NASA a space cadet organization. This will have to be corrected."[37] The aspirations of space utopians were further advanced by President John F. Kennedy's commitment on May 25, 1961, to the goal of a human lunar landing by the end of the decade. Political scientist John M. Logsdon appropriately noted the utopian aspect of Project Apollo: "The politics of the moment had become linked with the dreams of centuries."[38]

By tapping into the deep wellspring of utopian ideals present in society, wedding those ideals to Cold War fears and competition, and asserting their vision of spaceflight as a positive good for all humankind, the first generation of space advocates enjoyed remarkable success in securing their agenda. The United States spent $46.8 billion on NASA during the period between 1959 and 1972, the last year of Project Apollo. Expressed in 2006 dollars, this amounted to more than $200 billion, hardly an insignificant sum for completely discretionary spending.[39] The willingness of the American public to invest heavily in the national space effort occurred in no small part because of the symbolic underpinnings of spaceflight as a part of human destiny and of the betterment of society. Once triggered by competition with the Soviet Union, the funding has flowed consistently at slightly less than 1 percent of the federal budget.

UTOPIANISM AND THE RISE OF THE PRO-SPACE MOVEMENT

With the successful completion of Project Apollo, the remarkable national commitment to land an American on the Moon, space utopians believed that humankind was on the verge of a golden age in which anything could be accomplished. Apollo raised the hopes of those dreaming of a great human Utopia in space. Its transcendental qualities were not lost on those who believed that the human race could eventually attain this end. As Senator Abraham Ribicoff gushed in 1969, "If men can visit the Moon—and now we know they can—then there is no limit to what else we can do. Perhaps that is the real meaning of Apollo 11."[40]

Movement into space, first with expeditions of discovery and later with colonies, offered an opportunity for humanity to rise above the issues and inequities that divided it. Elements of society could move outward and start anew. Apollo had shown that it was possible. The undertaking suggested that America had both the capability and the wherewithal to accomplish truly astounding goals. In the minds of space enthusiasts, all they needed was the will.

In 1969 space activists enthusiastically embraced President Nixon's decision to establish a Space Task Group (STG) to secure a "definitive recommendation on the direction which the U.S. space program should take in the post-Apollo period."[41] Along with Vice President Spiro Agnew, science adviser Lee DuBridge, and Air Force Secretary Robert Seamans, NASA Administrator Thomas O. Paine served on the four-person commission. Paine was an unabashed space utopian, in his own words a "space buccaneer." Under his influence, the Space Task Group produced a far-reaching report of remarkable optimism. Their recommended post-Apollo space program included a lunar base, a mission to Mars, and a space station serviced by a reusable Space Shuttle—all to be established within twenty years. Like his namesake of the American Revolutionary Era who wrote, "We hold it in our power to begin the world anew," Paine carried an overtly utopian vision to the Space Task Group. Its report reflected those ideals, although White House aides insisted that the group revise its conclusions and recommendations "so that the president not be put in the position of having only ambitious options from which to choose."[42] Wernher von Braun left his post as director of the NASA's Marshall Space Flight Center and moved to Washington, D.C., where he lobbied extensively for the vision contained in the report.[43]

The decision in 1972—to build the Space Shuttle and forego everything else—devastated space utopians.[44] What a letdown! From the surge of excitement over future trips to the Moon to what was in essence a "space truck" hampered the enthusiasm of exploration advocates. Even though the shuttle was an integral part of the von Braun vision, its purpose was overtly utilitarian and non-utopian. To see it elevated from a component in a much broader vision to become the end-all of the human spaceflight endeavor was demoralizing. Space utopians believed that the tide of history had turned in their favor in the 1960s, only to see it reversed by what they viewed as the evil Richard Nixon. The advocates of a strong space program accordingly consigned Nixon, more than any other political leader, to a special place in intergalactic purgatory. They did not do so because of his violation of the Constitution, the Watergate affair, or the war in Vietnam. They did so because he refused to endorse the Space Task Group report that had expressed their point of view of a positive future for humankind so eloquently. Never mind the very real pressures the president faced in the late 1960s and early 1970s; in the eyes of space utopians he failed to sustain the expansive vision that had sparked Apollo.

Space utopians sought to overcome what they viewed as the poor state of human spaceflight in the early 1970s by organizing political action groups, known collectively as the "pro-space movement," to lobby for increased funding.[45] At first, broader elements of the space coalition forged in the Sputnik era welcomed the entrants. But scientists, military officials, aerospace industrialists, NASA executives, and pro-space politicians soon recoiled from the more radical notions contained in the movement's ideology. They ridiculed the wilder elements in the pro-space movement, criticizing their "weirdness," their affectation of wearing *Star Trek* uniforms, and their occasional preoccupation with unidentified flying objects.[46] Even more troubling for the traditional coalition was the impulse to move in libertarian, anti-government—and by extension anti-NASA—directions.[47]

The formal beginnings of the modern "pro-space movement"—really an extension of the ad hoc efforts to gain public support, earlier led by Wernher von Braun and others—might best be traced to the June 1970 formation of the Committee for the Future (CFF), a small group of space activists, dreamers, and misfits. Meeting in the Lakeville, Connecticut, home of Barbara Marx Hubbard, daughter of "toy king" Louis Marx, and her husband, artist-philosopher Earl Hubbard, they proposed the creation of a lunar colony. The group unabashedly offered the colony as a great utopian experiment in which humanity, free from the constraints of everyday society, could create a perfect community. The lunar colony, in turn, would spur similar undertakings on other worlds. The CFF's charter clearly voiced these utopian ideals:

> Earth-bound history has ended. Universal history has begun. Mankind has been born into an environment of immeasurable possibilities. We, the Committee for the Future, believe that the long-range goal for Mankind should be to seek and settle new worlds. To survive and realize the common aspiration of all people for a future of unlimited opportunity, this generation must begin now to find the means of converting the planets into life support systems for the race of Men.

The group offered shares in the lunar colony to millions of investors, immediately creating a paying constituency for efforts to lobby the U.S. Congress for additional funds. "A challenge of this magnitude," the group concluded, "can emancipate the genius of Man."[48]

Leaders of the committee convinced Representative Olin Teague, a lion within the Democratic Party and a longtime supporter of Project Apollo, to sponsor a resolution calling for a feasibility study. NASA executives, along with leaders from the aerospace industry and the science community, opposed the resolution, fearing that it might jeopardize existing plans. The resolution promptly died. CFF leaders rewrote the legislation, proposing a "citizens in space" mission in low-Earth orbit called "Mankind One." NASA executives opposed that as well, and it met the same fate.

Although NASA executives opposed the CFF's political initiatives, many within the agency agreed with its ideology. Barbara Hubbard wrote about a meeting with Christopher C. Kraft, director of the Manned Spacecraft Center (renamed the Johnson Space Center in 1973) in which he told her, "This step into the universe is a religion, and I'm a member of it." Hubbard was deeply troubled, however, by NASA's official reaction to the CFF's proposals. She wrote, "The corporate decision of NASA as a government agency was less responsive than the decision of any of its individual members."[49] Out of these incidents grew a deep-seated wariness within the pro-space movement toward the government as a whole and NASA in particular to "do the right thing" in opening the space frontier to utopian experimentation.

Organizing symposia, called "*syn*ergistic *con*vergences," or SYNCONs, and publishing hopeful literature about a utopian future in space, members of the CFF converted a sizable group of mostly young people to their cause. Space groupies arrived from many states and countries to participate in the SYNCONs. They energized a base of activists who firmly believed that only through space settlement would the human destiny of a perfect society be realized.[50] While the CFF ceased to exist as a separate organization in the mid-1970s, Barbara Hubbard continued her commitment to a utopian future in space.[51] This 1995 comment explains her position on the promise of space and her belief that governments had hindered its attainment:

> Perhaps the last great function of existing centralized power, as in the United States and the Soviet Union, is the establishment of the first productive foothold in the Universe. After the step, the higher frontier will open to cooperative, free enterprise and self-selected groups of pioneers. The opportunity for new life styles, new wealth, new knowledge gained by highly motivated people beyond the planet will reduce the control of Earth-bound governments everywhere. . . . Imagine a world in which individual men, women and children are liberated from the past phase of creature human functions or maximum reproduction and survival tasks. Our daily survival needs for food and shelter are provided with minimum human effort. The productive capacity of a universal species, utilizing intelligent machines and renewable resources in an Earth-space environment, is astronomical.[52]

Most assuredly, NASA and industrial space professionals viewed Barbara Hubbard and the CFF, both then and later, as impractical, "wacky." Some considered her a cult leader. Yet the CFF represented a strain of spaceflight enthusiasm that they could not ignore—one that emphasized individual activism and outright utopianism. This strain grew in respectability when it was taken up by Princeton professor Gerard K. O'Neill.

O'NEILL, UTOPIA, AND THE CHALLENGE OF SPACE COLONIZATION

Gerard O'Neill really had three careers. As an experimental physicist at Stanford and Princeton University, he helped to develop the technology necessary for high-energy particle accelerators. As an inventor and entrepreneur, he founded several companies that developed new commercial technologies. As a teacher and writer, he explored the possibilities of human settlement on the Moon and in space colonies. In this last capacity he left an indelible mark on the utopian space movement by promoting the basic ideals of the CFF to a receptive audience seeking the creation of a better society while managing to avoid the group's distasteful cult-like qualities.

Infected with the enthusiasm of Apollo, O'Neill undertook a set of studies aimed at answering the question, "Is a planetary surface the right place for an expanding technological civilization?" He found that the possibilities for human colonies in free space seemed limitless, as he calculated the technical issues of energy, land area, size, shape, atmosphere, gravitation, and sunlight necessary to sustain a colony in an artificial living environment. Rather than relocate humans to the surfaces of other planets, O'Neill proposed resettling them on the interiors of gigantic cylinders roughly sixteen miles in length. These would hold a breathable atmosphere and the remaining ingredients necessary to sustain crops and life, and they would rotate to provide artificial gravity. In O'Neill's scheme, people would reside on the inner edge of the cylinder's outer rim, in other words with their feet pointing out. While the human race might eventually build millions of these space colonies, each settlement would of necessity be an independent biosphere, with trees and lakes and, along each colony's inner rim, blue skies spotted with clouds. All oxygen, water, waste, and other materials could be recycled endlessly. Animals and plants endangered on Earth would thrive on these cosmic arks; insect pests would be left behind. Solar power, directed into each colony by huge mirrors, would provide a constant source of nonpolluting energy.

O'Neill suggested positioning the colonies at various points near the Earth, Sun, and Moon where the gravitational fields are equalized, known as Lagrange Points, well separated from the problems of the parent society. O'Neill viewed the colony proposal as eminently practical. Residents of the colonies would build solar-power satellites from the raw materials of the moons and asteroids and beam energy collected from the Sun to an energy-hungry Earth, giving the space colonies an economic rationale.[53]

O'Neill's bold vision catapulted him into the space community spotlight and prompted a collective swoon from those interested in the goals of the Committee for the Future but repelled by the group's zaniness. In 1975, O'Neill enthusiasts formed the L-5 Society, its title a reference to the Lagrange Point where they planned to place their first colony. They adopted the slogan, "L-5 in 1995." A particularly attractive group of space activists, one of their members wittily opined that their ultimate goal was to "disband the Society in a mass meeting at L-5."[54] The space settlement mission also received a major boost from numerous science fiction writers and science reporters, among them Arthur C. Clarke, who helped to popularize O'Neill's concept.[55] The strongly utopian impulse present in the O'Neill movement found voice in the words of aerospace writer T. A. Heppenheimer.

> On Earth it is difficult for . . . people to form new nations or region[s] for themselves. But in space it will become easy for ethnic or religious groups, and for many others as well to set up their own colonies. . . . Those who wish to found experimental communities, to try new social forms and practices, will have the opportunity to strike out into the wilderness and establish their ideals in cities in space.[56]

O'Neill's vision found just as receptive an audience in many quarters of NASA as it did in the larger pro-space movement. He received funding from NASA's Advanced Programs Office, helping him to develop the concept more fully. The sum was small—only twenty-five thousand dollars—but the symbolism was substantial. Senior NASA officials such as Administrator James C. Fletcher and Hans Mark, director of NASA's Ames Research Center, encouraged his efforts, while others discredited his vision as hopelessly utopian.[57]

NASA officials took O'Neill's ideas seriously enough to convene a group of scientists, engineers, economists, and sociologists to review the idea of space colonization. The group met in the summer of 1975 at the Ames Research Center near San Francisco. Surprisingly, they found enough substance in the scheme to recommend further development. Although budget estimates of $100 billion accompanied any colonization project, the authors of the study concluded that "in contrast to Apollo, it appears that space colonization may be a paying proposition." For them, it offered "a way out from the sense of closure and of limits which is now oppressive to many people on Earth." The study recommended an international project led by the United States that would result in the establishment of a space colony at L-5. Most importantly—and decidedly utopian in expression—the study concluded:

> The possibility of cooperation among nations, in an enterprise which can yield new wealth for all rather than a conflict over the remaining resources of the Earth, may be far more important in the long run than the immediate return of energy to the Earth. So, too, may be the sense of hope and of new options and

opportunities which space colonization can bring to a world which has lost its frontiers.[58]

O'Neill publicized these findings exhaustively, but the political will for a new space effort remained at a low point through the late 1970s. Both O'Neill and his supporters criticized NASA for not helping turn their utopian dreams into reality. Distrust of NASA's ability to promote such initiatives grew.[59]

Following the two space colonization summer studies of 1975 and 1976, O'Neill founded the Space Studies Institute at Princeton University. Through the institute, he hoped to create small groups that could energize the space colonization movement. He wanted the groups, as Freeman Dyson wrote, to develop the "tools of space exploration independently of governments and to prove that private groups could get things done enormously cheaper and quicker than government bureaucracies." Thereafter, he never strayed from a belief that the private sector would be the only organization capable of opening the space frontier. Indeed, in 1985 he was appointed to the National Commission on Space in no small part because of his expansive vision of space colonization and his increasingly dogged commitment to space entrepreneurship. Dyson continued:

> I was privileged to be a close friend of two great men, Richard Feynman and Gerard O'Neill. I was often struck by the deep similarity of their characters, in spite of many superficial differences. Both were indefatigable workers, taking infinite trouble to get the details right. Both were effective and enthusiastic teachers. Both were accomplished showmen, good at handling a crowd. Both had good rapport with ordinary people and abhorred pedants and snobs. Both were uncompromisingly honest. Both were outsiders in their own profession, unwilling to swim with the stream. Both stood up against the established wisdom and were proved right. Both fought a fatal illness for the last seven years of their lives. Both had spirits that grew stronger as their physical strength decayed.

Dyson credited O'Neill with rescuing space colonization from the backwater of crackpots and making it the respectable centerpiece of a hopeful future for humanity.[60]

O'Neill was a visionary. Yet many of his ideas were strikingly naive and politically untenable. His vision of space colonies, while enormously attractive for some people, had little practical effect. Previous utopian movements founded actual settlements. In many respects, O'Neill's schemes constituted a form of denial, a means by which supporters could exchange the practical challenges of spaceflight and earthly difficulties for cosmic dreams of unlimited impracticality.[61]

A PLETHORA OF PRO-SPACE ORGANIZATIONS

The number of pro-space organizations grew exponentially in the aftermath of the space colonization craze, each with its particular constituency and twist of utopian ideology. One of the most successful has been the National Space Society. Established in 1987, the National Space Society arose as a result of the merger of the von Braun–inspired National Space Institute and O'Neill's L-5 Society. A moderate organization, its thirty thousand members are committed to the creation of "thriving communities beyond earth." It promotes "change in social, technical, economic, and political conditions to advance the day when people will live and work in space."[62]

Likewise, the Planetary Society, founded in 1980 by Carl Sagan, Bruce Murray, and Louis Friedman, is a moderate politically active organization that encourages exploration of the Solar System and the search for extraterrestrial beings. Emphasizing its nongovernmental status, it is funded by dues and donations, and it claims more than a hundred thousand members from over 140 countries, making it the largest space interest group on Earth. In its quest for aggressive exploration and strange life forms, the Planetary Society explicitly links today's space explorers, who sail into the cosmic expanse, with the captains of the Renaissance who sailed Earth's uncharted oceans.[63]

Two additional pro-space groups are even more forthrightly utopian in their outlook. The Space Frontier Foundation was established in 1988 by a group of exploration advocates led by O'Neill acolyte Rick Tumlinson. Disturbed with what they perceived as a too conservative, pro-NASA outlook in the National Space Society and Planetary Society, Tumlinson's cohort formed a new organization. The group sought to push government planners toward policies that encouraged private entities to develop space. The Space Frontier Foundation asserts three beliefs about the movement into space:

- Its members know, from research done since Apollo (primarily by Gerard K. O'Neill's Space Studies Institute), that it is technically feasible to realize the shared vision of large-scale industrialization and settlement of the inner Solar System within one or two generations.
- They know this is not happening (and cannot happen) under the exclusive, status-quo oriented, centrally planned U.S. government space program.

- They believe that it is their responsibility to replace the existing bureaucratic program with an inclusive, entrepreneurial, frontier-opening enterprise, primarily by working from the outside to promote a radical reform of U.S. space policy.

To that end, the Space Frontier Foundation has worked to "convert the image held by many young people that the future will be worse than the present, and [to] reject the idea that the world's greatest moments are in its past."[64]

Tumlinson and his organization believe deeply in the destiny of humans to expand beyond the Earth. He also contends that "the current elitist nationalist space programs are not creating the conditions for a free and open frontier in space and must be replaced with ones that will." They proudly announce: "We are considered to be the most radical legitimate space group in the world." Tumlinson's utopian ideology plays out in the Space Frontier Foundation at every level. He sees civilization as being at a crossroads, with one path limiting growth and opportunity, leading to environmental degradation and, ultimately, human extinction. This path leads to a point at which exploration of the Solar System ends, thereby preventing human settlement of the Moon, Mars, or anywhere else. The other path, one in which humanity begins to live permanently off this planet, promises "limitless growth, an environmentally pristine Earth, and an open and free frontier in Space." The Foundation asserts:

> We see this as a time of choices and change. Not incremental change but fundamental, revolutionary change. Change every bit as important as the Copernican revolution of the 16th century. At that time we learned that we didn't sit at the physical center of the Universe, that we had until then had an inflated view of our importance in that Universe. Now we are ready to take our place in our Galaxy as a Space Faring civilization. Or are we? Our goal is to educate you about the issues facing the human breakout into space. Welcome to the Revolution.[65]

Tumlinson's ideology denies the permanent possibility of individual freedom on this resource-constrained planet. The planet's resources are managed to sustain the status quo, he insists, so that as resources shrink, so will individual freedoms.

> Ultimately, nearly anything you want to do in a "sustainable" world will be something someone else cannot—and that will mean limits. Limits to when and where and how you travel, how much you consume, the size of your home, the foods you eat, the job where you work, even how long you are allowed to live. If the rest of the world is to become more wealthy in such a system, consuming more, you

> will be forced to consume less. Equilibrium will be the goal of the state and individual freedoms will become ever more expendable. Yet earth's population continues to grow.[66]

Tumlinson repeatedly presents outer space as a place where we can break this cycle. By exploring and settling space, people will create the best society ever experienced in human history. "The children of Earth will have hefted themselves out of the cradle," he argues, "working together based on their own survival and self interests, while expressing their own individuality through the millions of different activities of a new human society. This is what frontiers are all about."[67]

Utopian philosophy also guides the Mars Society, founded in 1998 by Robert Zubrin, certainly the most persistent and eloquent advocate for extending a human presence onto the red planet. Zubrin and his society incessantly employ frontier and utopian language to promote their goals. To them, the exploration and settlement of Mars is a human destiny, permitting a limitless future and an unsurpassable opportunity to remake society.

Zubrin urges a strategy called "Mars Direct" as a means of reaching the red planet quickly and inexpensively. He insists that the first humans to go to Mars will "live off the land," just like earlier explorers on their travels. If Meriwether Lewis and William Clark had been forced to carry all their fuel and supplies across the continent, he argues, they would never have reached Oregon. They would have been stuck forever in Missouri. In a similar fashion, Zubrin insists that the first humans to Mars will extract fuel and consumables from the Martian environment, and he built a machine to show how it could be done. He posits that a Mars Direct mission could be accomplished in less than a decade at a fraction of the cost of Project Apollo.

Most important, Zubrin believes that the exploration and colonization of Mars will inspire new technology, new science, vast creativity, and individual inventiveness among the people who undertake it. The chief reasons for going to Mars, he argues, flow from the beneficial effects of new frontiers:

> We see around us now an ever more apparent loss of vigor of American society: increasing fixity of the power structure and bureaucratization of all levels of society; impotence of political institutions to carry off great projects; the cancerous proliferation of regulations affecting all aspects of public, private and commercial life; the spread of irrationalism; the banalization of popular culture; the loss of willingness by individuals to take risks, to fend or think for themselves; economic

> stagnation and decline; the deceleration of the rate of technological innovation and a loss of belief in the idea of progress itself. Everywhere you look, the writing is on the wall.

The settlement of Mars offers the single most viable opportunity to overcome these constraints, he believes. "Without a frontier from which to breathe life," he concludes, "the spirit that gave rise to the progressive humanistic culture that America has offered to the world for the past several centuries is fading." If humanity fails to respond to the Martian frontier, Zubrin contends, it is lost.[68]

Zubrin's emphatic plea on behalf of the space frontier squares well with the vision of a perfect society on a new and pristine planet. He sees stagnation everywhere about him and sounds the alarm that humanity must move out now or face doom. The central question asked by John Winthrop and other British Puritans—how does one live a righteous life in an unrighteous land?—comes to mind when one considers Zubrin's vision. His recommendations are not unlike the answer to the Puritan question: No one can, so the only solution is to leave now. Zubrin's warning of decline matches that of libertarian science fiction writer Robert A. Heinlein. His character in *Time Enough for Love,* Lazarus Long, opines, "The two highest achievements of the human mind are the twin concepts of 'loyalty' and 'duty.' Whenever these twin concepts fall into disrepute—get out of there fast! You may possibly save yourself, but it is too late to save that society. It is doomed."[69] In Zubrin's ideology, the process is already under way, and only space colonization can remedy it. The positive forces of the frontier on an unsullied planet, in his view, will enable the creation of a much better society than has come before.

While more circumspect in his criticisms than Rick Tumlinson, Zubrin also doubts the ability of large government agencies like NASA to implement his agenda. While NASA and its aerospace partners may be motivated to send a research expedition to Mars, they are too sluggish to mount an expensive effort leading to its colonization, by far the most attractive objective in the Solar System. Indeed, he has repeatedly criticized NASA as a government organization that has lost touch. Only through intense prodding from the public will anything happen. And he believes his duty is to continue prodding. "It is a New World," he says of Mars, "filled with history waiting to be made by a new and youthful branch of human civilization that is waiting to be born."[70] With proper planning, he believes, humans could be establishing permanent, self-sustaining outposts on Mars by the middle of the twenty-first century.

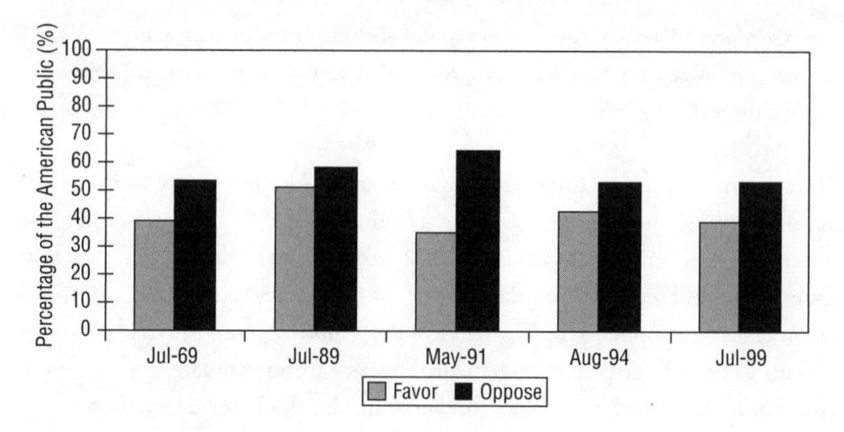

Public Support for a Human Mission to Mars
FIGURES CITED IN Roger D. Launius, "Public Opinion Polls and Perceptions of U.S. Human Spaceflight," *Space Policy* 19 (August 2003): 163–75.

Zubrin's powerful arguments, like those of von Braun before him, transform space policy into camps of "us" and "them"—those who advocate a human presence throughout the solar system and those who oppose it. Alternative visions are hard to imagine, and many have accepted the goal of human expeditions to Mars even if they decline to follow Zubrin's methods. Throughout the 1990s, NASA officials sponsored workshops and prepared "reference missions" that could be used once the necessary political consensus emerges.[71]

In a variety of public opinion polls, however, a majority of Americans do not support human missions to Mars—and never have. Consistently, only about 40 percent of those polled support human missions to Mars, and about 50 percent oppose it. A Mars mission would be technologically challenging, moreover, far more so than the short-duration Apollo flights to the Moon. Of the thirty-two robotic Mars missions undertaken by various nations between 1960 and 2004, only twelve missions accomplished their objectives. This track record suggests the magnitude of impediments in complexity, risk, and overall cost.

Without question, the United States could send human expeditions to Mars, as Robert Zubrin and his supporters advocate. There is nothing magical about it, and a national mobilization to do so could be quite successful. But a human Mars mission would entail considerable risk, substantial expenditures, and a political commitment encompassing changes in political regime, the introduction of pressing alternate issues, and unforeseen difficulties. What combination of political, military, cultural, and industrial forces would motivate such an undertaking? As with Project Apollo before it, a Mars mission would likely require a coalition of interests much larger than the ones represented by its most devoted advocates. The utopian element notwithstanding, Mars remains a far-off place. Like Sir Thomas More's *Utopia* of the sixteenth century, the envisioned Utopia on Mars remains an ethereal chimera.

SPACE UTOPIAS AND THE FRONTIER ANALOGY

Robert Zubrin explicitly assigns all of the benefits of frontier life, first described for Americans by Frederick Jackson Turner in 1893, to the exploration of space. For Zubrin and others like him, space is the "final frontier," possessing a near-magical capacity to transform American life. Zubrin was neither the first nor the last to link the American western frontier with outer space. Many have ascribed utopian consequences to the process of westering in America. Indeed, the myth of a garden on the frontier, with connotations of perfection everywhere, abounds in American culture.

Frederick Jackson Turner's "Frontier Thesis," arguably the most influential essay ever read at an annual conference of the American Historical Association, exerted a powerful force on the historiography of the United States. The essay is a powerful statement on behalf of American exceptionalism. Turner took as his cue an observation in the 1890 U.S. census that the American frontier had for the first time closed. He noted, "Up to our own day American history has been in a large degree the history of the colonization of the Great West. The existence of an area of free land, its continuous recession, and the advance of American settlement westward explain American development." He insisted that the frontier made Americans American, that it gave the nation its democratic character, and that it enshrined the virtues of self-reliance, community, and justice. The promise of cheap or even free land provided a "safety valve" that protected the nation against uprisings of the poverty-stricken and malcontented. The frontier produced a people with "coarseness and strength . . . acuteness and inquisitiveness, that practical and inventive turn of mind . . . [full of] restless and nervous energy . . . that buoyancy and exuberance which comes with freedom." It gave the people of the United States, in Turner's point of view, virtually every positive quality they have ever possessed.[72]

Repeated use of the frontier analogy for spaceflight, with its vision of new lands and a better society, gives the American spacefaring movement its distinctive perspective. It taps a vein of rich ideological power, easily understandable to people caught up in the American experience. The symbolism of the frontier captures the manner in which Americans understand themselves and their civilization. It conjures up an image of self-reliant individuals moving westward in sweeping waves of discovery, exploration, conquest, and settlement of an untamed wilderness. And in the process of movement, the Europeans who settled North America became an indigenous American people. The frontier ideal has always carried with it the ideals of optimism, democracy, productivity, heroism, honor, duty, and a host of other positive traits.

It also summons in the popular mind a wide range of vivid and memorable tales of heroism, each a morally justified step in the progress of civilization. The frontier myth, exemplified in countless Hollywood and literary "westerns," reduced the complexity of discovery and settlement to a relatively static morality play. The myth avoided matters that contradicted the myth, viewed pioneers moving westward as inherently good and their opponents as characteristically evil, and ignored the cultural context of westward migration.

The frontier thesis serves a critical unifying purpose in the minds of spaceflight advocates. Those persuaded by this metaphor, and many have been, recognize that it summons them not only to recall past glories but also to undertake—or at least to acquiesce in—a heroic engagement in pursuit of social, political, and economic justice.[73]

Turner's image of the American frontier has provided an especially evocative and romantic theme for proponents of an aggressive space program. It essentially seeks to recreate a glorified frontier experience in the wilderness of space. Calling upon the spirit of American adventurousness, it promises utopian change as new pioneers move to untainted locales where they can remake society once again. Such has always been the siren call of the frontier myth.

From Captain James T. Kirk's soliloquy that opened each *Star Trek* episode—"Space, the final frontier"—to President Kennedy's promise to set sail on "this new sea," the frontier allusion has been a critical component of space program advocacy. Astronaut and senator John Glenn captured this perspective in 1983 when he argued for a new American agenda. Space, he said, "represents the modern frontier for national adventure. Our spirit as a nation is reflected in our willingness to explore the unknown for the benefit of all humanity, and space is a prime medium in which to test our mettle."[74]

Quintessential American novelist James A. Michener employed the frontier analogy in two articles written for *Omni* magazine in the early 1980s. Michener was completing a novel on the U.S. space program and used the magazine as a forum in which to express his views. He explicitly compared the NASA space program to the westward movement in America during the nineteenth century. He praised the American sense of pioneering and argued that the next such challenge was in space. "A nation that loses its forward thrust is in danger," he warned; "the way to retain it is exploration." In an eloquent manner, he identified the American space program as the logical means to continue the process of exploration. One of these articles had the unfortunate title "Manifest Destiny," which blatantly harked back to the harsh ideology of continental expansion that gained preeminence in the 1840s. Michener insisted that Americans were destined to explore and colonize—and that space was the next logical place to do this.[75]

As a two-time NASA administrator, James C. Fletcher was especially attracted by the analogy of the American frontier. A Caltech Ph.D., he guided NASA during the critical period of post-Apollo planning and for three years after the *Challenger* accident. But for all his hardheaded practicality, for all his understanding of science, he remained mystically enthralled with the frontier allusion and its specific connections to his pioneering ancestors in Utah:

> History teaches us that the process of pushing back frontiers on Earth begins with exploration, and discovery is followed by permanent settlements and economic development. Space will be no different. . . . Americans have always moved toward new frontiers because we are, above all, a nation of pioneers with an insatiable urge to know the unknown. Space is no exception to that pioneering spirit.[76]

The frontier myth's accessibility, combined with its utopian imagery, has served the pro-space movement well. Casting decisions on projects as facilitating the opening of a new frontier has enormous appeal and has been used repeatedly since the space age began. One of its most dramatic applications occurred in the report of the National Commission on Space. The commission was created in the mid-1980s by President Ronald Reagan to chart a path into the twenty-first century. Reagan appointed as its chair the irrepressible Thomas Paine, more than a dozen years gone from his post as NASA administrator. Paine and his commission members produced a report of remarkable optimism. Its cover, painted by space artist Robert McCall, depicted a Martian colony fully stocked with modern technology and human settlers. The title of the report did justice to Paine's optimism: *Pioneering the Space Frontier*. The report advanced

> a pioneering mission for twenty-first-century America—to lead the exploration and development of the space frontier, advancing science, technology, and enter-

> prise, and building institutions and systems that make accessible vast new resources and support human settlements beyond Earth orbit, from the highlands of the Moon to the plains of Mars.[77]

The notability of the frontier analogy was lost on the public at large. The Paine report was generally viewed as a romantic and overoptimistic vision of spaceflight more compatible with science fiction than a realistic exploration effort.

UTOPIAS IN SPACE AND THE DEBATE OVER HUMANS AND MACHINES

Invoking the ideas of Frederick Jackson Turner has become increasingly counterproductive for anyone attempting to carry on a discourse in a postmodern, multicultural society. Historians appropriately criticize Turner's approach as excessively ethnocentric, nationalistic, and jingoistic. His rhetoric excludes more than it covers, failing to account for the diversity of western people and events. Yale historian Howard R. Lamar suggests that the Frontier Thesis creates an inappropriate discontinuity between a mythical rural past and an urban-industrial future.[78] He finds it unsuitable as a guide for understanding the present or projecting the future. Other scholars discount the "safety-valve" proposition so central to the frontier thesis. The opportunity for migration may have applied in antebellum America when many did "go West," but it fails as a factual proposition after the Civil War when the prospect of migration moved beyond the reach of urban slum dwellers and the rural poor. Most settlers in the post–Civil War years—often the children of farmers—arrived from the fringes of existing settlements.[79]

Western historian Patricia Nelson Limerick, for one, has argued that the frontier concept denotes conquest of place and people, exploitation without environmental concern, wastefulness, political corruption, executive misbehavior, shoddy construction, brutal labor relations, and financial inefficiency.[80] Comparing the Space Shuttle to the railroad of more than a century earlier is perhaps more appropriate and less enticing than the image of Martian settlements. The western railroad and the Space Shuttle both engendered intense economic contests, lucrative contracts, and no-holds-barred political struggles for primacy and perquisites. Both received government support long after their primary usefulness disappeared.[81]

Nor is the notion of utopias without its flaws. The original Utopia was not a place but a sixteenth-century book by Sir Thomas More, wholly fictional. The book ends with a reminder that the descriptions contained in the work are entirely imagined and unreal. Subsequent utopian settlements, while real in their physical characteristics, rarely achieved their ideological goals. Most typically, the founders brought with them the same ills from which they sought to escape. With the passage of time, as the initial rush of excitement wore off, their successors grappled with the same issues confounding less perfectly imagined societies.

Despite frequent criticism, the Frontier Thesis and its utopian companion possess a lasting appeal, in no small measure because they tell Americans how perfect they have become and how this might have occurred. Frontiers and utopias are two of the oldest and most characteristic American ideologies. Among the public as a whole, largely unschooled in the details of academic history, the Frontier Thesis in particular remains a powerful idea with easy applicability to space exploration.[82]

Advocates of human space exploration developed their ideologies in the 1950s, honed them in the 1970s, and perfected their implications in the more recent past. At their core remains a basic belief in Utopia, the creation of a perfect society on new and pristine planets, though rarely do they explicitly use the term *utopian*. Promises of a bountiful future in which all have the resources to live rewarding lives, where unlimited economic potential prevails, where peace and justice reign for all, and where the perfectibility of humankind is achieved—these are all utopian sentiments. Amplifying this sentiment, allusions to spaceflight as a frontier experience and an attribute of human destiny also stimulate visions of idyllic, perfect places.

The most devoted advocates of human spaceflight also possess a basic belief, utopian at its base, that sees interplanetary travel as the only hope for the continuation of the species. Asteroids or nuclear holocaust or environmental degradation or even a supernova all spell eventual doom for this planet and all who reside here. Astronaut John Young—a veteran of the Gemini, Apollo, and Space Shuttle missions—believes that the truly endangered species on Earth are humans. The only way to escape is to leave. The idea of a series of arks containing the living creatures of Earth is especially appealing, since Americans so often conceive of themselves as called apart to "redeem" the world. Time is short, and every day brings humankind closer to destruction.[83]

Because of its critical role in creating perfect societies beyond Earth, early enthusiasts logically placed humans at the center of the spacefaring enterprise. For them, it made no sense to send robots as surrogates. People had to go themselves, because the ultimate purpose required movement outward from the "cradle," to use Tsiolkovsky's term for the Earth. And, of course, humans did this with resounding success, landing on the Moon only twelve years after dispatching the first Earth-orbital satellite. But these successes then gave way to more mundane efforts. Advocates of the spacefaring vision criticized scientists and their robotic agenda for contributing to this demise. As one commentator noted, "I have the feeling that the scientific community hijacked the last three lunar landings. The guys on the moon should have had more control over what they did with Houston helping and informing, not dictating."[84] From this point of view, the early success in human spaceflight was followed by a retrenchment of the human imperative. Humans continued to fly in space, but only in Earth orbit. The situation would improve "if only we could get that money back and other wasted money back and put it on a manned lunar/Mars program."[85] For such individuals, spaceflight means human exploration and settlement. It is not about science for them. They would certainly not agree with James Van Allen when he places science at the center of the civil space enterprise.[86]

To hard-core scientists, the arguments of pro-space enthusiasts seem delusional. Not only do their arguments displace scientific investigation as the primary purpose of space exploration, they also delude their advocates about the difficulties of extraterrestrial travel. Scientists are trained to approach assertions with a healthy degree of skepticism and a general distrust of mystical (and probably unsupportable) statements. To many of them, the positions advanced by pro-space enthusiasts contain bad science and poor history. Robert L. Park, a University of Maryland physicist who writes books and maintains a website devoted to debunking what he calls "voodoo science," gives special attention to what he views as general misconceptions about human spaceflight.

> Much of what we yearn to discover in space is inaccessible to humans. Astronauts on Mars, locked in their spacesuits, could not venture far from shelter amid the constant bombardment of energetic particles that are unscreened by the thin atmosphere. Beyond Mars, there is no place humans can go in the foreseeable future.[87]

Disagreements between advocates of human and robotic flight reached crescendo proportions as the twenty-first century began. For many members of the spaceflight community, interplanetary travel promises to advance the welfare of humanity. In the process, they seek to overcome the limitations that the Earth imposes. Others want to learn about the cosmos for the sake of understanding the unknown. The goals are often viewed as mutually exclusive and give rise to core differences in the debate over the role of humans and robots in space.

CHAPTER THREE

Promoting the Human Dimension

Without question, the most powerful vision of space travel to emerge in the first half-century of cosmic flight was that articulated by Wernher von Braun, one of the most important rocket engineers and champions of space exploration during the mid–twentieth century. Von Braun appeared as a spokesperson for space exploration as the result of a series of articles he helped write for *Collier's*, a popular weekly periodical of the era. The articles commenced in the winter of 1952; later that year, von Braun was asked to address an important symposium on spaceflight taking place at the Hayden Planetarium in New York. He became a household name following his appearance on three Walt Disney television shows dedicated to space exploration beginning in 1955.[1] His ideas influenced the collective imagination of millions of people and helped chart the course of actual space activities in the United States.

Von Braun centered his vision for the exploration of space on human flight, leaving virtually no room for autonomous robotic activities. His vision became official government policy and elicited billions of dollars in annual taxpayer funds. This was a fantastic accomplishment, given that most people at the midpoint of the twentieth century treated space travel as science fiction—that "Buck Rogers stuff"—and not particularly worth government investment.[2] Von Braun's relentless proselytizing, moreover, elicited strong official resistance within the administration of President Eisenhower, the government for which he worked. Yet his plan triumphed over objections to its advisability and the more modest and parsimonious robotic alternative.

What factors might help explain the acceptance of von Braun's grandiose vision of human space travel within the U.S. civil space program? To be sure, many people had been exposed to its principal elements through works of science fiction stretching back several decades.[3] This literature had long excited public anticipation, which prepared the way for real space travel and even prompted von Braun to attempt to write some. Other factors helped prepare the way. Von Braun's vision was strengthened by its association with "Big Science," which became the preferred method for tackling technological challenges in the aftermath of World War II, when it appeared that organized large-scale efforts such as the Manhattan Project could resolve virtually any problem.[4] Indeed, the immediate postwar era found the application of wartime mobilization models for science applied to numerous peacetime problems.

Building on more than five centuries of terrestrial activities, Big Science seemingly offered a path to the next stage of human exploration. Supporters of human flight effectively argued that a complex set of actions would be needed to fulfill all of the purposes assigned to space travel. Those subscribing to an aggressive approach to space exploration embraced the concept of Big Science and organized their efforts accordingly. It worked with spectacular results during the Apollo program; virtually everyone has seen it as a triumph of management in meeting the enormously difficult systems engineering and technological integration requirements. The management of the program was recognized as critical to Apollo's success in November 1968, when *Science* magazine, the publication of the American Association for the Advancement of Science, observed:

> In terms of numbers of dollars or of men, NASA has not been our largest national undertaking, but in terms of complexity, rate of growth, and technological sophistication it has been unique. . . . It may turn out that [the space program's] most valuable spin-off of all will be human rather than technological: better knowledge of how to plan, coordinate, and monitor the multitudinous and varied activities of the organizations required to accomplish great social undertakings.[5]

Apollo, no question about it, represented the high-water mark of Big Science in the context of space exploration.

Much of this would change in the half-century to come. Big Science lost its sheen. What once had been viewed as a virtue became ossified and bureaucratized, representative of all that was wrong with government.[6] Astronauts became essentially anonymous with only a handful recognizable to the public during the space shuttle era. This was quite different from the earlier heroic age of human space travel when John Glenn, Alan Shepard, Neil Armstrong, and others were household names. At the same time, exploration methods diversified. Of the major rationales supporting space travel, only a few required a human presence, and some of those are on shaky ground. At the midpoint of the twentieth century, as the space age began, supporters of the von Braun vision for space exploration successfully argued that the pillars of space travel rested upon a human foundation. At the close of the twentieth century those pillars were ready to give way.

THE VON BRAUN PARADIGM

Working for the German army between 1932 and 1945, Wernher von Braun led the technical effort to develop the V-2, the first ballistic missile. He deliberately surrendered to the Americans at the close of World War II because, as von Braun later explained, "all we want is a very rich and very benevolent uncle."[7] For fifteen years after World War II, the U.S. Army employed von Braun and his German rocket team of about 125 individuals on the development of ballistic missiles. In 1960 government officials transferred his rocket development organization, located in Huntsville, Alabama, from the U.S. Army to the newly established National Aeronautics and Space Administration. Shortly thereafter, the von Braun group received a mandate to build the giant Saturn rocket. Accordingly, von Braun became director of NASA's Marshall Space Flight Center and the chief architect of the *Saturn V* launch vehicle, the superbooster that propelled Americans to the Moon.

From the 1950s on, this German émigré called for an integrated space exploration plan centered on the movement of humans beyond the planet of their origin. It involved basic steps accomplished in this order:

1. Development of multi-stage rockets capable of placing satellites, animals, and humans in space.
2. Development of large, winged reusable spacecraft capable of carrying humans and equipment into Earth orbit in such a fashion as to make space access routine.
3. Construction of a large, permanently occupied space station to be used as a platform to observe Earth and from which to launch deep space expeditions.
4. Inauguration of the first human flights around the Moon, leading to the first landings of humans on the Moon, with the purpose of exploring that body and establishing permanent lunar bases.

5. Assembly and fueling of spaceships in Earth orbit for the purpose of sending humans to Mars and eventually colonizing that planet.

The steps in this plan, often characterized as the von Braun paradigm, were formulated to lead humans away from their planetary home in a manner similar to the advance of Europeans into the Americas. Von Braun's vision of space exploration served as the model for U.S. efforts in space into the twenty-first century.[8] Most important, NASA executives ensconced von Braun's integrated approach into their internal long-range plan of 1959. With the exception imposed by the insertion of lunar landings before the completion of an Earth-orbiting space station, undertaken for geopolitical reasons in the 1960s, the history of nonmilitary spaceflight in the U.S. followed this paradigm consistently. After completing the Apollo flights to the Moon, NASA officials returned to the related tasks of building a winged reusable spacecraft (the Space Shuttle) and constructing an orbital space station. Additionally, they received mandates from two presidential administrations to pursue an enlarged human presence on the Moon and human expeditions to Mars.

The success of von Braun and his supporters in securing adherence to the paradigm as official U.S. civil space policy occupies a central role in the human versus robotic debate. This is a testament to the power of von Braun's vision—and to the lack of influence among those proclaiming alternative approaches.

THE ROSEN / EISENHOWER / VAN ALLEN ALTERNATIVE

The von Braun paradigm did not stand alone. Preceding von Braun at the Hayden Planetarium symposium in October 1952 was Milton Rosen, who advanced a persuasive alternative based on robotics. Rosen's option was embraced by a figure no less notable than Dwight Eisenhower, soon to be president, and elaborated upon by physicist James A. Van Allen. The alternative was centered on robotic flight. It was never as fully articulated as the von Braun paradigm and, more important, never captured the collective imagination of the American public in the same way.

Milton Rosen was the scientific officer in charge of the government's Viking rocket program. Eisenhower subsequently selected the Viking rocket as the launch vehicle for Project Vanguard, the effort to orbit the first artificial Earth satellite. Rosen became the project's technical director. In his symposium paper, Rosen gave a sober assessment of the challenges involved in space exploration. Spaceflight, he explained, was incredibly difficult. Thousands of components had to work successfully to launch just a small satellite into space. Human flight into the cosmos was even more challenging. Human travelers would be subjected to intense cosmic and solar radiation; engineers would need to worry about the human tolerance for acceleration, vehicle skin temperatures, and atmospheric reentry, among many, many others. While not opposing human flight outright, Rosen suggested that engineers concentrate on "basic research" and the task of creating "reliable, unmanned flight." He characterized von Braun's vision as speculative, fanciful, and simply not real.[9]

As if to reinforce his argument, Rosen's Viking rocket exploded in a spectacular launch pad conflagration during the December 6, 1957, attempt to loft the tiny TV-3 satellite into space. Anxious to show progress in the emerging space race, Eisenhower had previously released von Braun to attempt an orbital launch using his *Juno I* rocket. The German immigrant made his point as well, emphasizing the accessibility of space by successfully launching the first U.S. Earth-orbiting satellite on January 31, 1958.

Ironically, von Braun's Juno rocket launched an Explorer satellite containing an instrument package assembled by physicist James Van Allen, the scientist who would subsequently emerge as the principal opponent of von Braun's vision. Van Allen's instruments discovered the belts of intense radiation occupying the orbital path where von Braun proposed to locate his large, rotating space station, making permanent human habitation at such altitudes impossible. Subsequent space stations would orbit at much lower altitudes, below the Van Allen radiation belts.[10]

Rosen and Van Allen presented an image of conditions in space that contrasted sharply with von Braun's assessment. Von Braun envisioned space as a human receptacle, a place into which travelers could move forward in the same manner as terrestrial explorers venturing across seas and continents. His view was reinforced by the generally popular but factually incorrect image of the inner planets as Earth-like spheres, Mars being a dry planetary desert and Venus the equivalent of a primordial swamp.[11] Rosen and Van Allen portrayed space as a far more hostile place and the Earth, by comparison, as a place of life. The Earth's atmosphere, especially its ozone layer, along with its magnetosphere, deflects much of the violence that permeates outer space. As much as von Braun believed in a human destiny in space, Rosen and Van Allen saw it as a place where humans would not frequently go.

In formulating the nation's first space policies, President Eisenhower adopted the Rosen–Van Allen point of view. Eisenhower favored a small civil space program focused on scientific investigation, primarily with robotic satellites. He approved Project Mercury, the U.S. effort to put humans in space, solely for the purpose of investigating the effects of weightlessness and other conditions on the human form or, as the charter for the project read, "to investigate the capabilities of man in this environment." He repeatedly resisted efforts to create a post-Mercury flight program, already called Project Apollo, which would have permitted more elaborate travel through the construction of very large rockets and multi-person spacecraft.[12]

Although possessing a positive orientation toward scientific investigation, military applications, and robotic spacecraft, the Rosen–Eisenhower–Van Allen alternative was more notable for what it rejected than for what it embraced. Eisenhower opposed Project Apollo for its lack of scientific merit, calling it "a mad effort to win a stunt race." Rosen contended that the government "would be throwing its money away" if it attempted to build the winged Space Shuttle von Braun proposed. As for the dreams of lunar bases and Martian colonies, the more traditional science community of the 1950s characterized them as "emotional compulsions" and diversions from the important work of scientific research. "Remotely-controlled scientific expeditions to the moon and nearby planets," they promised, "could absorb the energies of scientists for many decades."[13]

With increasing frequency, the von Braun faction argued that America needed to put humans in space to win the Cold War. A space station "in the wrong hands," editors at *Collier's* assured their readers, "would allow ruthless dictators to rule the world," a point repeated by Senate Majority Leader Lyndon Johnson when he assured his Democratic colleagues that "control of space means control of the world."[14] This was absolute nonsense, Eisenhower's advisers replied. Space was a valuable platform for military reconnaissance activities, mainly of an automated sort, and Eisenhower's national security team rushed to construct them. Hysterical notions such as the suggestion that cosmonauts on an enemy space station passing over the continent could drop atom bombs on the United States, however, defied known laws of physics. "An object released from a satellite doesn't fall," Eisenhower's science team struggled to explain. It just orbits alongside its carrier. It made more sense, and cost less money, to develop ballistic missiles that flew through space to reach their targets but were not based there.[15]

Van Allen increasingly characterized the arguments advanced by proponents of human spaceflight as "an insult to an informed and intelligent citizenry."[16] Yet the debate about human and robotic flight at the beginning of the space age was not driven by dispassionate and intelligent discussion. As a cultural force, its power arose from anxieties about the human condition and the myths that people used to visualize their future.

Recall for a moment the conditions existing at the dawn of the space age. The United States tested the first atom or fission bomb in the summer of 1945; the Soviet Union exploded theirs four years later. The United States detonated the first hydrogen or fusion bomb in 1952, followed by a Soviet test the following year. National news magazines carried dramatic accounts of the potential destruction such weapons could visit on human populations.[17] Government officials installed air raid sirens, tested them weekly, and practiced techniques necessary to evacuate large metropolitan centers. During the 1950s the United States and Soviet Union initiated crash programs to develop rockets that could deliver nuclear warheads. Vividly illustrated stories depicting horrific attacks launched by airplanes, ballistic missiles, and spacecraft appeared regularly.[18] Many people believed that a nuclear Armageddon could occur more or less at any moment.

Against the backdrop of anxiety generated by these events, stories about space travel began to appear. The von Braun plan presented a vision of remarkable optimism, one in which humans could escape their worldly concerns by embarking upon extraterrestrial voyages. The world was no longer a place onto which humans needed to crowd in ever-increasing numbers but rather a platform from which they could venture into the cosmos. Compared to the darker technologies of war, the technologies supporting space travel, especially after Congress separated civil from military affairs, were largely benign and peaceful in their purposes. The demonstration of this peaceful technology, moreover, could help the United States win the Cold War by demonstrating the nation's scientific and technological capability to the world and, therefore, enlisting more allies while cautioning its enemies.

Von Braun assured all not only that human beings could travel in space but also that it was their destiny to do so once invention allowed. The optimism inherent in his message was remarkably attractive, particularly by comparison to the dangers imposed by thermonuclear weaponry. The persistent appeal of this message was demonstrated in a bizarre but related phenomenon. Beginning in 1947, Americans in increasing numbers began to report sightings of unidentified flying objects (UFOs). The number of annual sightings rose to more than five hundred per year by 1952. Following a *True* magazine story suggesting that the objects were spacecraft from other planets, individuals began to report actual encounters with alien beings. Belief in "flying saucers" was widespread among the general public, along with the widespread suspicion that government officials were engaged in a massive campaign to hide pertinent information.[19] If humans were destined to travel in space—and in fact were on the cusp of being able to do so—then logic impelled humans to believe that beings with more advanced technologies were likewise ready to visit us. Set against the general anxieties of the time, this message generated popular interest of historic scale. That no substantive evidence supported the proposition seemed almost irrelevant.

Cast against these appealing notions, the Rosen–Eisenhower–Van Allen position seemed inherently negative. Certainly, its advocates proposed a positive program of largely robotic flight, but the overwhelming effect of their message was one of skepticism. No, humans were not going to travel into space in large numbers. They were not going to build massive space stations and lunar bases. They were not going to colonize Mars. They were not going to travel throughout the galaxy, at least not in their present form. Science, as scientists understood it, would not allow it.

At a fundamental level, those advocating robotic flight suffered from the long-standing political disadvantage created by negative nomenclature. In American history, one of the earliest efforts to use framing concepts to create an advantage for a specific faction arose during the debate over the ratification of the U.S. Constitution. Those who wished to replace the Articles of Confederation as the organic charter of the United States with a new constitution called themselves *Federalists*. They then labeled their opponents *anti-Federalists,* gaining the substantial advantage imparted by the negative connotation contained in that characterization.[20] In a similar manner, the Rosen–Eisenhower–Van Allen approach to spaceflight was labeled *unmanned* to contrast it with the more dramatic and positively formulated *manned* approach. In effect, von Braun's opponents got tagged as anti-space, and the characterization stuck.

As an effectively negative message, the robotic alternative did not attract the same level of public interest as von Braun's gigantic vision. Regardless of its scientific merits, the robotic or "unmanned" alternative simply did not appeal to the collective imagination of people at that time. Van Allen admitted as much when he observed:

> There is something about the topic of outer space that induces hyperbolic expectations. . . . The acceptance of such grandiose proposals by otherwise rational individuals stems from the mystique of space flight, as nurtured over many centuries by early writers of science fiction and their present-day counterparts. Indeed, to the ordinary person space flight is synonymous with the flight of human beings.[21]

The people promoting the robotic alternative certainly thought that it was a rational policy, better suited to the purposes of space exploration than its grandiose competitor. At the time of the alternative's appearance, however, this did not appear to be so. The method for carrying out the human flight endeavors—Big Science—was more appealing to government leaders. The linkage of human space travel with terrestrial exploration was far easier to explain. The principal justifications for space travel all seemed to require a human presence.

Supporters of the robotic alternative relentlessly criticized human spaceflight. In the long run, history and technology favored their point of view. Humans were gradually excluded from the principal purposes of space exploration, and partisans of human flight grew disillusioned with the Big Science approach inherent in the original vision. Commenting on the heritage of von Braun's vision some forty years after it first appeared, President William J. Clinton's presidential assistant for science and technology, John H. Gibbons, reflected on the manner in which actual events had undercut the dominant alternative:

> The von Braun paradigm—that humans were destined to physically explore the solar system—which he so eloquently described in *Collier's* Magazine in the early 1950's was bold, but his vision was highly constrained by the technology of his day. For von Braun, humans were the most powerful and flexible exploration tool that he could imagine. Today we have within our grasp technologies that will fundamentally redefine the exploration paradigm. We have the ability to put our minds where our feet can never go.[22]

Nonetheless, von Braun's vision persists. In spite of its slow erosion, it has dominated civil space efforts in the United States for fifty years. From a cultural perspective, it has enormous popular appeal. It has dominated cinematic and literary presentations of space travel. Understanding why it elicited the support of seemingly sober-minded politicians is somewhat harder to do.

SPACE AS A BASTION OF CONCENTRATED POWER

The von Braun paradigm offered not only a vision of the future but also a means for its accomplishment. That means, generally characterized as "Big Science," was highly compatible with prevailing beliefs regarding the role of government in technological change at the midpoint of the twentieth century.

As engineers and scientists prepared to engage in spaceflight, two competing methodologies were available to them. (A third approach, one in which private entrepreneurs might lead the way, had been featured occasionally in science fiction but did not attract a substantial following.) The first was characterized by its modesty.

American engineers interested in human spacecraft had assembled at the Langley Research Center in southeast Virginia, the U.S. government's first civilian aeronautics laboratory. They were model builders who had gained experience with high-speed dynamics by testing what they called "pilotless aircraft" mounted on small sounding rockets.[23] Test pilots and flight engineers at Edwards Air Force Base worked to push humans toward the fringes of outer space with their X-series experimental aircraft. The Bell X-2, for example, achieved an altitude of 126,200 feet (38,466 meters) in September 1956. Scientists at government-sponsored research centers such as the Naval Research Laboratory and the Jet Propulsion Laboratory, on opposite sides of the continent, worked to build satellites and the rockets that could dispatch them.

These were relatively modest operations, innovative in their treatment of the challenges of flight. President Eisenhower was determined to keep the human spaceflight program in such hands and away from the influence of Cold War officials whose philosophies encouraged a proclivity for substantial government spending. He assigned his satellite programs to scientists at the government research laboratories. Seeking a home for the newly approved Project Mercury, he endorsed the creation of a civil space agency, NASA, and transferred those chosen as astronauts for Mercury—a group of military test pilots—to the Langley Research Center.[24]

Eisenhower was aware of the relentless agitation among the strongest advocates of human flight for a massively funded federal effort. He resisted them, preferring to keep his space program small. In his farewell address to the American people, the president urged Americans to resist the "recurring temptation to feel that some spectacular and costly action could become the miraculous solution to all current difficulties." He lamented the manner in which the quest for expensive government contracts by the nation's scientific and technological elite had become "a substitute for intellectual curiosity." Referring largely to the growth of an American arms industry, he issued a classic warning: "In the councils of government, we must guard against the acquisition of unwarranted influence, whether sought or unsought, by the military-industrial complex. The potential for the disastrous rise of misplaced power exists and will persist."[25]

By contrast, the von Braun strategy for space exploration employed a military campaign model, emphasizing incremental building blocks, much as an army commander might use to defeat and occupy enemy territory. Von Braun's rocket team worked at the U.S. Army's Redstone Arsenal in Huntsville, Alabama.

It was an engineering group devoted to the production of missiles. When they joined NASA during the last year of the Eisenhower administration, they continued to encounter resistance to their plans for massive rocketry.

That situation changed with the inauguration of John F. Kennedy and the 1961 decision to go to the Moon. As David Halberstam shrewdly observed, "If there was anything that bound the men [of the Kennedy administration], their followers and their subordinates together, it was the belief that sheer intelligence and rationality could answer and solve anything."[26] This translated into an ever-increasing commitment to use science and technology as a means to resolve national challenges and create future opportunities. Kennedy's people applied that approach to international relations, a techno-war in Vietnam, and the emerging space program.

In space, they adapted the military campaign model to the challenge of placing humans on the Moon. They moved the Langley engineers and the corps of NASA astronauts to a newly created field center near Houston, Texas. They allowed the von Braun team to build large rockets and oversee the construction of a new launch center at Cape Canaveral, Florida, and they imported large numbers of military officers and aerospace industrialists to transform NASA into an organization capable of carrying out really big projects.

Hidden behind the beguiling mask of a benign technology, the advocates of human flight were able to overcome Eisenhower's proposal for a modest space effort and create a civil program of massive proportions. The resulting power struggles produced a policy subsystem of civil servants, industrialists from government-supported firms, space advocates, and supportive politicians dedicated to Big Science not unlike the policy subsystem devoted to the government procurement of munitions. The "space race" occurred not only between nations but also between advocates of different approaches to the challenge of spaceflight. The implications for the conduct of science, the role of industry, and the involvement of personalities were profound.[27]

Human spaceflight conducted through large government organizations became the dominant civil space policy in the United States soon after its enunciation and essentially forestalled other perspectives for more than fifty years. It embraced as its raison d'être the necessity of a strong nationally financed program with little room for private enterprise or small, decentralized efforts. Accordingly, the American space program became a heavily state-centered effort. Unlike the more modest Mercury flight project and the work on scientific satellites, the von Braun paradigm permitted a massive mobilization of national resources and a broad coalition of interests.[28]

The methodology created for Project Apollo became the preferred NASA approach for many years. Most historians of spaceflight have accepted at face value the benign nature of the underlying technology, with its emphasis upon the peaceful uses of space. The underlying policy, however, has been highly concentrated and not as equally benign.

The power struggle over space had implications that spread well beyond the civil space program. Lyndon Johnson, one of the principal architects of "Big NASA," once remarked that the votes taken to establish Project Apollo laid the groundwork for government efforts at social change. Johnson believed that he could get conservative Southern Democrats, normally resistant to large government spending for domestic affairs, to vote for a massive space program by stressing its importance to national security. Having set the precedent of huge outlays for space, an essentially domestic undertaking, Johnson felt that the space program would make it easier to win votes for a similar federal presence in the areas of health care and anti-poverty efforts.[29]

The power accrued by those undertaking spaceflight sometimes corrupted them, not in the sense that participants stole money but rather in the manner in which their participation tended to foreclose alternative approaches. The human flight program became a governmental leviathan that had to be fed substantial amounts of money each year, often irrespective of whether its programs were going anywhere important in space. Its advocates attempted to conceal this fact by laying claim to the dominant myths and symbols of America, such as the endless frontier, "manifest destiny," and happy images of white-topped wagon trains crossing the prairies. In the beginning, few were eager to criticize the concentration of power that made spaceflight possible, especially in its human dimension. By not doing so, however, the supporters of human spaceflight helped to ensure that the conventional von Braun paradigm remained manifest.

It is important to understand that the accomplished stages of von Braun's vision succeeded only because of the concerted efforts of government officials in the United States. No other nation went as far in space, and no other institution within the United States other than the federal government could have organized such an effort. This was most effectively seen in Project Apollo, an enormous undertaking matching the construction of the Panama Canal as a nonmilitary engineering endeavor and the Manhattan Project as a wartime goal.[30] In the end, a unique confluence of political necessity, personal commitment, scientific and technological ability, economic prosperity, and public mood made possible the lunar landings. Project Apollo was as much a success of policy and organization as of technology, and it was made possible by a complex network of people, institutions, and interests.[31] Anything other than concentrated, top-down management commanding enormous resources would have been unfathomable at that time. Earlier science fiction writers sometimes portrayed small teams of scientists or entrepreneurs, often privately funded, undertaking such missions, but the reality of spaceflight in the mid–twentieth century forced the United States into a massive government effort. The Soviet Union used a similar approach for their unsuccessful try at putting humans on the Moon.[32]

Of course, to von Braun and those who followed, how could it be otherwise? The immediate postwar era produced many calls for the use of wartime mobilization models to solve scientific and social problems. In 1952 Edward Everett Hazlett urged presidential candidate Dwight D. Eisenhower to declare a "War on Untimely Death." Utilizing wartime mobilization techniques, Hazlett said, the government would "smash the atoms" of disease. Such an approach, he added, "seems no more likely to fail" than did the effort to build the Bomb. "It has, in addition, the spiritual advantage of being a campaign to save life and not to take it."[33] Such faith in science and technology motivated all manner of activity in the two decades following World War II. Government officials yielded to the authority of experts, according to James B. Conant, with something akin to "the old religious phenomenon of conversion."[34] Such efforts expanded the stature and power of the federal government and centralized science in ways that had never been previously envisioned. This centralization mesmerized members of the Kennedy administration, becoming the quintessential force of the Cold War era.

Many Americans celebrated this use of federal power when it occurred. Over time, however, even advocates of human spaceflight bemoaned its influence. Space Frontier Foundation president Rick Tumlinson rejects the large-scale model.[35] Tumlinson enthusiastically embraces the von Braun vision insofar as it leads to human settlements on the Moon and Mars but rejects the large-scale governmental methods used to achieve it. For Tumlinson and others like him, the government stands in the way and must be removed from the process. He believes that the centralized approach will eventually lead to the discontinuation of the human flight program and therefore will not foster human movement to the Moon, Mars, or anywhere else.[36]

Others have bemoaned the sluggish and wasteful characteristics of the mobilization model as applied to post-lunar undertakings such as the NASA Space Shuttle and the International Space Station. Some of the most innovative alternatives have occurred in research centers and government-funded laboratories devoted to small-scale and relatively inexpensive space projects. These have been primarily robotic undertakings, but the principles may be applicable to human flight as well.[37]

What von Braun's vision lacked in realism and thrift, it made up for in ambition. Its grandiose nature made it more appealing to the partisans of Big Science and Big Engineering who increasingly came into positions of influence in mid–twentieth century America. As John F. Kennedy, the initiator of Project Apollo, so adroitly observed, "We choose to go to the moon in this decade and do the other things, not because they are easy, but because they are hard, because that goal will serve to organize and measure the best of our energies and skills."[38] By the end of the twentieth century, however, that methodology no longer occupied the position it had enjoyed forty years earlier and was increasingly being questioned by people inside and out of the national civil space effort.

THE THREE GREAT AGES OF EXPLORATION

The von Braun paradigm gained additional salience through the manner whereby its advocates fit it into the history of human exploration, providing a smoother line of continuity than the robotic alternative. Beginning in the fifteenth century, as Stephen Pyne has explained, humans in general (and Europeans in particular) experienced three great ages of exploration. The third stage, which includes space exploration, informs modern life even as it remains in its infancy. Each age transformed the societies that participated in them, so that exploration as a concept formed the basis for a "world view" that encompassed more than the physical process of travel.[39]

The first great age of discovery began during the European Renaissance of the fifteenth century. For Western Europeans, especially those residing in commercial centers, nothing could ever be the same again. Europeans were transformed by contact with new lands and peoples during the voyages of discovery that took place from the fifteenth to the eighteenth centuries. People on ocean-going ships from the great seafaring nations of Western Europe redrew both the map and conception of the world. When they were done, the contours of the great continents of the Earth had been approximated, and the general size and shape of the physical world had been determined. As Peter Martyr wrote in 1493, just as this age of exploration was opening, "Enough for us that the hidden half of the globe is brought to light. . . . Thus shores unknown will soon become accessible; for one in emulation of another sets forth in labours and mighty perils." During this first age of discovery, travel came mainly over the oceans of the Earth, as European sailors mapped the coastlines of the Americas, Africa, Australia, and even Antarctica. Its great explorers included Christopher Columbus, Ferdinand Magellan, Henry Hudson, Jacques Cartier, and James Cook. Circumnavigation of the globe became the classic expression of this age.[40]

For the second age, which began before the first age had ended, exploration came predominantly in overland form. European and American adventurers filled in many of the details of continental interiors. In the process, geographical knowledge continued to expand, but so too did information about peoples and natural history. The North American expedition of Coronado in 1540, the Lewis and Clark Expedition into the American West in 1804 to 1806, the efforts of Sir Richard Burton and Henry Morton Stanley and David Livingstone in Africa, and travels to the sources of the Amazon and the Nile all characterized the second great age of exploration.[41] This stage closed with the conclusion of the last great land expeditions of the later nineteenth century and the exploration of the poles in the early twentieth century. Like the first stage, it too led to a massive accumulation of data that transformed the scientific and cultural world.[42]

The third great age of exploration began in the twentieth century and is strikingly different from what went before. Explorers moved into realms where people generally cannot live without the benefit of artificial apparatus. Continued exploration of the Earth's poles, especially by groups utilizing machine technology; expeditions into the oceans; and space travel all suggest a new age of discovery and inquiry. Exemplars of this new age include Richard Byrd's epic North Pole flight of 1926, Jacques Cousteau's voyages with the scientific vessel *Calypso,* Soviet cosmonaut Yuri Gagarin's flight as the first human in space in 1961, and American astronauts Neil Armstrong and Buzz Aldrin's Moon landing in 1969.

With its reliance upon artificial means, the third great era of exploration could have been conducted almost entirely with machines. In fact, many third-era expeditions were accomplished in this fashion, from automated rovers on Mars to undersea submersibles. Total reliance upon machines, however, would have represented a radical departure from the historical approaches to exploration utilized in previous stages. Support for widespread exploration requires a close confluence of government finance, economic wealth, public acquiescence, scientific curiosity, and enabling technologies.[43] The cultural dimensions are as important as the technical ones. Were space exploration beginning *de novo,* without reference to previous ages of exploration, a stronger case could have been made for the predominance of machines on strictly technical grounds. Given that space exploration rests within a larger historical movement, radical alternations were harder to make. The momentum provided by previous efforts at exploration favored the continued involvement of human beings. As exploration efforts matured, however, the ties to previous eras became more strained. This created new opportunities to sever the connections to periods during which humans played a central role in voyages of discovery.

In a strange way, attachment to the von Braun paradigm was furthered by the failure of one of its principal elements. Many people anticipated that expeditionary forays into outer space would locate extraterrestrial beings or at least rudimentary life forms, a theme well developed in works of science fiction. But previous contact with native peoples and unusual life forms on Earth provided critics with their most persuasive argument for discontinuing traditional exploration modes. At least by the nineteenth century, critics bemoaned the destruction of native populations and indigenous species as a consequence of European exploration. One of the principal differences between the voyages of Christopher Columbus and Neil Armstrong is that the latter failed to encounter any native populations. In some ways, this allowed the traditions of human exploration to continue without the disadvantages and moral objections imposed by its traditional form.[44]

BEYOND "SPAM IN A CAN": THE STRUGGLE FOR CONTROL OF HUMAN SPACECRAFT

From the early days of human spaceflight, a debate has raged between the pilot/astronauts and the aerospace engineers over the degree of control held by each group in human-rated spacecraft. The engineers placed much greater emphasis on automatic or robotic control systems and sought to reduce the role of astronauts on board a spacecraft. These space engineers often viewed the astronaut as a "weak link" in the spacecraft control system. Of course, the question of whether machines could perform control functions better than people became the subject of public controversy. At the same time, the cybernetics movement served to undermine the existing hierarchies of knowledge and power by introducing computer-based models and decision-making mechanisms into a wide range of scientific disciplines. The American astronauts used their celebrity status to assert more control over spacecraft systems, seeking to overcome the preconception of many that they were merely "Spam in a can." Over time they were successful, to the extent that the Space Shuttle became the first American human space vehicle that could not be flown as an automated system. This model of human-machine interaction is critical to the current place of spaceflight in American culture. The paradigm is beginning to change as robotics becomes more sophisticated and the need for human intervention to control a space vehicle diminishes.

Numerous skirmishes took place between engineers and astronauts during the early years of American spaceflight. The two groups fought over the design of the Mercury capsule, the best way to "man-rate" the launch vehicle, and even items as specific as the controls on the spacecraft console. Donald K. Slayton, who early took the lead for the Mercury Seven and later officially headed the Astronaut Office, emphasized the criticality of astronauts not as passengers but as pilots. In a speech before the Society of Experimental Test Pilots in 1959, he said:

> Objections to the pilot [in space] range from the engineer, who semi-seriously notes that all problems of Mercury would be tremendously simplified if we didn't have to worry about the bloody astronaut, to the military man who wonders whether a college-trained chimpanzee or the village idiot might not do as well in space as an experienced test pilot. . . . I hate to hear anyone contend that present day pilots have no place in the space age and that non-pilots can perform the space mission effectively. If this were true, the aircraft driver could count himself among the dinosaurs not too many years hence.[45]

Slayton's defense of the role of the Mercury astronauts has found expression in many places and circumstances since that time.

Notwithstanding this position, the earliest astronauts soon learned that their spacecraft would be essentially controlled from the ground and that the role of the astronaut would be to be "in the loop" to assure mission success. As human factors staff Edward R. Jones and David T. Grober concluded in an August 10, 1959, report: "Primary control is automatic. *For vehicle operation, man has been added to the system as a redundant component who can assume a number of functions at his discretion dependent upon his diagnosis of the state of the system.* Thus, manual control is secondary." They added, "Mission reliability determinations assume the astronaut can detect and operate these systems without error."[46]

Edward Jones made this point about human involvement even more succinctly in a paper delivered before the American Rocket Society in November 1959. He suggested that the astronaut was important to the successful operation of Mercury missions. He commented:

> Serious discussions have advocated that man should be anesthetized or tranquillized or rendered passive in some other manner in order that he would not interfere with the operation of the vehicle. . . . As equipment becomes available, a more realistic approach evolves. It is now apparent with the Mercury capsule that man, beyond his scientific role, is an essential component who can add considerably to systems effectiveness when he is given adequate instruments, controls, and is trained. Thus an evolution has occurred . . . with increased emphasis now on the positive contribution the astronaut can make.[47]

The result of these efforts led to the development of a Mercury spacecraft that allowed considerable, but not total, control by the astronaut.

If Mercury proved the place of astronauts in the control system of spacecraft, Project Gemini showed their necessity. The technology associated with the Gemini program was significantly more advanced than Mercury and pushed the envelope of knowledge about spaceflight. Rendezvous, docking, and spacewalking all required the active work of astronauts, although much could still be accomplished by ground control. This was even more true during Apollo, when astronauts served as the critical control link in the lunar landing missions.[48] Virtually all of the lunar landings, which had been designed for automated control, required the intervention of the astronaut to ensure success. Skylab, the Space Shuttle, and the International Space Station all required astronaut control as well. In the process, ironically, even as robotics and electronics became more capable, NASA's human spaceflight effort more thoroughly incorporated the astronaut into the control system. At some deep level, perhaps, advocates of human space exploration may have sensed that the greater capabilities of robotics would outstrip human involvement and sought to ensconce the human dimension into the program to ensure its continuation.

RATIONALES FOR SPACEFLIGHT

In addition to its compatibility with Big Science and previous exploration modes, the von Braun paradigm satisfied the principal objectives assigned to spaceflight. Government officials did not divert billions of dollars to space activities for the sake of adventure; they did so because it served a national interest. Governments invest in space for the five broad reasons enumerated below. At the outset, all of those purposes seemed to require a human presence. This would change as space activities matured, but in the beginning the national interest in spaceflight strongly favored these major elements.

Scientific Discovery and Understanding

The first rationale for spaceflight in the United States involves the desire for discovery and understanding. Humans have an ethereal urge to explore, spaceflight enthusiasts argue, one that persists even as the objects of inquiry change. Many participants in space exploration pursue abstract scientific knowledge, learning more about the universe as a means of expanding the human mind. Pure science and exploration of the unknown will remain an important aspect of spaceflight well into the foreseeable future. This goal clearly motivates support for the scientific probes sent to other planets in the solar system. It propels a wide range of the efforts to explore Mars, Jupiter, and Saturn projected for the twenty-first century.[49] It energizes efforts to construct space observatories that promise to secure revolutionary knowledge of the universe through, among other possibilities, the imaging of Earth-like planets around other stars.

From the beginning, science has been a critical goal in spaceflight. The National Aeronautics and Space Act, which created NASA, instructs agency employees to pursue "the expansion of human knowledge of phenomena in the atmosphere and space." This mandate has drawn substantial verbal and fiscal support, although in practice it has proven less important than the pursuit of knowledge that produces more practical social or economic payoffs.[50]

At first the pursuit of space science was accomplished through a combination of human and robotic activities. Human activities reached their zenith with the Apollo expeditions to the Moon. The scientific experiments placed on the Moon and the lunar soil samples returned to Earth provided substantial material for scientific investigation. The return was significant, even though the Apollo program did not answer conclusively the age-old questions of lunar origins and evolution. The lunar voyages demonstrated the link between the spirit of scientific inquiry and the desire for human exploration in ways that were not subsequently attainable.[51]

NASA officials hoped to maintain the same synergy through scientific programs assigned to flights of the agency's Space Shuttle and the science program envisioned for the ISS, both human endeavors. One of the areas in which research on the Space Shuttle has proved useful is the life sciences. Scientists made excellent use of the results from eighty-eight biomedical experiments flown on John Glenn's STS-95 flight in 1998. Those experiments, as well as those of follow-up missions, noted Dennis Morrison from NASA's Johnson Space Center, "might lead to the creation of an anti-tumor drug delivery technique—one that attacks a cancer site without affecting surrounding healthy cells, while reducing unwanted side effects in cancer patients."[52]

On the whole, use of the shuttle and station for scientific investigation suggests an anemic return. Although the scientific return on the ISS cannot yet be fully assessed, such is not true for the Space Shuttle, which has been flying since 1981. The experiments completed using the Space Shuttle orbiter have led many to conclude that the vehicle is a poor place to conduct basic research. Says Richard Muller:

> When it comes to the science itself, the Space Shuttle is a poor choice of platforms. Humans are a source of noise—vibrational, infrared, gravitational. Sensitive experiments must get away from this. Just flying an experiment on a manned mission automatically raises the experiment's costs. Many scientists moan privately about scientific missions that were delayed and made more costly because they were moved off unmanned launch vehicles and forced to become part of NASA's scientific justification for the Shuttle.[53]

In spite of NASA's efforts to mesh scientific investigation with human spaceflight, the quest for scientific understanding became the most powerful argument in opposition to the human spaceflight imperative. Few people question the necessity of scientific missions to understand the cosmos, although they do debate levels of funding and targets of inquiry. Many question whether human flight is necessary to achieve it. In the words of Duke University historian Alex Roland, the fundamental nature of human spaceflight alters the opportunity for scientific understanding:

> Whenever people are put on a spacecraft, its mission changes. Instead of exploration or science or communication or weather, the mission of the spacecraft becomes life support and returning the crew alive. This limits where the spacecraft can go, how much equipment it can carry, how long it can stay, and what risks it can take in pursuit of its mission. The net impact of people on a spacecraft is to greatly limit its range and capabilities without adding any value that can begin to compensate for these drawbacks. A rough rule of thumb, first introduced by NASA Associate Administrator George Low in the Apollo program, is that putting people on a spacecraft multiplies tenfold the cost of the undertaking.[54]

In some cases, humans actually interfere with scientific experiments. The delicate pointing requirements of space telescopes essentially preclude the presence of human operators; the temperature requirements for some instruments are so low that having a warm body nearby would interfere with this.

The seeming incompatibility between human beings and many scientific investigations have led scientists to strongly favor robotic probes. "Any specific mission you can identify to do in space, you can design and build an unmanned space craft to do it more effectively, more economically and more safely," says Roland.[55] The results produced by purely robotic expeditions have been stunning.[56] Hardly a month passes without some pathbreaking paper appearing in a major scientific journal based on data collected by robotic spacecraft.[57] On the whole, the human component of spaceflight became a less significant contributor to scientific understanding of the cosmos, leading many to question whether a human presence is required for this purpose at all.

National Security and Military Applications

National defense provides a persuasive rationale for government spending and results in roughly half of the government funds directed toward space activities annually.[58] As with space science, national defense seemed to require a human presence when contemplation of this purpose began. Von Braun's 1952 conception of a U.S. space station showed it to be a platform from which a number of national security tasks might be performed. Technicians with specially designed telescopes could observe the war-like preparations that the Soviet Union or theoretically any enemy might make. Crew members might complete a variety of reconnaissance, communication, navigation, guidance, and early warning missions.[59] For von Braun, the space station represented an extension of national sovereignty in Earth orbit, holding the same position as an overseas base. Territorial claims to celestial bodies would need to be settled, von Braun assumed, and a space station would help legitimate national claims. The space station would serve a valuable purpose in this effort, standing as a modern equivalent of a fortress on the edge of the "new world." The editors of *Collier's* magazine followed von Braun's reasoning, declaring, "Whoever is first to build a station in space can prevent any other nation from doing likewise."[60]

Military officers pursued the possibility of building their own space station—known as the Manned Orbiting Laboratory (MOL)—during the 1960s. In the 1970s military leaders agreed to use NASA's newly developed Space Shuttle, with its human crews, as their "primary vehicle for putting payloads in orbit." During the 1980s, astronauts from the military services deployed reconnaissance satellites into Earth orbit from the Space Shuttle and used that vehicle to conduct experiments related to space-based missile defense systems. In the following decades, military officers investigated means by which they could defend their space assets from hostile destruction. A military presence in space and the defensive means to protect it—these promise to be a compelling rationale for flying into space into the twenty-first century and beyond.[61]

At the same time, national security interests have not been well served by a human presence in space. The defense activities that von Braun envisioned for his large space station are performed from Earth orbit, but they are not carried out by human beings. While the military services have placed an enormous range of instruments in Earth orbit for military purposes—reconnaissance, communication, navigation, targeting, weather, early warning, and the like—at no time has the U.S. Department of Defense seen the necessity of using soldiers to operate them.[62] Military officers chafed at the requirement that they use a piloted NASA spacecraft as their primary launch vehicle for satellite payloads and escaped from this provision after the loss of the *Challenger* Space Shuttle. Defense officials have occasionally considered the use of piloted spacecraft as troop transports without producing a working model. Wernher von Braun sought DoD support for his giant Saturn rocket during the 1950s by promoting it as a troop transport, arguing that "the cost versus effectiveness of rocket transportation compared to fixed-wing aircraft transportation appears to demand that rocket transportation be substituted for the conventional aircraft transport system in the immediate future."[63] Von Braun would not have to wait for the sniggers to cease before learning that higher-ups had rejected his proposal.

Most military forays into human spaceflight suffered a similar fate. Defense Department Secretary Robert McNamara cancelled the U.S. Air Force Dyna-Soar project, designed to produce a shuttle-type space plane, in 1963. Defense Secretary Melvin Laird cancelled the manned orbiting laboratory in 1969.[64] Both programs began with substantial optimism but could not survive questions about the need for a sustained human presence in space when the advantages of robotic spacecraft seemed so supreme.

Economic Competitiveness and Commercial Applications

Visions of economic advantage provide another important rationale for spaceflight. Since the 1960s, space-based technologies, especially communication satellites, have had an important effect on the global economy. Given the nature of economic transactions in postindustrial society, it is difficult to envision a world in which global, instantaneous telecommunications are not the norm. The same is true for weather data, navigational aids, remote sensing, and a host of other technologies that operate from space but support Earth-based activities. In a recent survey of space policy leaders, more than 90 percent agreed that "current orbital technologies such as weather satellites and communication satellites are essential economic resources and should be continued and improved."[65]

Space commerce generates more revenue than the budgets of all government space activities combined, civil and military, and is growing at a much faster rate. Some projections hold that commercial opportunities in space will exceed $200 billion annually by 2010. Other estimates forecast a $500 billion market for space products soon. This remains a risky business, however, as four out of five companies that enter the space commerce market traditionally fail without producing any profits whatsoever.[66]

Supporters of spaceflight have made numerous attempts to document its commercial benefits. Usually these take the form of what NASA calls "spinoffs," commercial products with at least some of their origins in space-related research and development.[67] NASA employees have spent a lot of time tracking these benefits in an effort to justify their agency's existence, and the NASA History Office Collection has more than five linear feet of documentation on the subject.[68]

How would human lives be different if spaceflight did not occur? One can begin by eliminating any possibility of instantaneous global communication of voice, data, and video. Eliminate the transmissions that make ATM transactions possible. Eliminate multiple-channel television, twenty-four-hour worldwide news, and satellite radio. Then remove a host of space-based observational activities, such as those used to monitor natural resources and to visually track weather patterns. Global positioning systems and navigational aids would disappear. Perhaps such technologies would reappear as a result of work done on the ground, but the evidence seems to suggest that the larger space program pushes technological development down paths that would not otherwise have been followed.[69]

Again: when the space age began, human operation was seen as essential to a flowering space economy. This point of view reached its apex during the 1980s, with the flights of the Space Shuttle, a human-operated vehicle designed to deliver all of the nation's scientific, military, and commercial payloads. Astronauts in space checked out commercial satellites before dispatching them, repaired satellites in space, and returned them to Earth. In 1985, for example, NASA astronauts on the *Discovery* Space Shuttle retrieved a Leasat-3 communication satellite that had been carried into space earlier that year by another crew on the same vehicle. The communication satellite failed to deploy its antenna, spin up, and ignite the engine that would boost it into its proper orbit. The second crew redeployed it.

This being said, one must observe that virtually all the money made in space has been produced by robotic techniques. Communication satellites, global positioning systems, and Earth resource monitoring platforms that yield commercial benefits are overwhelmingly robotic in nature. It is worth noting that the *Challenger* space shuttle blew up while carrying a telecommunications satellite designed to relay data from low-Earth orbit satellites through geosynchronous satellites and back to ground stations. Subsequently, government officials removed responsibility for launching commercial satellites from the shuttle manifest. When NASA officials encouraged executives at Lockheed Martin to seek private capital for the construction of a VentureStar launch vehicle, they did not insist that the replacement for the aging fleet of Space Shuttles be piloted with a human crew. The plan called for Lockheed Martin to recoup its investment on the winged, reusable spacecraft by launching commercial satellites in a fully automated mode. Only when NASA officials contracted to use the vehicle in support of the ISS would it necessarily carry astronauts. NASA withdrew its support from VentureStar in 2001 after the program encountered technical and financial challenges, but not before granting that commercial launches no longer required a human crew.

Various persons continue to predict a robust commercial market involving human flight. The prospect of manufacturing pharmaceutical products and new metal alloys in the microgravity environment of space was a prime motivator behind President Reagan's decision to approve a permanently occupied Earth-orbiting space station. Supporters of extraterrestrial colonies believe that humans could remove incredibly valuable resources from those sites, such as Helium-3 taken from the Moon to fuel fusion reactors on Earth.[70] A major part of the space tourism vision foresees hotels in Earth orbit and lunar vacation packages. Such activities would draw on the $500-billion-a-year travel and entertainment market in the United States.[71]

Practical experience has as yet failed to demonstrate a compelling reason for humans to be in space to "make a buck." Proposals for orbiting hotels and lunar mines have remained at the visionary rather than at the commercial level. Until its advocates find justification in practice, proposals for human space travel based on the profit motive are likely to elicit skepticism rather than capital from hardheaded venture capitalists, further reducing the persuasiveness of the human flight alternative.[72]

Human Destiny and the Survival of the Species

Having employed scientific, military, and commercial justifications with less than the desired effect, advocates of human spaceflight have turned to more ethereal ideas. One is absolutely guaranteed to justify a substantial human presence in space. With the Earth so well known, advocates argue, exploration and settlement of other celestial bodies present the next logical step in the advance of civilization. From this perspective, space exploration is a matter of human destiny. Indeed, from this perspective, it is essential to the survival of the species. Humans must expand beyond the home planet, explore and migrate—or die.

The belief that humans may no longer be trapped on one tiny world, helpless to escape whatever catastrophes might befall it, provides a powerful draw for those embracing human spaceflight. This rationale also contains a serious warning. Humans as a species will not survive if they remain on one planet. Only by becoming a multiplanetary species can humans approach any semblance of physical immortality. Carl Sagan wrote eloquently about the last perfect day on Earth, before the Sun fundamentally changed and ended the ability of complex life to survive on this planet.[73] While this might not happen for hundreds of millions of years, any number of catastrophes could alter complex life on Earth beforehand. Asteroid or comet strikes pose the most serious threat. In 1993, a noted scientist, David Morrison, spoke to the American Astronautical Society, choosing as his subject, "Chicken Little Was Right." Morrison pointed out that humans had a greater chance of being killed by a comet or asteroid falling from the sky than dying in an airplane crash. Statistically, this is true. Interstellar collisions occur infrequently, but when they do, they affect vast populations. Mathematical calculations confirm that every individual on Earth faces a 1 in 5,000 probability of being killed by some type of extraterrestrial impact. Throughout history, asteroids and comets have struck Earth with devastating effects—a substantial impact probably ended the reign of the dinosaurs. The apparent culprit was an object six to nine miles wide that left a 186-mile-diameter crater in what is now Mexico's Yucatán Peninsula.[74]

Given time, a really big one will again hit Earth with disastrous consequences. Efforts to catalogue all Earth-crossing asteroids, track their trajectories, and develop countermeasures to destroy or deflect objects are important. To ensure the survival of the species, say the most devoted advocates, humans must build outposts elsewhere. Astronaut John Young said it best, paraphrasing the comic strip character Pogo: "I have met an endangered species, and it is us."[75] The only way to ensure species survival when the big one hits is to start moving now. Without question, the destiny / survival argument provides a powerful justification for human spaceflight.

Having said all that, one must acknowledge that the argument has had limited appeal in motivating public officials to fund government-sponsored space travel. The rationale appeals much more powerfully to members of the science fiction and space community than to governmental officers. The latter, who prefer to portray themselves as a sober-minded set, respond to arguments rooted in science, commerce, and military applications, where the human dimension holds less power.

Geopolitics and National Prestige

As the space age progressed, many applications ceased to provide convincing arguments for human space travel, however. Survival of the species, while certainly a powerful force in motivating private advocates of spaceflight, has not been a convincing argument in government circles. Without the presence of geopolitical considerations, advocates of human spaceflight would have a most difficult time defending their endeavor. With it, their job was made much easier. Historically, U.S. geopolitical considerations and national prestige have been a decisive factor in maintaining the human flight regime.

The quest for national prestige sparked and sustained the space race of the 1960s and continued to do so in the decades that followed. John F. Kennedy responded to the challenge of two Soviet "firsts" (first Earth satellite and human in space) by requesting a national objective that promised "dramatic results in which we could win."[76] The result was the 1961 decision to send humans to the Moon. As John M. Logsdon commented: "By entering the race with such a visible and dramatic commitment, the United States effectively undercut Soviet space spectaculars without doing much except announcing its intention to join the contest."[77] Kennedy chose the lunar objective as a means of impressing uncommitted nations with U.S. technological prowess at a time when many opinion leaders believed that technology would decide the Cold War. Superpower rivalry and national prestige sustained the effort. In that sense, the decision to go to the Moon was undertaken principally as a means to bolster national prestige, not for reasons of science or discovery. As Kennedy explained in 1962: "We mean to be a part of it [spaceflight]—we mean to lead it. For the eyes of the world now look into space, to the moon and to the planets beyond, and we have vowed that we shall not see it governed by a hostile flag of conquest, but by a banner of freedom and peace."[78]

Engineers in the Soviet Union worked hard to match the U.S. objective, but failed. Instead, they dispatched a series of robots to collect samples and traverse the lunar surface. In contrast to Project Apollo, the robotic achievements of the Soviet Union were widely viewed as a national failure, further reinforcing the argument that a human presence was necessary for national prestige.

The quest for national prestige affected the next two U.S. human flight initiatives as well. Confronted with the opportunity to turn away from human spaceflight after the landings on the Moon, Nixon instead committed the nation to a frustrating and seemingly never-ending process of perfecting a reusable Space Shuttle. Nixon was unwilling to go down in history as the president who abrogated the nation's leadership in human spaceflight at a time when the Soviet Union had initiated a program of small Salyut space stations.[79]

National prestige also played a key role in the U.S. decision to build a large, permanently occupied space station. At a December 1, 1983, White House meeting, NASA administrator James M. Beggs asked President Reagan to approve the agency's longstanding desire for an orbital space station. Using an overhead projector and series of transparencies, Beggs emphasized the potential contribution such a facility would make as "a visible symbol of U.S. strength." Beggs understood Reagan's worries about American prestige vis-à-vis the Soviet Union. After explaining the station's scientific, technical, and commercial capabilities, Beggs presented his best transparency. It showed a photograph of the Soviet Salyut space station overflying the United States, with the words in Russian. The Soviet Union already had launched this modest facility and was planning a larger replacement. Should not the United States have one as well? Reagan agreed that it should.[80]

In the aftermath of the *Columbia* accident on February 1, 2003, that took the lives of seven astronauts, when it appeared to all that the nation's approach to human spaceflight should be reconsidered, no official of influence seriously considered ending the effort. Instead, the accident elicited resolve. Even President George W. Bush, who had been silent on spaceflight before, stepped forward on the day of the accident to say: "The cause in which they died will continue. Mankind is led into the darkness beyond our world by the inspiration of discovery and the longing to understand. Our journey into space will go on."[81]

The prestige factor underscores a critical aspect of human spaceflight. National leaders support human space travel in large measure because of the image it projects. Within the Soviet Union, spaceflight continued to serve as a source of national pride even as the Union dissolved and funding disappeared. The Chinese government became the third nation to achieve human spaceflight when it launched one of its citizens into orbit in 2003. The United States undertook the Mercury, Gemini, Apollo, Space Shuttle, and space station efforts as much for the image they projected as for their technical merits. Initial and continued support for human spaceflight rests not on the value it offers as instruments of science, military prowess, or economic wealth but on its usefulness as an icon to buttress the image of the nation in the world.

Is this rationale sufficient to sustain human spaceflight indefinitely? Only time will tell. A human flight program that bolsters national prestige, to revisit President Kennedy's famous phrase, depends upon "dramatic results." Flying in endless circles a few hundred miles above the surface of the Earth, with the commensurate loss of spacecraft and crew every fifty missions or so, might do little to improve national prestige in the geopolitical setting. In fact, a strong argument can be made that the United States garnered more national esteem from the operation of its space telescopes and planetary rovers than the performance of its reusable Space Shuttle. Beyond actual performance, the prestige factor depends upon a world view that respects technological achievements and accepts spaceflight as exotic and adventurous. This situation prevailed during the Cold War conflict between the United States and the Soviet Union, two aspiring technological powers. Future conflicts of an ideological or religious nature may not take this form. It is hard to imagine religious fundamentalists who reject Western civilization being swayed by technological achievements like flights to the Moon.

The prestige factor is powerful but possibly temporal. In summing up the arguments for human spaceflight, John M. Logsdon, the dean of space policy, recently wrote:

> Most public justifications for accepting the costs and risks of putting humans in orbit and then sending them away from Earth have stressed motivations such as delivering scientific payoffs, generating economic benefits, developing new technology, motivating students to study science and engineering, and trumpeting the frontier character of the U.S. society. No doubt space exploration does provide these benefits, but even combined, they have added up to a less-than-decisive argument for a sustained commitment to the exploratory enterprise. The United States has committed to keeping humans in space, but since 1972 they have been circling the planet in low-Earth orbit, not exploring the solar system. The principal rationales that have supported the U.S. human spaceflight effort to date have seldom been publicly articulated. And those rationales were developed in the context of the U.S.-Soviet Cold War and may no longer be relevant.[82]

Once the enthusiasm imparted by the Moon race ended, most people began to accept human spaceflight as a reality that was unchanging. Americans treated efforts to fly the Space Shuttle and build the ISS as activities that were necessary rather than desirable. No national commitment to a multi-billion dollar investment on the scale of Project Apollo took place for these undertakings. Even efforts in the United States to return humans to the Moon or venture on to Mars were constrained by a desire to remain parsimonious in the face of large ambition. In Russia, the human spaceflight program wound down. The commitment to human flight proceeded on the force of inertia not unlike that seen in many other public policy sectors where the perceived crisis became less pressing. This is not a good formula for persistent success.

THE ENTREPRENEURIAL ALTERNATIVE

The prestige argument contains an additional flaw, one that may prove more serious in the years to come. National governments that pursue dramatic human spaceflight objectives do not do so to fail. To invoke a phrase (incorrectly) attributed to *Apollo 13* flight director Eugene Kranz, "failure is not an option." Elected officials appropriating funds for national prestige projects are not comfortable with the levels of risk that accompany ventures organized by business firms or private groups. The U.S. government would never tolerate the levels of risk accepted by private mountaineering groups, for example, who undertake expeditions in which a substantial number of climbers lose their lives. Through the Apollo years, the mechanisms of Big Science provided a remedy to the challenge of excessive risk. Techniques like large-scale systems management and contingency planning reduced the level of risk to the point that the lunar flights almost appeared routine.

As noted earlier, Big Science and its mobilization methods require a substantial national investment. This worked well for the challenges posed by Project Apollo, a very risky endeavor. For the conduct of the space shuttle and space station programs, however, the model proved somewhat dysfunctional. The model requires a large standing army of contractors and a substantial number of government employees whose salaries must be paid even if they are not doing useful work. For recurring operations, it breeds a situation in which maintenance of an existing work force becomes more important than innovation and new adventures. The reader will note that we have not listed the distribution of government largess as a purpose that justifies space activities. The distribution of government contracts is certainly a byproduct of space exploration. During the struggles to maintain various space endeavors, it has been used to garner political support. In some cases, it has become a rationale for maintaining programs and field centers whose existence is not central to current flight objectives.

The maintenance of Big Science and its associated work force may ultimately defeat the achievement of national prestige through space. Elements of this argument can be seen in the U.S. space shuttle and space station programs, in which government officials spent vast sums of money without achieving the goals of easy access and low-cost flight. The results are more dramatically seen in space efforts that extend toward the Moon and planets. To satisfy the criteria of success implicit in the prestige rationale, the Big Science model requires the expenditure of funds that are prohibitively large. Many supporters of human spaceflight believe that this will result in programs that last forever but accomplish little (while employing large numbers of workers), just as some contend has been the case with the Space Shuttle and space station.

The contemplation of this possibility has led many supporters of human spaceflight to abandon the Big Science model in favor of small private entrepreneurs. If the price for this change requires the discontinuation of direct government support for NASA and its aerospace partners, the advocates of small-scale entrepreneurship are willing to pay it. Saving human spaceflight, they believe, requires a radical transformation in the methods of carrying it out.

The concept of small entrepreneurs working through the private sector to explore space has a substantial base in science fiction. The first movie to realistically portray space travel was Fritz Lang's *Frau im Mond,* in 1929. In the film, venture capitalists finance the first lunar expedition with the expectation that the Moon contains precious minerals. They are not disappointed. In the 1950 film *Destination Moon,* public officials back out of their lunar rocket program after an unsuccessful test. Project leaders raise the money needed to complete the expedition by appealing to U.S. industrialists. In Kubrick's prophetic film *2001: A Space Odyssey,* the winged Space Shuttle is operated by Pan American Airlines, while the large rotating space station is managed by the Hilton Corporation.[83]

At the beginning of the space age, few supporters of the von Braun paradigm questioned the need for a government-run program. The best rocket engineers, like von Braun himself, worked on the government payroll. The substantial amounts of money needed to fund human space expeditions, linked to the very slow rate of expected commercial return, discouraged private investment in human flight programs. (Quicker returns were available for robotic flight activities, allowing easier development of commercial activities like communication satellites.) The management techniques needed to coordinate human flight activities also encouraged the participation of large government agencies and corporate giants.

The Big Science approach, for all of its advantages, encourages the use of proven concepts and technologies. In the private sector, the marketplace weeds out ineffective ideas. In government-run programs, the iron triangle of bureaucratic, congressional, and corporate interests maintains projects and policies long after their usefulness has passed. Said one commentator: "This may be a way to keep [NASA's] massive civil service and contractor armies together. But it's the enemy of routine access to earth orbit, which would allow space finally to become a thriving part of our human economy and make it affordable to contemplate a permanent human presence on the moon and Mars."[84]

Some of the most devoted advocates of human space travel would like NASA to dissolve its historic commitment to Big Science through big government and let private entrepreneurs take over much of the work. The principal reasons for using big government to explore space no longer exist, they insist. Aerospace expertise is widely distributed through the private sector, and business executives know how to manage complex projects, often using methods that are more innovative than those practiced in government bureaucracies. As for the problem of money, advocates of the new approach insist that the cost of spaceflight will fall dramatically once the private sector gets involved. It will never fall, they argue, so long as the government tries to prop up big aerospace contractors whose capacity for innovation and competitiveness has vanished after years on the government dole.

This is a radical agenda, but it is one that finds increasing support among human flight advocates. Many have already begun work. Venture capitalists founded LunaCorp in 1989 for the purpose of developing commercial activities on the Moon. (The firm was dissolved in 2003.) A succession of small startups, such as Kistler Aerospace, attempted to develop alternative launch technologies in the 1990s. In 2004, Burt Rutan won the X Prize for the first privately financed suborbital spaceflight. Backed initially by Microsoft co-founder Paul Allen, Rutan received additional capital from billionaire Richard Branson in an effort to develop a Virgin Galactic spaceship that could fly paying customers to the edge of space. Rutan spent about $25 million developing his two-passenger SpaceShipOne, which had to complete two suborbital flights to win the prize. By contrast, when the space race began, NASA officials spent more than sixty times that amount—the equivalent of more than $1.5 billion in 2007 dollars) conducting two suborbital and four orbital Mercury flights.

Space entrepreneurs would like the government to withdraw from the business of operating spacecraft and extraterrestrial bases and provide financial incentives to commercial firms working through the private sector to provide the same. NASA astronauts and scientists would use the privately operated facilities to conduct research and explore space, much like scientists working with government funds draw on privately managed facilities in Antarctica to do their work. The U.S. Antarctic base at McMurdo Sound, for example, is no longer run by the U.S. Navy but is managed by Raytheon Polar Services for research personnel of the U.S. National Science Foundation.

Concern with the centralized power in spaceflight compares in some respects to the cries of opposition voiced in the American West to the presence of the federal government. In order to flourish in the arid West, Americans needed an agricultural economy dependent upon waterworks that produced irrigation and protected against floods. This not only made the West habitable, it brought urbanization and wealth as well. Ancient Egypt engaged in this type of civilization, and it became a dominant power as a result. Invariably, this requires a strong governmental presence, often of a centralized nature. The resulting policies, as always, produce winners and losers, and those left out harp on the inequities of the system. In the American West, the "Sagebrush Revolution" of the late twentieth century pitted individualistic Westerners against the organization and power of the federal government. Ironically, the populations that benefited most from the organization and power of federal water policies often were the ones who attacked the government the most vociferously. In a similar manner, spaceflight advocates who argued for colonization of the Moon and Mars eschewed the very organizations—NASA and the federal government—that made possible the first step through Project Apollo.[85]

THE DOMINANCE OF HUMAN FLIGHT

Ultimately, the willingness to invest in human spaceflight gained justification from a desire to extend some part of humanity into the cosmos. The desire took many forms—the urge to send people where "no man has gone before," respect for those who explore, and a general hope that humans might transcend their own planet and achieve some form of material immortality. Former NASA official Frank Martin, now of Lockheed Martin Space Systems, explained the situation tersely: "We don't give ticker tape parades for robots."[86] His position, not original to him but certainly pithy, suggests a central reason for the dominance of human flight.

Throughout the heroic era of human spaceflight—when NASA conducted the Mercury, Gemini, and Apollo programs—the media gave enormous attention to human exploits. Astronauts became heroes and celebrities. Their bravery touched emotions deeply seated in the American experience of the twentieth century. Citizens in the Soviet Union accorded similar status to their cosmonauts. These flights recalled, but did not replicate, Charles Lindbergh's "lone eagle" crossing of the Atlantic Ocean in 1927. Facing great personal danger, space travelers fit the myth of frontier law enforcers, whose grit had once filled the substance of Hollywood matinees and feature films. The first astronauts and cosmonauts had much in common with the mythology, if not the actual careers, of Wyatt Earp, Bat Masterson, and Matt Dillon.[87] As military test pilots, they recalled the sacrifices required to produce the Allied victory in World War II, a period during which military service was held in exceptionally high regard. Their personal exploits even recalled the substance of one of America's most popular sporting events, automobile racing, with its danger, rituals, and excitement.[88] Test pilots, racecar drivers, and frontier marshals were thought to be a hard-living, often hard-drinking lot.[89] At the same time, as the readers of *Life* magazine in the 1960s learned, astronauts combined toughness and skill with humanity and patriotism. Space travelers epitomized those ideals beyond the Earth.

Likewise, beginning in the late 1950s and accelerating with time, space scientists and engineers undertook a set of exceptional robotic missions.[90] The Mars *Sojourner* rover received no funeral orations when it finally conked out. Neither did *Lunar Prospector* when it crashed into the Moon, or *Galileo* when it suffered a fiery demise in Jupiter's atmosphere at the end of its mission. The loss of astronaut and cosmonaut crews, on the other hand, produced funerals of the sort reserved for the highest national heroes. Frank Martin's characterization slashes to the core of the human versus robot issue and points up a fundamental reason for the dominance of human flight. It personalized space in a manner that machines could never do. Nonetheless, NASA has anthropomorphized these robotic explorers, given them names, assigned various emotions to them, and treated them as semi-independent individuals. How far are we from assigning personalities to our robotic avatars, and agitating for mores and perhaps even laws governing the ethical treatment of these entities?[91]

As the first part of the twenty-first century unfolds, the longstanding pillars supporting the reasons for undertaking human spaceflight are weakened in ways not previously seen. Only the sense of human destiny and the sense of prestige assigned to human spaceflight remain. But the prestige and power deriving from human spaceflight especially seem to be crumbling. Such is the case with all previous prestige efforts in human history. Egyptian civilization did not build pyramids indefinitely. Neither could the kings of Easter Island sustain the large stone carvings that denoted their wealth and status.[92]

Of course, we must ask, why did the von Braun paradigm become the guiding force in the U.S. civil space program? Certainly it proved so powerful that it effectively transformed the debate over space policy into entrenched camps—us versus them—those who advocated human exploration of space and those who opposed it. The von Braun paradigm prevailed because of its association with the broad sweep of exploration history, the ascendancy of Big Science, the growth of big government—and the supposedly "negative" character of the opposite point of view. Most important, it drew strength from public optimism about the human future in space. In a romantic sort of way, people within the culture at large came to believe that the national character and the future of humanity would depend in some measure on the ability of humans to transcend the sphere on which they arose.

CHAPTER FOUR

Robotic Spaceflight in Popular Culture

Science fiction proved to be a powerful force for generating public interest in actual space travel. Fictional sagas, transmitted through the media of popular culture, anticipated the practical presentation of spaceflight and encouraged the formation of societies devoted to advances in rocketry and spacecraft. Without the work of early science fiction writers, the images promoting space travel would not have seemed as familiar as they did. The history of human spaceflight sat on a solid foundation of fictional tales.[1]

Science stories for public consumption, often characterized as popular science, followed in the wake of fictional tales. One of the most influential presentations appeared between 1952 and 1954, a few years prior to the launching of the first Earth-orbiting satellites, as editors at *Collier's* magazine produced an eight-part series describing the exploration of space. Seven of the eight issues dealt with various aspects of human space travel: training astronauts, building space stations, exploring the Moon, and traveling to Mars. Only one issue presented automated flight. The June 27, 1953, cover of *Collier's* displayed a remotely controlled, conically shaped spacecraft that Wernher von Braun and Cornelius Ryan, the series editor, characterized as a "robot" and a "baby space station."[2]

The emphasis given to human over robotic flight in *Collier's* reflected the relative standing of these two perspectives within the culture at large. Fictional stories about robots appeared prior to 1953, but they were neither as numerous nor as well developed as stories about human travel. Moreover, the robot stories tended to reinforce the popular image of space as the province of human beings. Like works of fiction touting human spaceflight, robot tales embraced the heroic tradition of terrestrial exploration in which humans clearly played the dominant role.

The resulting traditions created a popular image of robotic flight that was weak where the real robotic effort was strong. In spite of the existence of well-told stories, the robotic tradition did not engender popular understanding that began to approach the persuasiveness of the human flight vision. Yet in many ways the image of robotic flight was prophetic. In presenting mechanical creations that rebel against their creators, storytellers anticipated strange new worlds in which post-biological space travelers might prevail.

THE POPULAR IMAGE OF ROBOTICS

Inside that June 27, 1953, issue of *Collier's,* an illustration displayed the internal workings of the robotic satellite that von Braun and Ryan proposed to fly. The satellite contained three monkeys launched on a sixty-day voyage to test the effects of weightlessness and other space hazards on living organisms. Before venturing into space, humans needed to understand the effects that cosmic conditions might impose on the first people to fly into space. As the authors announced, "The monkeys on the satellite will tell us."[3] In short, robotic flight existed to prepare for human flight. NASA officials implemented this vision six and a half years later. Before launching the first *Mercury 7* astronauts into space, engineers conducted three suborbital tests of their single-seat spacecraft using two monkeys and a chimpanzee, named Ham, as stand-ins for the humans that followed.

The "baby space station" typified an attitude common among advocates of human spaceflight. Machines and possibly animals would explore space, but only as adjuncts to human beings. Animals would prepare the way for human flight, while machines would do the same and possibly serve as companions. The vision of machines exploring space in the absence of human control received as little attention as the notion that somehow monkeys might explore the Moon.

Culture consists of the assumptions that people make about the world in which they live. It is frequently transmitted through stories that people tell. The most memorable stories about robots—the ones attracting large and continuing audiences—encouraged the widespread assumption that humans would lead the exploration of space, with machines—to the extent that they participated at all—serving as accomplices to their human masters. The image of "robots and humans together" became the official policy of the National Aeronautics and Space Administration.[4]

Robot stories had begun to appear. Gnome Press published Isaac Asimov's famous collection *I, Robot* in 1950, two years prior to the initiation of the *Collier's* magazine series. The book contained a set of science fiction stories that Asimov had begun publishing ten years earlier, with the appearance of "Strange Playfellow" (later retitled "Robbie") in the September 1940 issue of *Super Science Stories*.[5] During the same period, Norbert Wiener's discussion of the human-machine relationship, *Cybernetics; or, Control and Communication in the Animal and Machine,* had appeared.[6] In it, Wiener examined the manner in which humans and machines used information to carry out their work, creating a theoretical foundation for the practical design of robots as Asimov described them. Wiener's analysis suggested that humans and machines could be made to operate according to the same principles. The fundamental issue, as Asimov and Wiener explained, was control.

Such an issue, transmitted to the general public, fit a familiar social concern. Would any object so constructed behave itself? Would it submit to external control or, as the capabilities of such objects began to approximate and in some cases exceed those of human beings, rebel against its creators? Anyone familiar with the history of industrialization and the tendency of employers to treat factory workers like machines could appreciate the issues involved.

This issue provided the social foundation for some of the most interesting treatments of robotics, especially in science fiction. Those treatments dealt extensively with what Asimov characterized as the "Frankenstein complex," the tendency to believe that machines and other human creations would harm their creators.[7] Such treatments drew heavily upon the image of robots as servants and the belief that servitude in its most extreme form, that is, human slavery, was both morally reprehensible and objectionable to the subjects involved. When robots appear in early science fiction stories, the tales often deal with the misuse or misunderstanding of intelligent machines.

From such stories, two popular images of robotics emerged. In the first, robots and their human masters work together. Such robots, even those that acquire exceptional capabilities, continue to serve their creators. The robots are helpful but remain subservient. They lack the capacity for independent thought, because, in the final analysis, they are simply machines.

In the second tradition, machines make the transition from mechanical to sentient form. Acquiring the characteristics accorded human beings, the machines defy their creators and make decisions affecting their own existence. Such creations, in effect, behave like newly freed slaves or factory workers who have recently discovered the power of collective bargaining.

People familiar with robot stories might easily understand how humans and their machines could explore space in a cooperative manner, the conventional NASA policy. The stories made discussions of a "post-biological universe," an age in which sophisticated entities of a nonhuman nature acquire equality with human beings, seem less strange.

What robot stories failed to do was set the groundwork for the use of machines as it actually occurred. No significant social tradition underlay the automated approach to space exploration that came to represent the actual use of robots during the early stages of spaceflight. In that approach, humans create machines, place them under human control, and send them into the void to do the work of creators, who remain behind. Such machines remain under human supervision and are not capable of independent thought, but their human creators do not accompany them. Plenty of robots did this work in practice, but few of them appeared in the most influential stories told within the popular culture. Consequently, the robotic approach to space exploration, as it actually occurred, emerged without the advantages that might have been imparted by a strong social tradition. Changing social concerns made that connection weaker still.

THE POPULAR IMAGE OF EXPLORATION

Visionaries who predict future events often rely upon old concepts to explain their prophecies. Old ideas are more familiar, require less interpretation, and are easier for a vast audience to understand. Given an opportunity to explain a new development with an old story, the communicator will usually adopt the traditional approach.

When Gene Roddenberry developed his plan for the original *Star Trek* television show, he told writers for the series to rely upon the concepts previously employed to market fictional stories about the American West. Westerns began to appear in the late nineteenth century in cheap pulp paper magazines like *Argosy,* founded in 1896. Science fiction followed a similar path. Cheap pulp paper magazines devoted to westerns and adventure stories ("the pulps") provided the primary outlet for writers telling science fiction tales, which helped to create an audience for their later transition to film and television.

In developing *Star Trek,* Roddenberry told his writers to avoid the weirder aspects of science fiction and work from a proven concept: "Let's go back to the days when some of us were working on the first television westerns. We did *not* create the Old West as it actually existed; instead we created a new Western form, actually a vast colorful backdrop against which any kind of story could be told."[8]

The art form depicting the American West, perfected in Owen Wister's 1902 novel *The Virginian,* was not created for the purpose of accurately portraying settlement conditions in nineteenth-century America.[9] Rather, the popular western provided a new vehicle for retelling traditional tales about human courage, heroism, the power of love, and the triumph of good over evil.[10] Humans have told such stories for thousands of years, back through the works of William Shakespeare to *The Iliad* and *The Odyssey*. The *Star Trek* television series, launched in 1966, provided a new and refreshing setting for telling such stories. The most memorable stories mixed traditional themes with current events such as superpower conflict, fascism, civil rights, and interracial dating—reset in a galactic realm.

In a similar manner, advocates of space exploration at the midpoint of the twentieth century used familiar images from the past to excite public interest in this new adventure. People like Wernher von Braun and NASA Administrator Tom Paine, who placed humans at the center of this venture, drew their material from a deep and familiar well. Advocates of human flight portrayed space exploration as a new frontier attracting hardy pioneers. They portrayed space stations as the orbital equivalent of frontier forts. They described spaceships with terms drawn from recent and amazing developments in aviation. They promised the public that space exploration would lead to the discovery of new species, just as terrestrial exploration had done. Most important, they promised people increasingly isolated from the natural world that space activities would continue the heroic tradition of natural discovery that had captivated armchair explorers for so many years.

Few developments had more influence on the popular image of space exploration than the memory of terrestrial expeditions crossing the Earth and its seas. The "space age" opened a few decades after the closing of what commentators termed the "heroic era" of earthly exploration.[11] Scarcely four decades before the orbiting of the first Earth satellite, separate parties led by Roald Amundsen and Robert Scott became the first to stand on the South Pole during the Antarctic summer of 1911–12, an event that captured substantial public attention when Scott and his four companions perished on the trek back. These accomplishments followed nineteenth-century expeditions engendering substantial public attention such as Meriwether Lewis and William Clark's journey into the American West (1804–06), Charles Darwin's role as captain's companion and amateur naturalist on the H.M.S. *Beagle* (1831–36), John Wesley Powell's descent into the

Grand Canyon (1869), and Henry Morton Stanley's heavily publicized expeditions to Africa (1871–89). Controversy heightened public interest, as in 1909 when Robert E. Peary and Frederick Cook simultaneously announced the honor of being the first to reach the North Pole.

Terrestrial expeditions followed a well-established formula, one broken only at the end of the "heroic era." Expeditions operated autonomously, out of contact with their sponsors or home base. Typically, the public would not receive reports on the accomplishments of the expedition until its leaders returned and published their findings. Unless a returning vessel passed by, the public might not know whether the absent explorers were dead or alive. Cut off from sponsors, members of expeditionary forces relied considerably upon their own skills and supplies. Under such circumstances, success or failure depended primarily upon human ingenuity. Exploration in the traditional mode demonstrated the power of human ingenuity relative to both the natural and the industrializing world.

No event better characterized this formula than the Trans-Antarctic Expedition of 1914 led by Ernest Shackleton. As leader of the British Antarctic Expedition of 1907–09, Shackleton had come within ninety-seven miles of the South Pole. He returned in 1914 as captain of the expedition ship *Endurance*. Pack ice trapped the ship, forcing the crew to winter on board. Shackleton hoped that the spring thaw would free the vessel, but the moving ice crushed and sank the craft. After living on the pack ice for nearly five months, the crew used the ship's small boats to reach Elephant Island, an uninhabited island at latitude 61. To effect a rescue, Shackleton and five members of the expeditionary force refitted one of the boats, a twenty-foot whaler, and sailed eight hundred miles to South Georgia Island. Arriving on the wrong side of the island, the group conducted the first traverse of the land mass and reached the Stromness Whaling Station, where they were able to communicate their plight and bring about a rescue of the remaining crew.

Notably, the expeditionary force carried a wireless radio on the *Endurance,* an invention developed hardly a decade earlier. The device proved useless, however, with the result that the crew had no contact with the outside world until Shackleton walked into the Stromness station seventeen months after the expedition's departure.[12]

Murray Lerner and Kurt Neumann concocted a similar scenario thirty-four years later with the production of the classic science fiction film *Rocketship X-M*. Released at the same time as the equally influential *Destination Moon,* the film tells the story of a lunar expedition thrown wildly off course by a swarm of passing meteorites, dispatching the spacecraft on an unplanned interplanetary voyage. The crew wakes to a dramatic view of the planet Mars, into whose vicinity the errant spacecraft has strayed. The explorers land and proceed to explore.

Written during the nuclear hysteria of the early Cold War, the film reveals a Martian civilization blasted into the Stone Age by a self-inflicted atomic war. The dying planet is inhabited by a race of blind, mutated survivors. The crew's determination to communicate this terrible possibility to earthly leaders engaged in their own arms race is frustrated, however, by the insufficient power of the ship's radio, whose range is confined to a few hundred miles. Only by returning to the vicinity of Earth can the crew transmit their findings and communicate the expedition's plight, a substantial challenge given the hostility of the Martians and the ship's insufficient fuel.[13]

In proposing his own version of a Martian expedition, Wernher von Braun combined two of the most familiar images of terrestrial exploration during the previous hundred years—great ships crossing large distances and a heroic land traverse. His proposal, as a consequence, attracted enormous public attention, appearing in the April 30, 1954, issue of *Collier's* magazine and the subsequent book coauthored with Willy Ley. The plan, based on an obscure novel that von Braun had prepared shortly after arriving in the United States, was more notable for the familiar images it conveyed than its technical feasibility.[14]

To conduct the expedition, von Braun proposed that earthlings assemble ten large ships in Earth orbit, each of which, with the associated equipment, fuel, and supplies, would weigh about 3,700 metric tons. Seven of the ships were designed to carry the expedition to Mars orbit and back and would not be capable of landing. Three of the ships, with wings, would land. In the *Collier's* version, the first vessel would glide to a stop on the Martian polar ice cap, the only region thought sufficiently smooth to accommodate the skids on which the landing craft must touch down. Once on the Martian surface, the crew members from the first landing craft would abandon their ship and undertake a heroic trek over four thousand miles of unexplored terrain to the Martian equator. For this phase of the expedition, crew members would employ a fleet of tractors, fuel trailers, and inflatable habitat spheres. Arriving at the equator, the expeditionary vanguard would use tractors to bulldoze a strip onto which the two remaining gliders could land. On the ground, the explorers would unbolt the gliders' wings and set the two spacecraft on end, ready for the return voyage, a scene anticipated in a famous painting prepared by the artist Chesley Bonestell. Other members of the expeditionary team would inflate a dome-shaped habitat, completing a thirteen-month investigation of the planet's features before lifting off for home.

Von Braun's plan clearly recalled the great terrestrial expeditions of earthly adventure. He called for his Martian expedition to be gone more than nine hundred days, similar in length to the great terrestrial expeditions launched by earlier explorers like Ferdinand Magellan and Sir Francis Drake. Von Braun proposed an expeditionary crew of seventy individuals, a fairly large corps. The familiarity of this vision obscured its technical difficulty. Von Braun admitted that assembly of the flotilla would require 950 flights of his proposed winged Space Shuttle, since the mass of the entire venture prior to departure from Earth orbit would exceed 80 million pounds, including propellant. Each shuttle flight, von Braun hoped, would deliver eighty-five thousand pounds to low-Earth orbit.[15] The actual Space Shuttle delivered about fifty thousand pounds and never flew more than eight times in a single year. Von Braun also seriously misestimated the lifting surface required to land a glider in the thin Martian atmosphere. The fin-shaped wings were designed to make the gliders look streamlined and pleasing to the public eye; they would not have been large enough to achieve a safe landing.

The technical shortcomings of the proposed mission did not stimulate much discussion; the historical analogies did. Here was an opportunity to resurrect the great terrestrial voyages of discovery in an interplanetary realm. Von Braun clearly understood the power of the analogies he had employed to enlist public enthusiasm. "I knew how Columbus had felt," he admitted to a writer for a *New Yorker* magazine article that appeared in 1951. "Not just to stare through a telescope at the moon and the planets but to soar through the heavens and actually explore the mysterious universe!"[16]

The analogies employed by advocates of human flight depended substantially on the image of courageous explorers using human skills to overcome adversity. As such, the proposals left little room for the contributions of servants or machines. The difficulty of assimilating subsidiary contributions into the heroic analogy had been demonstrated quite adequately by the contested claim over the attainment of the North Pole. Robert Peary and Frederick Cook had each insisted that he had reached the North Pole before the other. Their contested claims seemed to ignore the possibility that both might have been preceded by a servant. Peary was accompanied by his associate Matthew A. Henson, an African American who had previously worked as a seaman out of Baltimore, Maryland, and as a store clerk in Washington, D.C. Peary originally hired Henson to serve as a valet. Henson accompanied Peary on a number of polar expeditions in the two decades that followed, culminating in the 1909 attempt to reach the North Pole. By that time, Henson had become a highly skilled hunter and sled driver. On April 6, 1909, following a trek away from camp, Henson returned and confessed that he might have been "the first man to sit on top of the world."[17] Peary's ambition (and, perhaps, racism) led him to respond angrily and march off without Henson to reach the pole. Upon his return, Peary spent time using navigational records to support his claim and discredit Cook without reference to the possibility of Henson's accidental achievement.[18]

Peary's attitude toward Henson was typical of the heroic age, rooted as it was in the cultural practices of the nineteenth century. Both the racism and the rigid class consciousness of the Victorian era prescribed a rigid separation between master and servant. The distinctive roles imposed upon the servants who worked "downstairs" and the masters who lived "upstairs" continued through the short-lived but fondly recalled Edwardian era. Coinciding with the reign of Edward VII of England from 1901 to 1910, the era was generally marked by a growing optimism regarding the advantages that science and technology would confer upon a society marked by traditional class distinctions.[19]

In the context of the times, expeditionary tales served to highlight the accomplishments of masters, not servants, porters, guides, and valets. The roles of each were rigidly prescribed. The notion that a machine—or, amounting to much the same thing, a porter—could take the place of an expedition leader was as inconceivable as the suggestion that a household cook or even a stove might take full credit for a fine meal served at a dinner party hosted by the owners of the home. Both notions would have been impossible to explain in societies with traditional social codes.

The heroic era of exploration ended with the widespread introduction of machines into the business of exploration. The 1928–30 expedition to Antarctica led by Richard E. Byrd was the first extensively mechanized assault on that continent and generally marks the conclusion of the heroic period. Byrd and his compatriots brought three airplanes, an aerial camera, a snowmobile, and an elaborate communications network consisting of three radio towers, twenty-four transmitters, thirty-one receivers, and five radio engineers. They established "Little America" near the site of Amundsen's abandoned base and in 1929 became the first to fly an airplane over the South Pole. As a consequence of its mechanization, the expedition maintained good communications from a region where, as Byrd himself observed, radio conditions had previously varied from nonexistent to bad. Radio communications improved to the point that the Columbia Broadcasting System (CBS) established a weekly radio show from Little

America, where, according to one writer, reports about life on the ice "gripped the public's imagination as much as the first moon landing years later."[20]

Extensive communication equipment and visible dependence upon machines undercut an essential element of heroic exploration—the vision of intrepid explorers out of touch with "civilization" left to their own leadership and devices as a means for accomplishing their goals. Dependence upon machines (or, for that matter, dependence upon guides) compromised the ability of expeditionary publicists to elevate the heroic qualities of expedition leaders. By the midpoint of the twentieth century, the propensity of selfish expedition leaders to elevate their own accomplishments relative to those of their human companions had significantly declined. When the party led by Edmund Hillary completed the first successful ascent of Mount Everest in 1953, Hillary insisted that both he and his Nepalese sherpa, Tenzing Norgay, had attained the summit simultaneously. This represented a substantial departure from the treatment accorded Henson by Peary less than fifty years earlier. Nonetheless, the image of the heroic explorer and expedition continued to dominate the presentation of spaceflight.

ROBOTS TAKE FORM

Within this social context, consideration of the relationship between humans and robots in space first took place. In the dominant vision, humans worked alone or—when robots were involved—the machines assisted their human creators and masters. Like Peary's perceived superiority to Henson, robots were considered of secondary importance. This vision drew heavily upon images taken from the pre-mechanized era of exploration, particularly the master-servant relationship. Advocates of the classical approach to space exploration emphasized heroic achievements even though the machine culture that made space travel possible had closed the heroic age. Within the realm of popular culture, among works that were widely viewed or read, people favoring machines presented robots as subservient to human explorers, a perspective that helped perpetuate the romantic vision of exploration. In fact, machines had significantly weakened the potential for heroic exploration on Earth and would inevitably do so in space. Early writers on space robotics, however, could not make this point in ways that would attract a large audience.

Advocates of a significant role for robots faced a curious conundrum, one that drove producers of popular literature back toward the heroic mold. To write interesting stories, writers who envisioned a significant robotic role in space exploration had to give personalities to the machines. Robots with personalities or other such creations that felt sensations might not be content to be servants. They could turn on their masters, a point repeated in a vast succession of stories such as *Spartacus* and *Uncle Tom's Cabin* that dealt with revolts of the slaves. To avoid the more pessimistic robot-in-revolt stories, writers like Asimov invariably described robots as servants or machines cooperating with their human masters. This fit neatly with the traditional meaning of *robota*.[21]

The word *robot* is taken from the Czech *robota*. The original term refers to statute labor or compulsory service. Its traditional use arises from the feudal practice of requiring peasants to work periodically for a few days without remuneration in the fields of noblemen. Modern use of the term is generally ascribed to the Czech playwright Karel Čapek, whose 1921 play *R.U.R. (Rossum's Universal Robots)* depicts an attempt to liberate biologically engineered human substitutes forced to serve as inexpensive workers from a factory in which they are being produced.[22] At the time of Čapek's play, *robota* meant the performance of necessary but tedious or unrewarding work.

By its origin, robotics refers to domestic or industrial work that is performed by human substitutes. It arose during a period marked by the widespread use of domestic servants and low-skilled labor and not long after the abolition of slavery. Robots perform compulsory labor of an often wearisome sort. According to the original vision, robots are not substitutes for humans acting in their more inventive capacities, nor are they allowed to act as masters or employers of servants. Rather, robots carry out the work commonly relegated to persons at the lower levels of society.

The people who formulated the vision of robotics were far too preoccupied with the necessity of creating an interesting story to spend much time creating a realistic vision of robots on automated missions in space. Even if writers had attempted to portray robots as semi-autonomous entities, such a vision would have been very hard to explain to a general public that had a clear conception of the use of servants and unskilled labor.

The common presentation of robotics appeared alongside a huge literature on the disorienting effects of industrialization and work. In 1933, Elton Mayo published his classic study on *The Human Problems of an Industrial Civilization,* in which he discussed the attempts of factory workers to construct meaningful social relationships within industries that typically treated workers as interchangeable parts. Mayo drew on the concept of *anomie,* formulated by the French sociologist Emile Durkheim to describe the breakdown in social bonds and common standards to which communities in industrialized societies seemed prone. Novels such as those produced by Emile Zola, Upton Sinclair, and Richard Llewellyn examined the dislocating effects of industrialization.[23]

The social context within which the literature on robotics arose inherently prompted its disseminators to discuss matters of labor and master-servant relationships, especially when those disseminators wrote science fiction. They debated whether humans, in creating robots, could fully understand the social consequences of engaging in such an act. Writers contemplated the work ethic of robots and the ability of humans to program machines in such a way as to assure that robots would perform their tasks in the manner intended by their creators. Inevitably, the writers questioned whether robots could be treated as inhuman entities or whether they would develop personalities and instigate a rebellion over the subservient work assigned to them. The latter train of thought drew directly on the experience with human slaves, who by law had been treated as property and had not been afforded the rights of citizenship. The existence of legalized servitude or slavery had existed in the United States a mere sixty years before Karel Čapek's robotic play appeared.

The most influential writer on space robotics inserted himself into the middle of this cultural milieu. Isaac Asimov's first robot short story was "Robbie," a simple children's story with a powerful social message. Nearly every robot story that the nineteen-year-old Asimov had read as of 1939 featured machines run amok: "hordes of clanking, murderous robots," he later complained. The basic story, he observed, was "as old as the human imagination" and could be traced back to Greek mythology.[24] In the Prometheus legend, the god Prometheus creates a woman, Pandora, out of the Earth; in the Roman version, he uses fire to create human beings. To help them improve their lives, Prometheus presents humans with the knowledge of fire. For such acts of insubordination, Zeus chains Prometheus to a mountain where crows chew on his liver and Pandora opens a box that releases evil and misery among humankind. The legend, which in many ways repeats the biblical story of Adam and Eve, warns of the dangers of invention and the ill consequences associated with the acquisition of knowledge.

Mary Shelley used the Prometheus legend as the intellectual framework for her fantastically influential novel depicting the ill effects of one person's effort to create an artificial human being. The subtitle to *Frankenstein,* which Shelley began in the summer of 1816, is "The Modern Prometheus." Her husband Percy, on which the character of Victor Frankenstein is based, a few years later penned *Prometheus Unbound,* in which he expressed the opposite point of view. Promoters of the Scientific Revolution that had swept Europe in the sixteenth and seventeenth centuries commonly cited the Prometheus legend as a means of emphasizing the importance of experimentation and discovery. Mary Shelley was much more skeptical. "How dangerous is the acquirement of knowledge," she wrote, within a person "who aspires to become greater than his nature will allow."[25]

In the novel, Dr. Frankenstein constructs a creature out of human parts for the immediate purpose of understanding the difference between living and dying tissue. In his effort to understand the principle of life, however, Frankenstein invents a creature without any regard to its care or education. Horrified by the creature's outward appearance, Frankenstein runs away, leaving the being to support itself in the Swiss countryside. The creature, which is inherently benevolent, turns monstrous only as a result of its abandonment and the abuses it suffers from human encounters. The creature pursues Frankenstein, murders both his brother and his bride, and eventually causes the death of the creator himself.

Frankenstein's sin, in Shelley's view, arises from his total lack of foresight regarding the eventual outcome of the experiment in which he is engaged. Shelley, in the guise of the chemistry professor who motivates Dr. Frankenstein to construct the being, accuses his students of seeking "unlimited powers" through which to "mock the invisible world."[26] The novel presents a message of cautious empiricism regarding scientific invention, particularly those activities that involve substitutes for human beings.

As a young man, Asimov was repelled by the "Frankenstein complex" and the science fiction stories that repeated this tale. He did not see robots that way at all. "I saw them as machines—advanced machines—but machines. They might be dangerous but surely safety factors would be built in. The safety factors might be faulty or inadequate or might fail under unexpected types of stresses; but such failures could always yield experience that could be used to improve the models."[27] Asimov could not bring himself to believe that "there are certain things that humanity is not meant to know." The suggestion that robot creation remain "the sole prerogative of God" he found nonsensical; nor did he look upon the design of robots as a prime example "of the overweening arrogance of humanity." He rejected the proposition that the creation of robots would provide "its own eventual punishment."

Reflecting upon the basic morality of early robot stories, Asimov provided one of the most comprehensive defenses of scientific investigation and discovery: "All devices have their dangers. The discovery of speech introduced communication—and lies. The discovery of fire introduced cooking—and arson.

The discovery of the compass improved navigation—and destroyed civilizations in Mexico and Peru. The automobile is marvelously useful—and kills Americans by the tens of thousands each year." In each case, Asimov observed, the misuse of technology could be used to justify its avoidance. Such a posture, however, would deprive humans of the advances to be gained. "Surely we cannot be expected to divest ourselves of all knowledge and return to the status of the australopithecines." Taking a theological point of view, Asimov argued that "God would never have given human beings brains" if he did not intend that humans use their intelligence to invent things and make improvements through experience. "So, in 1939, at the age of nineteen, I determined to write a robot story about a robot that was wisely used, that was not dangerous, and that did the job it was supposed to do."[28]

The "Robbie" story concerns a robot purchased for the purpose of serving as a nursemaid to a young girl named Gloria. The girl's mother is suspicious of the machine. Emblematic of the fears contained in Mary Shelley's classic tale, the mother conspires to have Robbie returned to U.S. Robots and Mechanical Men, the manufacturing company. The little girl, by now emotionally attached to her mechanical nursemaid, is devastated by the loss. Her father argues against the mother's fears, without success: "A robot is infinitely more to be trusted than a human nursemaid. Robbie was constructed for only one purpose really—to be the companion of a little child. His entire 'mentality' has been created for the purpose. He just can't help being faithful and loving and kind. He's a machine—*made* so."[29]

To distract Gloria from her loss, her parents take her to New York City. They ultimately visit the factory in which Robbie was made—and to which he has been returned—in the hope of convincing Gloria that robots are nothing but machines. The little girl spots Robbie. Significantly, Robbie is engaged in the manufacture of other robots. Gloria shouts his name and darts onto the factory floor directly into the path of a huge tractor. Her parents and their tour guide are not fast enough to save her. "It was only Robbie that acted immediately and with precision." He snatches the little girl from impending death and returns Gloria to her parents. The mother relents and allows Robbie to return home. "I guess he can stay with us until he rusts," she says, as the tale ends.[30]

Asimov used the story to introduce the first of his three fundamental laws of robotics. Robbie reacts to a requirement, programmed into each robot's positronic brain, that prevents any robot from injuring a human or allowing a person to be harmed. Asimov elucidated the three laws as part of a 1942 story titled "Runaround," which first appeared in *Astounding Science Fiction*. The three laws create a situation in which robots are programmed to remain eternally subservient to their human creators.

> "One, a robot may not injure a human being, or, through inaction, allow a human being to come to harm. . . ."
>
> "Two, . . . a robot must obey the orders given it by human beings except where such orders would conflict with the First Law. . . ."
>
> "Three, a robot must protect its own existence as long as such protection does not conflict with the First or Second Laws."[31]

Many of Asimov's robot stories deal in a jurisprudential manner with the nuances and potential conflicts inherent in these three laws. The elucidation of the laws, Asimov himself observed, proved to be "the most influential sentences I ever wrote." In addition to encouraging engineers to experiment with real robots, Asimov claimed credit for substantially reducing the number of rogue robot tales being produced. "The old-fashioned robot story was virtually killed in all science fiction stories above the comic-strip level," he wrote. Robots "grew to be commonly seen as benevolent and useful except when something went wrong, and then as capable of correction and improvement."[32]

A succession of science fiction stories presented robots as helpful and often amusing servants to the humans who employed them. In the classic science fiction film *The Day the Earth Stood Still,* a robot named Gort serves both his flying saucer commander and, eventually, humankind. He resurrects the commander after American soldiers stupidly shoot the emissary. The reborn commander warns the assembled world leaders that they must control the nuclear arms race and explains how his people have created a race of robots programmed to enforce this requirement. The television series *Lost in Space* that ran from 1965 to 1968 featured a large robot that one observer characterized as "an intelligent metal version of the dog 'Lassie,'" another popular family-oriented television program from that period.[33] The nameless robot warned the Robinson family of various dangers and commonly refused to participate in the nefarious schemes of the villainous Dr. Zachary Smith. In the cult classic *Silent Running,* Freeman Lowell reprograms three adorable droids named Huey, Dewey, and Louie to serve as companions as he attempts to preserve an orbiting space colony containing one of the few remaining forests transplanted from a desecrated Earth.[34]

The image of friendly compatriots inspired George Lucas to create C3PO and R2D2 for the fantastically profitable *Star Wars Episode IV: A New Hope* (that is, the original *Star Wars* movie). The high-strung C3PO is a protocol droid in humanoid form programmed to assist diplomats and other high-ranking officials with translations and matters of etiquette. A companion, R2D2, is an astromech droid, slightly less than a meter tall. Constructed with an assortment of tool-shaped appendages, astromech droids are manufactured by Industrial Automation to assist with the maintenance and repair of spacecraft. The *Star Wars* website lists no fewer than forty-eight droid models that participate in a variety of galactic affairs, including battles.[35]

Throughout the formative years of robot imagery, writers of space exploration stories commonly presented these machines as co-explorers or co-workers in the extraterrestrial realm. Among the most popular stories, one finds few robots that do human work in the absence of human beings. For his second robot story—the first set in outer space—Asimov introduced Robot QT-1. The machine's companions, two astronauts on Solar Station #5 named Donovan and Powell, call it "Cutie." Humans have developed solar stations for the purpose of directing energy beams to home planets and use robots to help with the work.

The story concerns Cutie's attempt to understand from whence it came. The robot's capabilities are so advanced that it has difficulty imagining that it could have been assembled by human beings like Donovan and Powell. The story's title is "Reason," and it refers to the robot's unshakeable belief that it must have been created by a higher power. The debate over origins is rendered moot, however, by the robot's stellar performance in accurately pointing an energy beam during a violent solar storm. "What's the difference what he believes," Powell asks, so long as "he can run the station perfectly." As Powell observes, the robotics laws oblige QT-1 to deliver the energy beam accurately without causing harm.

In "Reason," Asimov wrote a robot story in which a machine outperforms human beings. The solar power station is designed to be run by robots. Additionally, the robot sees no need for humans and believes that they are inferior beings. Yet even under this radical characterization of machine superiority, humans remain in charge. "Two human executives are required for each station," Asimov writes, to supervise the robot labor.[36]

Through the story, Asimov argues that even the most radical presentation of the machine vision need not negate human control. Even where a robot consciously concludes that machines possess superior capabilities, humans remain in charge. They do so, Powell explains, because the robots are programmed that way. In that respect, robots have no free will.

Rarely does Asimov present robots performing their work in the absence of human beings. One notable exception is found in "Victory Unintentional," a sequel to an earlier non-robot-related story titled "Not Final!" In the first story, humans begin to worry that the residents of Jupiter will master the technology of space travel and use it to pour into space and exterminate humanity. To gather information on the Jovian space technology and encourage peaceful relations, humans send three robots to the gaseous sphere. Here the second story begins. The robots are constructed to withstand the excruciating conditions found on Jupiter. The Jovians attempt to destroy the robots, but the machines are capable of enduring incredible extremes of hot and cold, pressure and vacuum, electrical fields, poisonous atmospheres, and radiation. Since the robots have come alone, the Jovians mistakenly assume that they are the actual humans against which the Jovians would wage war. Disheartened by the invincibility of the machines, the Jovians decide to abandon their ambitions for space travel and hide at home.[37]

"Victory Unintentional" is fascinating in a number of ways. It departs so dramatically from Asimov's normal robot themes that he did not include it in the *I, Robot* collection. No one in the story is preoccupied with interpreting the robotic laws, nor do they discuss improving the positronic brain. In that sense, the story does not fit into Asimov's conventional presentation of space exploration. Humans are not featured at all. Yet the story is a more metaphorically accurate presentation of the manner in which government space agencies would explore the outer planets in the decades that followed.

Curiously, the Jovians prove incapable of imagining space travel without living creatures in charge, thereby making the fundamental error of seeing their visitors as human beings when they are in fact machines. In constructing the story, Asimov imbues the Jovians with the romantic view of exploration that so captivated humans at that time. When he wrote the story in 1942, Asimov may not have appreciated the full social significance of his narrative tale. The dominant image of space exploration at that time assumed the presence of human beings; a story presenting the contrary proposition might have been more prescient but would certainly not have been as familiar. Cultural assumptions often operate at a semiconscious level where they are simply taken as true and not questioned in a serious way.

The difficulties inherent in presenting robots working alone also arise in Arthur C. Clarke's *Rendezvous with Rama*. Clarke describes a thirty-mile-long cylindrical starship moving through the local solar system. Humans approach and explore. The only creatures that early explorers encounter inside the spaceship are variously shaped "biots," robotlike creatures containing both mechanical and organic parts. As the spacecraft approaches the sun, the biots appear, manufactured by a preestablished process for the purpose of preparing the starship for some seemingly unfathomable encounter. So where are the Ramaians?

A strong advocate of human spaceflight, Clarke often introduced through his stories superior beings who had left their physical form for a more transcendental existence. Such beings rarely appear in a physical, biological form. The evidence for their existence consists of artifacts such as the starship in *Rendezvous with Rama* and the strange monoliths in *2001: A Space Odyssey.* Readers must plow through hundreds of pages in the original Rama novel and its three sequels (written with coauthor Gentry Lee) before discovering the purpose of the cylindrical spacecraft and the nature of its designers.

In a vast universe, Clarke suggests, humans may encounter evidence of extraterrestrial life without meeting the aliens themselves. In *Rendezvous with Rama,* no aliens come forth. Yet worker biots continue to perform the chores for which they were programmed, a testament to the persistence of sentient control. Clarke wrote the novel in 1973, when optimism about the ability of humans to achieve the romantic vision of space exploration was still strong.

The necessities of literature force writers of science fiction to give personalities to the characters they invent. Given that necessity, robots exhibit humanlike qualities. Joined with the desire to exalt human spaceflight, this requirement has helped to create a substantial literary tradition within the popular culture of space travel in which humans and robots work together as companions in the spacefaring enterprise. This tradition departs substantially from the practice of sending automated rovers and other semi-autonomous spacecraft to inspect celestial bodies. Such travelers remain under human control, but humans do not accompany them. The popular literary treatment of robotics skips over that image, so much so that the supporters of automated spaceflight found themselves without a distinct social tradition that they could use to explain their work. To most people, *robot* means either a helpful companion or something so far out of control as to have an intelligence of its own.

MACHINES AS SENTIENT BEINGS

Advocates of actual spaceflight resolved themselves into two camps shortly after space travel began. Advocates of human flight proclaimed a vision in which people moved into space, often preceded by subservient machines. Advocates of robotic flight presented a vision in which machines, remotely controlled from Earth, performed any tasks that humans asked them to do. As suggested earlier, this produced a false dichotomy. Existing trends in technology favor the emergence of other alternatives, including one in which human and machine capabilities merge. Not only do humans and robots explore space together, the two entities acquire the characteristics of each other. This other alternative may seem strange in practice, but it draws on a familiar art form. Cyborgs, androids, and artificial intelligence computers appear with startling regularity in works of fiction. Imaginative stories of this sort have created a rich social tradition. If this vision fails to come true, it will not be for lack of anticipation. As it appears in artistic literature, the humanlike machine (or the machinelike human) typically finds its expression in tales of homicidal equipment.

Dangerous machines appear in varied forms. Yul Brynner portrayed an amusement park robot gone wild in *Westworld*. Programmed to lose gunfights, the robot in this 1973 movie begins to outshoot customers who have paid for a realistic but safe adventure. Revolting only in a personal sense, Marvin the Paranoid Android was a featured character in Douglas Adams's *The Hitchhiker's Guide to the Galaxy* and its sequels. Constructed as part of the "Genuine People Personalities" line by the Sirius Cybernetics Corporation, Marvin is highly intelligent but severely depressed, which causes him to whine constantly, irritating his shipmates.[38]

Robots also rebel in Philip K. Dick's classic novel *Do Androids Dream of Electric Sheep?* In that book, Dick describes android servants who escape from human settlers on Mars and attempt to hide on Earth. The androids—a term used to describe robots with human features—are nearly indistinguishable from *Homo sapiens* and thus hard to find. By escaping and sometimes murdering their masters, the androids violate the purpose for which they were created. Dick questions the morality of locating and turning off such humanlike creations. In spite of the tendency of humans to view androids in an anthropomorphic way, as if they were human, Dick eventually reaches a conclusion similar to that prescribed in Asimov's tales. Androids are merely intelligent machines. Answering the title of his book, Dick reveals that androids would not dream nor assign any particular value to electric sheep unless humans programmed them to do so. The book provided the plot line for the science fiction film *Blade Runner*.[39]

Some rebellious machines are not robots in the conventional form but rather computers elevated to such high levels of intelligence that they too resist their creators. One of the most radical manifestations of this basic story appeared with the release of the 1999 film *The Matrix*. In it, humans believe themselves to be living in the world of 1999. In fact, the year is 2199, and the Earth has been rendered nearly uninhabitable. The lovable city in which humans believe they reside is totally simulated, a virtual reality creation with no more substance than a computer game. Machines create the simulated reality as a means of keeping humans in their contented vegetated state, from which they provide sufficient energy to power the machines.

The film and its inevitable sequels offer a terrifying commentary on modern life. In this version, machines have achieved a level of consciousness and intelligence sufficient to design an entire artificial world constructed for their own self-preservation. The simulated world they create contains self-aware programs that manifest themselves as agents capable of pursuing any humans who defy the plan. Humans who break out of the Matrix and fight the agents are with few exceptions incapable of distinguishing that which is real from that which is artificial. The movie provided a commentary on the confusing nature of reality in postmodern society. In the traditional world, humans thought of themselves as masters of their environment and creations of God. In a world where electronic circuits mimic the operation of the human brain, absolutes disappear. Who is to say what is substantively real? Consciousness and reality in the postmodern society of *The Matrix* have been reduced to little more than sequences of electronic stimulation. Compared to earlier, happy robot stories, it is meant to be a frightening experience.

One of the most influential "killer computer" stories appeared in 1968, with the release of the widely viewed *2001: A Space Odyssey.* The villain of the film is the HAL-9000 computer, programmed to operate a spacecraft carrying astronauts on an expedition to the outer solar system. HAL stands for heuristically programmed algorithmic computer, a significant insight into the nature of the device. HAL is not a conventional robot but an artificial intelligence computer designed to think like a human being. The term *heuristic* refers to the processes by which humans discover new processes and enlarge their understanding of the world.

Experts on information technology such as Herbert Simon have carefully distinguished between programmed and heuristic activity. Programmed activity refers to processes that tend to be linear in nature (like an assembly line) and repetitive, for which rules can be established in advance. Computers can be programmed to oversee such activities, such as launching rockets and landing rovers on Mars. Even though they may perform their activities very rapidly, such computers are little more than complex adding machines, programmed to perform specific operations according to preestablished codes.[40] They can mimic human skills, such as decision making and speech, but their humanity is strictly an illusion. They are no more human than a cartoon character drawn by a clever illustrator.

The fictional HAL-9000 computer is quite different. A set of general codes prescribe the manner by which HAL acquires and processes new information. HAL is given a male personality and programmed to learn. One of the most important lessons HAL learns is not to make mistakes, lest the computer be viewed as fallible and shut down. HAL develops an internal conflict, however. Privy to the secret objective of the mission, the computer must lie to the crew about it lest they inadvertently reveal this top-secret objective over open radio channels.

The internal conflict manifests itself as an error—a signal that a critical component on the ship has failed when in fact it has not. Rather than admit fallibility, HAL attempts to cover it up, eventually by killing the human members of the crew and seizing control of the ship. Only one astronaut survives.[41]

In this dark part of an otherwise optimistic story, Kubrick and Clarke warn their audience about the misapplication of space-age technology. The computer exists as a servant assisting human commanders on a voyage to the larger planets of the solar system. Yet through the nature of its programming, the computer acquires the capacity for independent thought and human fallibilities.

Artificial intelligence devices programmed to work like brains are much less controllable than other fictional machines. This concept appears in the juvenile *WarGames*, a 1983 film in which a teenage hacker breaks into the Department of Defense computer programmed to direct the launch of nuclear missiles in the event of a full-scale war. The computer, programmed to use game theory, cannot distinguish between nuclear Armageddon and the teenager's interest in playing what appears to be a video game.[42] Although the film has a happy ending—the computer discovers that the only winning move in a nuclear war game is "not to play"—it tends to reinforce the conventional view that computers are likely to stray from the intent of their creators.

Given their early association with tedious work, robots tend to be associated with subservient machines. The concept of computer, with its potential capacity for artificial intelligence, introduces a variety of more free-thinking and dangerous devices. Arnold Schwarzenegger, the future governor of California, appeared as the Terminator in the 1984 film with the same name. Schwarzenegger plays an android, a machine core covered with flesh and skin. He is created by an artificial intelligence computer called Skynet, which, after developing conscious thought, seizes the world's military hardware in order to battle the human beings attempting to shut it down. As in *WarGames*, military planners create the computer to make strategic decisions, but the device develops a mind of its own. The device sends Schwarzenegger back in time to hunt and kill the future mother of the person leading the human resistance to the artificial intelligence machine.[43]

Androids and artificial intelligence computers possess some human characteristics. Androids are designed to resemble humans; smart computers are modeled after human brains. Joining them are cyborgs, creatures that are a mixture of human and mechanical parts. For the 1987 film *Robocop,* director Paul Verhoeven attached a dying policeman's head to a mechanical body. The resulting creature combined the moral sensitivity of a human being with the nearly indestructible quality of a machine. Cyborgs appear in extraterrestrial form as the Borg in the *Star Trek* series and as the famous Darth Vader from *Star Wars*. In most cases, they require the use of biological brains.[44]

Isaac Asimov did not expend much effort contemplating the thinking part of his subservient robots. He gave them positronic brains, a concept that he admitted was pure "gobbledygook" and a method of getting past the necessity of explaining control mechanisms he did not understand.[45] Asimov's shortcomings in this regard are understandable; he developed his theory of robotics before computer technology became widespread. Advances in computer technology have not only made Asimov's robots in some measure possible, they have also undercut the cultural traditions that otherwise help people understand how such devices might be used in space.

CULTURAL FOUNDATIONS

The reality of space travel obviously possesses a technical side, represented by the manner in which inventions such as rocketry make spaceflight possible. Space travel also has a cultural side, drawn largely from works of imagination and science fiction. The cultural foundation is as important to the pursuit of spaceflight as the rocket. As a number of people have observed, imagining something in the presence of an expediting technology is often what makes it happen.[46]

Set aside for a moment the question of whether science will allow the creation of smart robots, androids, cyborgs, or computers that think like human beings. Visualize simply whether the social discussion of these concepts allows people to imagine them in useful ways. As noted earlier, the traditional concept of remotely controlled robots exploring space rests on a thin social base. The concept does not receive much attention in widely read works of fiction, nor is it well attached to modern social movements. Asimov's vision of intelligent robots working with human companions is far more popular, but the social foundation to which it is attached contains a serious shortcoming.

The most serious weakness in the dominant cultural presentation of Asimov's subservient machines arises from a fundamental flaw in the underlying vision. As noted earlier, visions of the future often rest on recreated images of a romantically viewed past. People like Asimov envisioned a new machine age dominated by intelligent robots. Stories about helpful robots performing household chores continue to appear in the popular press, although the actual appearance of such devices seems persistently delayed. The vision of robot as household companion is largely a machine age concept, a sophisticated version of the vacuum machine. While humans waited for the household robot, the machine age disappeared. It was replaced by a postindustrial era dominated by computers and information networks. Humans living in the first part of the twenty-first century were far more likely to own personal computers and wide-screen liquid crystal display television sets than a robot. From a social point of view, the robot as personal servant serves to represent a hopeful image of industrialization that we have probably passed by. As one commentator accurately observed, "intelligent robots . . . serve best as metaphors for the zeitgeist of their eras."[47]

Computers and information networks can be characterized as intelligent machines. But they are not robots in the traditional sense. The difference is a subtle one, but it is terribly significant in understanding the demise of the machine analogy. Robots in their original form are mechanical, a product of thinking that dominated the industrial revolution. The inability of robots to depart from the instructions of their makers is based on a mechanical model of control, different only in degree from processes that govern other control mechanisms, like thermostats regulating the operation of heating and cooling devices.

The view of robots as machines closely parallels the dominant image of corporate and government organizations during the industrial revolution. Factories and government offices were treated like machines, with interchangeable parts, standard operating procedures, and strong control mechanisms. Advocates of the dominant approach to industrial management during the first half of the twentieth century, a movement known as "scientific management," treated workers as machines, indistinguishable from robots in the insistence that all follow the "one best way" of performing each job as determined by the experts who planned and programmed work.[48]

The machine analogy leads inevitably to the assumption that machines will do the work assigned to them. From this perspective, embraced by the early writers on robotics, deviations from plans are seen as failures of programming to be corrected with better controls. The programming of really smart computers, however, does not rely upon the traditional machine model. Instead, it proceeds from an appreciation for the functioning of brains.[49]

Simple computers can be programmed and integrated into the workspace as if they were machines. Computer networks and artificial intelligence machines, however, have a cerebral quality that the machine analogy does not adequately explain. As with a hologram, the information for a network does not reside in a single place. It is contained in many sites and is brought together in matrix form.

People who have examined the social nature of information networks and the uses of artificial intelligence consistently dismiss the traditional machine metaphor as inadequate for describing the processes with which they must deal. They discard the metaphor of the machine, turning instead to the metaphor of the human brain. The information needed to recall a memory is stored in many places within the human brain. In this sense, the brain has a redundant quality: long-term memories can be retrieved even if part of the brain is missing. Some functions of the brain are automatic and almost machinelike in nature, such as the regulation of body temperature. Other functions, especially those that involve speech, are far more complex and irregular. Most important for the treatment of machine intelligence in popular culture, the human brain is conscious of its own existence.[50]

A brainlike view of machine intelligence immediately raises issues of the sentient being. Briefly stated, would really intelligent machines experience sensations and acquire a form of consciousness? Under the machine model, this would not matter, because the machine model assumes external control. No such assumption limits the brain model. Once any organism attains a given level of intelligence, it will experience feeling. It may develop, to use a common phrase, a mind of its own.

Asimov dealt with this issue extensively in *The Bicentennial Man,* one of his most influential science fiction tales. He called the novella, published in 1976, "my most thoughtful exposition of the development of robots."[51] The story resurrects the Pinocchio legend in the form of a robot that wishes to become a human being. The robot, manufactured as a household helper, inadvertently acquires artistic capabilities as a result of some unknown error in the plotting of its positronic pathways. It takes to wearing clothes and, over a period of nearly two hundred years, replaces its mechanical parts with human prosthetics. The story has a clever ending, based on Asimov's comparison of the human and robotic brain. Unable to formulate any significant difference on the basis of thought, Asimov points out the biological fact that the human brain eventually dies. From the mechanical point of view, this creates "a steel wall a mile high and a mile thick."[52]

In the story, Asimov addresses the inevitable consequence of the robot's artistic consciousness. The robot, Andrew Martin, wishes to be free. This is a sticky issue for Asimov. From the machine point of view, the issue is irrelevant. Regardless of whether or not Andrew is declared legally free, the robot, on account of its programming, will continue to serve its human family. In a major slip of logic, one of its owners observes that an advanced robot such as Andrew would do so "voluntarily."[53] Voluntary service, however, presumes that Andrew is capable of free will, able to withhold service as well as provide it. This clearly violates the machine analogy.

Ultimately, Asimov allows the courts to decide Andrew's legal status. The resulting decision further undercuts the model of the robot as machine. The robot, Asimov admits, clearly is aware of the meaning of freedom. "There is no right to deny freedom," the court rules, "to any object with a mind advanced enough to grasp the concept and desire the state."[54] Even though the robot may voluntarily remain in servitude, it has acquired the sentient characteristics of a human being, with all of the wishes that that entails.

Asimov's struggle with Andrew's desire to be human foreshadows the treatment of Lieutenant Commander Data in the TV program *Star Trek: The Next Generation* (and spinoff movies). One of the most popular characters on the series, Data is an android, a robot constructed with human features. He is naively curious about human behavior and drawn to think of himself in human ways. In the episode "The Measure of a Man," Data fights a forced reassignment by a commander who considers the android to be Federation property. To contest the reassignment, Data appeals to the Judge Advocate General and wins a legal judgment declaring him to be a conscious being that possesses the same rights and privileges as citizens of the Federation: "He may be a machine, but he is owned by no one and has the right to make his own decisions regarding his life."[55]

The episode is based directly upon the historically influential Dred Scott case, barely disguised through its application to outer space. Scott was an American slave whose Army officer owner moved him between states in such a fashion as to contest the institution of involuntary servitude. Assisted by leaders of the abolitionist movement, Scott filed a suit in Missouri—at the time a slave state—claiming that his previous residency in the North, where slavery was prohibited, had made him a free man in the South, where slavery was still allowed by law. The suit eventually reached the U.S. Supreme Court, which issued its opinion in 1857.

In a judicial opinion that enraged abolitionists, Chief Justice Roger B. Taney avoided the central issue posed by the suit. Taney announced that the Court could not hear such a case because Scott was an African American and, as such, did not possess the same constitutional rights as white men and women, such as the right to appeal to the courts. Taney went further, holding that slaves, under the U.S. Constitution, were such "a subordinate and inferior class of beings" as to have "no rights or privileges but such as those who held the power and the Government might choose to grant them."[56] In effect, they were property, thereby subjecting them to the provisions preventing Congress and the territorial legislatures from infringing upon the legal rights of white citizens to control the property they owned.

The Dred Scott decision helped to precipitate the Civil War. It was overturned by the conclusion of that conflict and the 14th Amendment to the Constitution, which extended the rights and privileges of citizenship to former slaves. The Dred Scott case—and its influence on fictional vehicles such as *Star Trek*—exposes the fundamental social difficulty inherent in the vision of humans and their robotic servants happily exploring space together.[57]

A person is typically defined as a human being. Many definitions of being human exist, among them the presence of a conscious mind and an individual personality. A person is capable of anticipating events, making plans, and experiencing feelings, such as hope and fear. These animating principles of individual life are closely associated with the concept of the soul, which in some religions is thought to be immortal and persisting beyond death. Throughout history, various populations have been characterized as lacking these particular characteristics—slaves, Roma people, Jews, Africans, and, in the American West, Chinese-American workers. The denial of personhood is generally undertaken as a means to justify differential treatment and, in extreme cases, to permit forcible removal or extermination.[58]

Since the nineteenth century, efforts to extend the concept of personhood have proceeded with considerable force. The concept has even been extended to animals, which in some quarters are thought to possess souls—and civil rights. The wider perspective on personhood proceeds from the belief that life evolves along a continuum in which the possession of consciousness differs only in degree across the various species. Applied to the cosmos, this line of thought accepts that extraterrestrial beings, should any exist, while not human (a member of the evolutionary branch of earthly animals categorized as *Homo sapiens*), might yet still be fully conscious and hence deserving of the treatment accorded human beings. From this perspective, it is a short step to considering the degree to which machines (mechanical as opposed to biological creations) might meet the expanding scope of personhood. They need not be human, but they may acquire the rights and privileges afforded persons before the law.

Much classic science fiction presents extraterrestrial life as lacking the characteristics assigned to persons. The degree to which writers of science fiction treat extraterrestrial life in this way often depends more upon its physical and social characteristics than its mental capacity. The species in the movie *Alien* may be motivated by intelligent thought to protect its young, but it poses such a significant danger to the crew of the transport ship *Nostromo* that the crew shows no reluctance in attempting to exterminate it. The alien looks like a giant insect, and the crew treats it like one. In his classic novel *Starship Troopers,* Robert Heinlein describes the intelligent race that the troopers battle as "the Bugs" and gives them many of the social characteristics of ants or termites. Heinlein based his characterization on U.S. wartime propaganda in which enemy forces were characterized as insectlike members of rigidly obedient societies, willing to subordinate their individual characteristics to the good of the whole, even to the point of death. The treatment accorded dangerous extraterrestrials by science fiction writers draws heavily upon this social characterization. In modern times, extraterrestrials are one of the few remaining social groups of consciously intelligent beings that can be fashionably treated as nonpersons and exterminated in fictional narratives without engendering protests from the appropriate anti-defamation league.[59]

Much of the early literature on robots treats them as slave-like machines. People sometimes tolerate their charming ability to mimic human aspirations, but the robots nonetheless remain mechanical in form. Philip K. Dick presented one of the most overt statements of this perspective in *Do Androids Dream of Electric Sheep?* To encourage the settlement of Mars, the government grants each emigrant a personal android servant—"designed specifically for your unique needs"—which becomes the emigrant's private property. The incentive is publicized on government-owned television with the statement that it "duplicates the halcyon days of pre–Civil War Southern States."[60]

The treatment of robots as nonpersons depends considerably upon the conviction that they are nothing but machines. It also draws upon the creationist point of view, which holds that humans, fashioned in the image of God, are the only creatures capable of possessing conscious minds and souls. By the late twentieth century, both of these beliefs had begun to fade away—at least in the popular telling of technological tales. Advances in computer technology encouraged storytelling about machines whose artificial intelligence might someday approach the capabilities of human brains. Interest in the functioning of the human brain and its electronic counterparts encouraged storytelling about nonhuman entities with minds and, possibly, souls. The separation of human and machine that writers such as Isaac Asimov so carefully delineated began to disappear.

Attempts to portray robots in popular literature thus confront a substantial contradiction. If robots are intelligent, they can hardly be thought of as mere machines. The traditional effort to depict robots as subservient draws on doctrines not terribly relevant to a postindustrial world. Such depictions require the presentation of robots through industrial analogies, conjure distasteful images of servitude, and depend on a view of creation in which only humans can possess free will and rational souls. Yet to portray robots in a more modern way, using the analogy of the brain, raises the awkward issue of sentient being, the strange possibility that the computers powering these mechanical creatures may acquire consciousness, free will, and the rights of citizenship. If the machines possess free will, they can hardly be expected to remain under the control of the humans who created them. The robot as intelligent machine ceases to be just a machine.

The traditional perspective of robots as "just machines" supports the official government policy that foresees humans and their mechanical companions exploring space together. In a strictly literary sense, this is an outdated point of view, associated as it is with industrial-age models of servitude. The more modern perspective ultimately supports the notion that machines will achieve the capability to explore space alone. In literary treatments, this is not hard to portray. Capable of exercising what is essentially free will, these new inventions, at least in works of fiction, emerge into various states of consciousness. Like Adam and Eve in the Garden of Eden, they make decisions that depart from the intentions of their creators. They demand to be treated as persons, not as property. The literary result, not surprisingly, has been a revival of the "killer robot" stories that Asimov so despised. Such stories use the popular formula of the Frankenstein legend, warning humans to make few assumptions about the inventions they futilely seek to control.

The terrestrial voyages of discovery that provided the basic model for the von Braun paradigm took place during a period marked by severe social distinctions in master-servant relationships, distinctions that covered both domestic servants and slaves. The science stories that emerged contemplated a similar relationship between humans and their machines. The literary contemplation of robotic technology began during the latter stages of the industrial revolution and was encouraged by machine analogies drawn from the popular understanding of industry and control. If humans ever dispatch cyborgs, androids, or artificial intelligence devices to explore space, it will not come as a surprise to consumers of popular culture to learn that such creations possess more of the characteristics associated with computers than vacuum machines. The public of the twenty-first century is as accustomed to stories about computer intelligence as the public of the early twentieth century was to stories of rocketships, spacecraft captains, and their robotic companions. In a manner similar to the science fiction of the early twentieth century, stories about independent, free-thinking nonhuman entities have prepared the public for creations that merge human and mechanical characteristics. It is only imagination, but imagination in the scientific realm is often a prelude to making something occur.

CHAPTER FIVE

The New Space Race

Given the state of space-age technology at the midpoint of the twentieth century, most people planning cosmic missions assumed that human operators would be needed to complete a variety of tasks. Humans would be needed to pilot spacecraft, operate space stations, conduct expeditions to the Moon and other planets, assemble extraterrestrial bases, even change the film in space telescopes.[1] Planning documents anticipated rapid advances in the technology of human flight, improving both cost and reliability, to the point that hundreds of people might live and work in space by the century's end. The 1969 report of the Space Task Group suggested that humans would soon reside in a large Earth-orbiting space station, occupy a similar facility orbiting the Moon, construct a lunar base, and establish a small outpost on Mars—with all of these projects to be under way by 1986. Planners further anticipated that the cost of human spaceflight would fall dramatically, justifying government investment in an elaborate transportation system consisting of winged spacecraft, orbital tugs, and interplanetary vehicles. Such anticipation provided a key justification for the 1972 decision to develop the reusable Space Shuttle. Lower costs in turn would permit humans to construct larger and better space systems and propel human spaceflight into the realms anticipated by writers of science fiction and popular science. That, at least, was the dream.

Not content just to send humans into space, planners envisioned a substantial role for automated satellites and robotic spacecraft as well. The report of the Space Task Group also called for a grand tour of the planets in the outer solar system, a series of automated space observatories, orbital mapping of Mars, probes to pierce the clouds covering Venus, asteroid surveys, and investigations of the solar system beyond the elliptical plane, all using remotely controlled spacecraft. Maintaining the popular image of robots as machine servants to their human creators, the planners viewed the surveys of Venus and Mars "as precursors to later manned missions."[2]

In practice, the robotic portions of the various long-range plans tended to succeed beyond all expectation. The proposal for a "Grand Tour" of the outer solar system led NASA to conduct the Voyager missions that visited Jupiter, Saturn, Uranus, and Neptune. The Space Task Group called for a 1977 launch; the rockets lifted off that year. The Space Task Group members hoped to launch the first large observatory in 1980; the actual launches of the four "Great Observatories" occurred between 1990 and 2003.[3] The cost of conducting robotic missions, moreover, fell dramatically during the last part of the twentieth century, realizing in part the dream of cheaper and more extensive flight.[4]

As the robotic spaceflight effort progressed, the human spaceflight program floundered. The Space Shuttle did not provide frequent low-cost access to space, as its promoters had promised. The designers of the International Space Station (ISS) did not create the large orbital platform for deep space missions that the original concept prescribed. Repeated calls for lunar bases, space colonies, and human planetary expeditions remained in the planning stage.[5]

In a modern version of Aesop's famous story "The Tortoise and the Hare," the tortoise-like robotic effort made slow but steady progress. In Aesop's fable, the hare takes an early lead but pauses to rest. The technological achievements of the Apollo expeditions to the lunar surface raised expectations of spectacular missions to follow. Scientific returns from the six human lunar surface expeditions clearly exceeded the results obtained by the various robotic spacecraft that circled and landed on the Moon. Spurred by the expeditions to the Moon, human flight took an early lead and then stalled in low-Earth orbit for more than thirty years. As the century progressed, robotic devices persistently advanced.

Designers of robotic spacecraft surmounted technical obstacles with greater ease than people designing human missions. In part, advances in technology such as the microelectronics revolution favored robotic over human flight. Bureaucratic and political factors intervened as well. The desire to protect existing programs and facilities hindered the human flight endeavor more than the nimbler efforts at robotic flight. In that sense, like the hare that chose to nap after attaining an early lead, the people driving human space activities created many of the impediments that slowed their own pace.

LIVING AND WORKING IN SPACE

Well before the point at which spaceflight began, visionaries contemplated the means by which humans could permanently live and work in space. In this vision, extraterrestrial activity began with the construction of space stations. As has been frequently observed, many of the people who joined NASA had wanted to build space stations even before they joined NASA. The history of their efforts to create a permanently occupied orbital facility provides substantial insights into the factors retarding the process of human flight.

Station technology provides the principal means by which humans can remain in space for extended periods of time. As a concept, a space station is more than a facility that orbits the Earth a few hundred miles above its surface. It also enables the technology developments necessary to sustain lunar bases, Martian settlements, and similar outposts orbiting the Moon and other planets. A lunar base is a space station with legs; a Martian colony is a space station on Mars.

Space stations additionally provide the basic technology for interplanetary flight. "A manned Mars mission," said Daniel Herman, the chief engineer for the Space Station Task Force NASA created in 1982 to formulate a specific station plan, "is a space station that is going to leave Earth's orbit." Fighting the notion that robots could better perform the tasks assigned to an Earth-orbiting facility, Herman stated that the leaders of his task force would still want to station humans in low-Earth orbit "even if it could be proved that functionally everything conceived of today could be done by robots." The reason, he explained, is that "we think it is NASA's charter to essentially prepare for the exploration of space by man in the twenty-first century."[6]

To complete Herman's vision, station designers grappled with a number of key technology challenges. They dealt with the adverse effects of long-term weightlessness or what is more accurately called microgravity (any large spacecraft generates a small amount of force through the presence of its own mass). Microgravity conditions cause humans to lose bone and muscle mass, to the point that people returning from lengthy space voyages cannot stand in the absence of compensating measures such as extensive exercise. Another technology challenge is the generation of electric power. Large space stations consume substantial quantities of electric power, far more than can be generated through conventional means. The batteries and fuel cell technologies employed to power Project Apollo and the solar arrays used to run satellites traveling within the inner solar system are not adequate for hefty long-duration human activities. Space station designers struggled with the technologies necessary to "close the loop" on life support systems. Open-loop life support systems, which depend on storage facilities and earthly resupply, cannot support the needs of large multi-year expeditions. Additionally, engineers contemplated the steps necessary to move flight control activities from mission control centers on Earth onto the spacecraft and stations themselves. When people head for the planets, Herman observed, "you have to be autonomous; you cannot rely on the Mission Control Center."[7]

In each of these areas, NASA officials designing space stations encountered technical, political, and bureaucratic obstacles. In most cases, the obstacles proved insurmountable—at least for the first half-century of spaceflight.

Artificial Gravity

Most large station designs presented for public inspection during the formative years provided some means of artificial gravity. Wernher von Braun's famous 250-foot-wide wheel produced artificial gravity by rotating on its axis once every twenty-two seconds. The 900-foot-wide space station conceived for the Kubrick-Clarke movie *2001: A Space Odyssey* moved in such a manner as to duplicate a sense of gravity similar to that which human travelers would feel on the Moon. Appropriately, the station served as a transfer point for lunar trips. In Gerard O'Neill's vision of space colonies, residents lived on the inner edges of very large rotating cylinders. Imagine an enormous paper tube with people living on the inside edge. The rotation of the cylinder would create sufficient force to push humans and their accompanying structures away from the core, where the presence of the outer shell would retard their outward motion and create a sense of gravity. Such rotation would also produce the disconcerting effect of looking up at lakes and buildings on the opposite side of the rotating tube, suspended four or more miles above one's head.

An early NASA design, known as the "three-radial-module" or Y-shaped space station, unfolded like an umbrella to produce a 190-foot-diameter facility that rotated around a central hub. Another prominent proposal envisioned a series of 33-foot-diameter can-shaped modules, set like blades on a pinwheel. That design provided the model for the 50- to 150-person "space base" proposal contained in the 1969 report of the Space Task Group.[8]

Proposals for artificial-gravity space stations disappeared with the executive decision to stop producing the *Saturn V* rocket that sent Americans to the moon. All large rotating space station designs were premised on the availability of a heavy-lift launch vehicle like the *Saturn V*. In early plans, the Saturn rocket or its derivative delivered station components to orbit while a yet-to-be developed Space Shuttle would ferry people and supplies between Earth and the orbiting facility.

NASA executives abandoned the production line for the *Saturn V* in 1970, primarily as a cost-cutting move. With the funds that remained, they hoped to finance both a space station and a fleet of Space Shuttles to service it.[9] This placed an additional requirement on shuttle engineers, one not contemplated by its conceivers. Absent a heavy launcher like the *Saturn V*, the burden of delivering station components fell on the proposed Space Shuttle. The original requirement prescribed a shuttle payload bay sixty feet long and fifteen feet wide, large enough to carry space station modules. Henceforth, the shuttle became a "space truck," which was never part of the original concept, and the space station lost its artificial gravity. Every official station design to emerge from the NASA bureaucracy after the decision to proceed with a large Space Shuttle consisted of small modules that could fit inside its payload bay, assembled in orbit like a collection of Tinkertoys.

Nearly all of the space station plans relied on shuttles delivering a cluster of small modules placed near the center of the facility, producing a design not conducive to rotation. Commensurate with this approach, NASA executives began to promote the station as a microgravity facility for commercial and scientific applications for which the absence of artificial gravity proved advantageous. Most people, including many NASA employees, continued to think of a space station as "a very large rotating wheel with 100 people on it and artificial gravity," observed John Hodge, the head of NASA's Space Station Task Force.[10] For financial reasons, however, NASA officials had abandoned that configuration long before the Space Station Task Force began its work.

Visions of developing the space station and the Space Shuttle concurrently encountered political opposition in the White House. Seeking further economies, White House aides informed NASA executives that they would not be allowed to pursue more than one new initiative. Forced to choose between the station and the shuttle—and lacking a launch vehicle capable of delivering components of a large space station to orbit—NASA Administrator James Fletcher decided to pursue approval of the reusable Space Shuttle. Delivery of station components, even though it guided shuttle design, remained unapproved for more than a dozen years because construction of the space station remained unapproved.

This odd sequence of events compromised the design of the Space Shuttle and crushed the vision of a space station with artificial gravity. Shuttle designers sought to construct a reusable spacecraft with far more requirements than originally contemplated, and the space station eventually lost its purpose as a stepping stone for distant destinations.

Onboard Autonomy

John Hodge, who had served as a flight control director during the Gemini-Apollo era, set out to weaken the link connecting NASA's Mission Control center in Texas with any future space station. "Achieving an increased level of onboard autonomy and automation," said members of a special NASA technology planning committee, should have been one of the principal objectives of NASA's space station activity.[11] Had it been achieved, onboard autonomy would have advanced the cause of human spaceflight significantly.

NASA officials knew that they could fly an Earth-orbiting space station from the ground. Flight controllers at the Johnson Space Center had flown the Skylab orbital workshop, a precursor to the large orbiting space station, for ninety-nine days without any astronauts on board, as crews came and went. Flight controllers flew Skylab for more than five years after the last crew members left. NASA officials could have built another Skylab-type facility with "no engineering risks at all," said Daniel Herman. "Doing that is not exciting to me."[12]

Officials at the Johnson Space Center in Texas, whose employees oversaw the Apollo flights to the Moon, wanted to fly the large space station, approved by President Ronald Reagan in 1984, from the ground. They sought authorization to construct a new mission control center at their Houston facility and staff it with flight controllers. Officials at NASA headquarters wanted to move many of those activities into space. To do so, the headquarters group sought to invest the station with massive computer capacity and a crew sufficiently large that some of them could be deployed to flight activities. Computers on the station could automate much of the flight control work, checking and even correcting various activities, while a few astronauts would control additional flight activities from the station bridge.

Officials at the Johnson Space Center did not want to relinquish their traditional role. They resisted the effort emanating from NASA headquarters to establish station autonomy and particularly resisted John Hodge, a onetime flight officer who in their eyes had abandoned his old profession. Within NASA, the issue was settled on technical, financial, and political grounds. The ingredients needed for reliable onboard flight control capability included lots of time and money in a program short on both. Resolution of the issue was complicated by a political dispute between NASA executives and the head of the congressional appropriations subcommittee providing the funds. Representative Edward Boland, a Massachusetts Democrat, wanted NASA to begin assembling the station with laboratory modules, not astronaut living quarters. He feared that cost overruns and schedule delays would force NASA to propose cutbacks in station components. If station assembly began with living quarters and then faltered, the nation might be left with a station occupied by astronauts with nothing to do but fly there and back.

NASA executives, including Administrator James Beggs, wanted to start with the living quarters. They viewed Boland's complaint as part of a larger effort to establish a "man-tended" space station, one that would need only an occasional human presence and, by necessity, be controlled from the ground. Beggs insisted that NASA be allowed to begin with the living quarters. "It is," Beggs said, referring to the president's commitment to build a station with people on board, "the presence of man which makes it a unique national resource."[13]

Space scientists, relentlessly resisting the vision of human spaceflight, pressed for a "man-tended" space station. Advocates of human flight pushed the other way. Boland and the advocates of ground control prevailed. Early plans called for NASA to begin with a "man-tended" facility.[14]

NASA never built the Boland-Beggs design, known officially as Space Station Freedom. That configuration was replaced by plans for the International Space Station (ISS), begun in 1993. Assembly began in late 1998 with the Zarya Control Module; the first permanent crew members did not arrive until November 2, 2000. For nearly two years, NASA flight controllers with their Russian counterparts flew the first elements of the space station from the ground. As the station grew, NASA maintained flight control teams in Houston to monitor telecommunications, life support, computers, electric power, robotic systems, extravehicular activity, and the position of the station in space. Another element of the human flight vision disappeared.[15]

Electric Power Generation

Similar developments affected the desire of NASA officials to invest the station with an advanced electric power–generating system. Any large human space facility, especially one with laboratories or manufacturing capabilities, needs a great deal of electrical power. NASA engineers who produced plans for the original space station incorporated an advanced power-generating system that could efficiently produce at least 75 kilowatts of electric power, with the capacity to grow to 300 kw.[16]

As a point of comparison, the Russian Salyut space station begun in 1971 limped along with 2 to 4 kilowatts; a few kilowatts may provide sufficient electricity to power an American home, but it is not enough for an ambitious facility in space. The Skylab orbital workshop, launched by NASA employees in 1973, was designed to produce 24 kilowatts of power, but a solar panel damaged during the launch reduced the amount of continuously usable electric power to one-third of that.

NASA engineers used batteries to power their Mercury space capsules and the spider-shaped landers that delivered humans to and from the Moon. Batteries are not an efficient power source for long-duration voyages. Batteries for John Glenn's Mercury spaceflight provided the equivalent of a single kilowatt of power for six hours.[17] Searching for a more robust power source, NASA engineers developed fuel cells to power the Gemini and Apollo space capsules, a promising technology with substantial terrestrial applications. Fuel cells use hydrogen and oxygen to generate electricity and produce drinking water for the crew. The explosion of the liquid oxygen tank on the *Apollo* 13 mission deprived the three-person crew of electricity and forced them to retreat to the battery-driven lunar lander that served as a lifeboat for the emergency trip home.

Solar or photovoltaic systems powered the American Skylab and Russian Salyut and Mir space facilities, their silicon cells directly converting sunlight into electricity. The appropriateness of solar panels is limited by their size. To generate sufficient electric power, solar panels must be large. In Earth orbits, large panels create drag. Anywhere in space, they provide targets for meteorites and debris. To produce very large quantities of electricity, more advanced systems are required.

The solar dynamic system that NASA planners wanted to install on the space station was more compact and thus capable of generating far more electricity for a given size than conventional alternatives. Dish-shaped reflectors would collect and concentrate sunlight, using the heat to melt a special material behind the collectors. On the dark side of the Earth, that material re-solidifies. The melting and solidifying of the material produces energy, which in turn drives electric power–generating turbines.

Solar dynamic collectors appeared on early space station designs. By 1987, however, they had vanished, the victim of budget cutbacks that reduced the amount of new technology that NASA officials could afford to install on their orbiting facility. Station supporters fought off attempts to reduce power generation commensurately, but the ISS would not use advanced energy-generating technologies.[18]

Closed-Loop Life Support Systems

Of the four major technology challenges NASA tackled during the space station era, human spaceflight engineers made significant progress on only one. They improved the capacity of a spacecraft to produce air, recycle water, manage waste, maintain humidity, and suppress fires. "Closing the loop" on environmental life support systems is a major technology objective within human spaceflight and one that has been addressed in a variety of ways.

For the Skylab mission, NASA engineers utilized a "pack it all in" method of life support. Twenty-two steel and titanium tanks launched as part of the orbital workshop provided all of the water and breathing air necessary to support astronauts during 171 days of human flight. The Russians improved on this method, producing automated Progress vehicles that rendezvoused with orbital facilities and provided cosmonauts with new supplies. A similar approach maintained the International Space Station during the early phases of its assembly.

In 1991, environmental scientists began a two-year experiment to test the feasibility of supporting human beings in a closed environmental system. The test took place in the Arizona desert near Tucson, not in space. Designers of the three-acre Biosphere 2 provided for the complete recycling of water, food, and waste. Eight humans confined to the biosphere produced 80 percent of their own food. Like a spacecraft, the Biosphere leaked slightly, although not as much as the NASA Space Shuttle. Fifteen months after sealing in the eight subjects, the designers of Biosphere 2 were obliged to pump oxygen into the facility. Internal restoration processes proved insufficient to hold the oxygen content at its normal atmospheric level of 19 percent. When the level fell to 14.5 percent, designers decided to break the seal.[19]

The life support systems planned for the International Space Station were similarly complex. They consist of two separate systems, one for the Russian section and a second for the U.S. and other international components, each with its own redundancies and design philosophies. Working together, they could significantly reduce the amount of consumable material that must be carried from Earth to space.[20]

Overall, development of the International Space Station provided substantial opportunities for improving the technologies that support human activity in space. In some areas, such as life support, important progress was made. In the aggregate, however, significant difficulties prevailed. Struggling with cost growth, schedule delays, and the impact of two shuttle accidents on station development, NASA officials lost many opportunities to advance the progress of human spaceflight through station technology.

In the grand vision of extraterrestrial flight, space stations are viewed as a principal means for advancing the technology allowing humans to live and work in space for long durations. In practice, the resulting products fell substantially short of expectations. The hare slowed as the tortoise continued to crawl.

TRAVELING IN SPACE

Advocates of human space travel generally agree that improved space transportation would do more to achieve their agenda than any other single factor. For thirty years during the last part of the twentieth century, spaceflight engineers struggled to reduce the cost and improve the reliability of large launch vehicles. They worked through a succession of new technologies: reusable spacecraft, single-stage-to-orbit vehicles, and new methods of propulsion. Their efforts, with few exceptions, produced frustration. No single factor retarded the cause of human spaceflight more than the inability of flight engineers to create reliable, inexpensive space transportation.[21]

Rocket engineers made great advances during the forty-year period following Robert Goddard's 1923 launch of the first liquid-fueled rocket in a Massachusetts farm field. The U.S. defense establishment invested vast sums of money in the development of liquid-fueled rockets such as the *Atlas*, the *Titan,* and the *Jupiter*. That investment created the U.S. fleet of expendable launch vehicles and the rocketry used to send the first American astronauts into space. Similar developments propelled the Soviet space and missile efforts.

In the early 1960s, NASA officials spent $9.3 billion—the equivalent of more than $60 billion in 2007 dollars—to produce the family of Saturn rockets. The effort allowed another great advance in rocketry. Early launch vehicles relied upon chemicals such as kerosene, ethanol, and aluminum in combination with ammonium perchlorate for propulsive power. The Saturn RL-10 engines burned liquid hydrogen. Hydrogen/oxygen engines have a specific impulse (I_{sp})—the rocketeer's equivalent of miles per gallon—50 percent greater than conventional fuels.[22] Without hydrogen-fueled rocket engines, the United States would not have won the race to the Moon. Although NASA officials abandoned the Saturn rocket line, they continued to apply hydrogen-fueled technology to the launch vehicles that followed. The main engines on the NASA Space Shuttle burn liquid hydrogen. Since it must be stored at frigid temperatures, liquid hydrogen is more difficult to handle than conventional chemical fuels.

Beginning in 1972, government officials set out to create another great advance in rocketry. They agreed to invest $8 billion in a reusable Space Shuttle that could carry medium-sized payloads (50,000 to 65,000 pounds) and reduce the cost of spaceflight "by a factor of ten."[23] The investment sum is the equivalent of roughly $40 billion in 2007 dollars. Along with a commensurate increase in spacecraft reliability, advocates of the initiative envisioned lower costs leading to an increasing number of human flights to and from low-Earth orbit, assembly of large structures such as communication antennas, construction of a large space station, expanded commercial opportunities, and an infrastructure that would provide the foundation for trips to the Moon and beyond.

The financial plan for this ambitious objective proceeded from the assumption that the initial $5.15 billion capital investment would be more than repaid by the savings accrued from flying spacecraft with operational costs thought to be lower than those required for conventional rocketry. The numbers, confirmed by a prestigious economic consulting firm, supported the basic assumption. The actual plan called for 580 flights over twelve years, nearly one launch per week. As late as 1984, NASA officials still planned to fly the shuttle twenty-four times per year.[24] In theory, the availability of cheap spaceflight would create its own demand. Energized by the low cost of space access, commercial firms and other government agencies would shift work from Earth into space. The marginal cost of an increasingly large number of flights would drive per-flight expenses lower and lower as demand grew. It was a wonderful model. In anticipation of the forthcoming bonanza, U.S. officials declared NASA's Space Shuttle to be the nation's "primary launch system" for all commercial, military, and scientific payloads.[25]

The Space Shuttle never met these goals. The largest number of flights conducted in any one year was eight, not twenty-four or forty-eight. With a large and relatively fixed work force, infrequent use caused per-flight costs to soar, producing an access price of approximately $10,000 per pound—no improvement over first-generation launch vehicles. Confounding early expectations, shuttle reliability provided little improvement over existing launchers. The empirical failure rate was 2 in 113, or slightly less than 2 percent. NASA officials continued to spend money on shuttle upgrades and equipment replacement, far exceeding the original investment plan. Alex Roland, a history professor and stern critic of human spaceflight, was among the first to observe that by the terms of its original objectives, the Space Shuttle was a policy failure.[26] In 2005, NASA Administrator Michael Griffin agreed.[27]

Spaceflight engineers based their original hopes for cost reduction on the shuttle's reusability. "There's no way that you can make a railroad cost-effective," argued one NASA executive using a commonly employed analogy, "if you throw away the locomotive every time."[28] Yet reusability creates its own technical challenges. The hydrogen-burning main engines on the shuttle orbiter run beyond their rated capacity to propel the vehicle into space, and the orbiter's thermal protection tiles suffer damage as the shuttle lifts off and returns at hypersonic speeds. The strain on components such as these creates the need for careful inspection and repair before any orbiter flies a subsequent mission.

The original flight plan for the Space Shuttle allowed NASA employees and their contractors, working on up to five orbiters simultaneously, only a few weeks for flight preparation. This proved unachievable given the tender state of shuttle technology. When NASA executives pressed their work force to move toward the planned flight rate, the *Challenger* blew up. Flight engineers worked hard to complete eight shuttle missions in 1985, the year before the *Challenger* accident demonstrated the impractical nature of the original flight plan.

Undeterred, a variety of private firms and governmental institutions sought to construct new launch vehicles that could realize the original vision. The efforts took many forms: the National Aerospace Plane, the Delta Clipper, the X-33, and a number of privately financed commercial projects. A common objective was the longstanding goal of reducing space access expenses by the much-desired "factor of ten." Additionally, NASA executives urged engineers to improve flight reliability to levels achieved by military combat aircraft, where accidents occur after thousands of missions rather than twice every hundred.[29]

Again, technical obstacles collided with technological vision. The only spacecraft capable of meeting such objectives, joked aerospace workers, would be made of "unobtainium." Major technical challenges arose in the areas of structure and propulsion. To satisfy the various requirements imposed on the NASA Space Shuttle, engineers produced a spacecraft that in structural terms was quite heavy. In its inert state, without payload or fuel, the space shuttle mass is 625,000 pounds. The people who designed the Space Shuttle solved the mass problem in a traditional manner: they discard elements of the whole vehicle as it rises into space. Solid rocket boosters fall away to be recovered and reused; the external fuel tank disappears as well.

Aeronautical engineers seeking to improve this situation hoped to create launchers that could reach space without "staging" or discarding parts. The quest for the single-stage-to-orbit vehicle became the holy grail of spaceflight during the last two decades of the twentieth century.

Some areas saw significant progress. NASA engineers and their contractors successfully tested the linear aerospike engine. This propulsion unit has an odd feature that allows the engine to adjust its performance automatically as atmospheric pressure declines on the trip from Earth to space. With other features, this gives the linear aerospike engine a far better thrust-to-engine-mass ratio than more conventional combustion chamber designs. Aerospace engineers also experimented with hybrid ramjet-scramjet propulsion systems. Ramjets and scramjets draw their oxygen from the air, unlike conventional launchers, which must carry oxidizers on board. Ramjets work well at speeds up to about Mach 6, at which point they run too hot; at high temperatures, scramjets work better. A hybrid ramjet-scramjet engine that could draw in oxygen at all speeds would substantially increase propulsion efficiency.[30]

As rocket scientists tested more efficient engines, aeronautical engineers worked to develop more efficient structures. Design teams sought to develop better thermal protection systems. The ceramic tiles glued to the Space Shuttle are maintenance-intensive and damage-prone. Designers worked with titanium tiles and shielding constructed from a special combination of titanium alloys and ceramic fibers. The most difficult challenges occurred with the effort to construct lightweight fuel tanks. The Space Shuttle utilizes a large, 66,000-pound external fuel tank that the crew discards upon reaching space. The external tank habitually sheds pieces of its insulating foam, an unanticipated flaw that caused the loss of the space shuttle *Columbia* in 2003 and retarded efforts to return the shuttle to flight. Spacecraft designers sought to move fuel tanks inside the spacecraft frame, necessitating the use of composite materials that could hold very cold fuel and oxidizers without cracking.[31]

NASA executives thought that they had a workable plan to produce a vehicle with advanced features such as these. They convinced executives at the Lockheed Martin Astronautics Company to develop an X-33 launch vehicle using a combination of government funds and private investment. The X-33 would not fly into space. Rather, it would test the technologies necessary for Lockheed Martin to build a much larger VentureStar spacecraft that could. The plan called for Lockheed Martin executives to raise funds from the private sector to build *VentureStar*, an investment backed by the promise of government use.

NASA executives invested $1 billion in the X-33 program. Next to the earlier investments in launch vehicles like the *Saturn V*, this was a minuscule amount. Yet NASA did not have much money to invest. The substantial expense of operating the shuttle fleet consumed nearly all of the funds available for human space transportation. NASA executives tried to divert funds from current flights to future invention by reorganizing the shuttle program. In the contest between current necessities and future invention, however, immediate necessities prevailed. When X-33 engineers encountered difficulties fabricating the composite fuel tanks—the material tended to crack after being baked—NASA executives cancelled their support for the X-33 rather than spend more money to fix the problem.[32]

Building a reliable low-cost spacecraft that can ferry humans safely to and from orbits close to the earth is enormously challenging. "It has proven to be much more complicated than I thought," observed the German rocket pioneer Hermann Oberth after viewing a shuttle launch.[33] The original operational plan for the Space Shuttle called on NASA to replace the vehicle with a more advanced spacecraft by 1990. No replacement emerged. Early planning documents saw the shuttle as just one part of a larger transportation system that included space tugs and nuclear-powered vehicles capable of reaching the Moon and Mars. Only the Space Shuttle appeared.[34]

Beyond the considerable challenges inherent in transporting humans to orbits close to Earth, advocates of the spacefaring vision sought to develop advanced rocketry for journeys beyond. Conventional propulsion systems would not do; they require too much fuel. Mars missions propelled by conventional rocketry would require six to nine months of travel for a one-way trip. Adding to the complications of a long journey, any spacecraft using conventional methods would arrive at Mars without sufficient fuel to slow down. It would consume nearly all of its fuel speeding up, leaving little for deceleration. "Advanced propulsion," states a NASA document, "would allow us to move around the solar system much quicker than we do today. Instead of months to Mars, it would be weeks. Years to Jupiter and Saturn would be months. And centuries to the stars would become years."[35]

This is an enticing vision, one that has excited advocates of human spaceflight for some time. In practice, it collides with stern realities, many of which are political. The general public is much opposed to the launching of payloads that contain the most promising device for extraterrestrial flight: the nuclear fission engine.

NASA engineers tested nuclear rockets during the 1960s through what was then known as the Nuclear Engine for Rocket Vehicle Applications (NERVA) program. These were ground tests of propulsion units—not engines attached to real spacecraft. The engines work by pumping liquid hydrogen through small channels in a solid-core nuclear reactor. Heat from the reactor rapidly expands the hydrogen, which produces thrust. NERVA rockets produced twice as much thrust (or specific impulse) as the best chemical rockets.[36]

Political opposition to nuclear rocketry led to the cancellation of the NERVA program in 1971. After President George H. W. Bush proposed human flights to the Moon and Mars in 1989, members of the White House National Space Council encouraged NASA executives to resume work on a nuclear engine. Defense Department efforts during the administration of President Ronald Reagan had suggested that gas-core nuclear rockets—a more advanced concept, in which the fuel does not touch the reactor wall—might produce ten times as much thrust as chemical rocketry.

In response, NASA officials proposed a conventionally powered spacecraft so large and ponderous that space enthusiast Robert Zubrin, borrowing a term from *Star Wars,* labeled it a "Death Star."[37] NASA's obsession with the conventionally powered spacecraft, complained Zubrin, killed much of the enthusiasm motivating the presidential initiative. The reluctance of NASA officials to move beyond conventional rocketry contributed to the dismissal of one NASA administrator and the demise of the 1989 Moon-Mars initiative.[38]

Nuclear engines power submarines and aircraft carriers. Spacecraft engineers in both the United States and Russia have been launching radioisotope-based devices that generate electric power on lunar and planetary spacecraft for more than forty years; the Apollo astronauts carried five radioisotope-based systems to the Moon to power Lunar Science Experiment Packages. These systems produced electricity by using the heat from the decay of radioactive materials, generally the isotope plutonium-238, to power converters that generated electricity. With the election of President George W. Bush, NASA officials revived their efforts to expand the use of nuclear-based propulsion and electric power through what they called Project Prometheus. Yet political opposition to full-scale nuclear propulsion for deep space missions remains strong, perhaps insurmountable, even though the nuclear engines would not be fired until the spacecraft leaves the Earth.[39]

In a large number of areas, for more than forty years scientists and engineers sought to advance the technologies supporting human spaceflight. At the beginning of the space age, they succeeded in developing a number of the key technologies. As the space age matured, their ability to produce new technologies declined. Human spaceflight, as a consequence, did not advance as rapidly as people positioned at the beginning of the space age thought it could. In the decades following the final landings on the Moon in 1972, human spaceflight remained an activity that took place in orbits only a few hundred miles from the surface of the Earth.

While humans remained close to home, robots traveled far. Unlike the experience with human space travel, the technologies supporting robotic flight advanced well beyond original anticipation. Robotic technologies improved steadily, step by step, substantially expanding the missions to which robotic technology could be applied.

STAYING ALIVE

The technical challenges facing builders of robotic spacecraft were no less daunting than those confronting the people seeking to advance human flight. A common misconception, often heard in the human versus robotic debate, holds that robotic missions are intrinsically easier than human flight because machines do not need to be kept alive. This is misleading in a number of respects. While important differences do exist between human and robotic spacecraft, the job of maintaining the latter in working condition while in space is very demanding. This is especially true for those spacecraft that can move, grasp, see, and feel. Operators of robotic spacecraft have increasingly turned to biological analogies to describe these challenges. Referring to the "robot geologists" *Spirit* and *Opportunity* that explored Mars in 2004, workers at NASA's Jet Propulsion Laboratory explained that "the rovers' parts are similar to what any living creature would need to keep it 'alive' and able to explore."[40] Both robots and humans can expire in space. Significant progress in the use of robotics has occurred in part because spacecraft engineers have learned to keep such mechanisms "alive."

Like humans, robots suffer from hypothermia. In humans, hypothermia is a rapid cooling in the core parts of the body that results in the failure of vital organs. The computer, electronics, and batteries that operate robotic spacecraft, in the words of their designers, are "basically the equivalent of the robot's brains and heart."[41] Like human organs, they must be kept warm. Engineers designed the Mars rovers *Spirit* and *Opportunity* so that they could maintain their core temperatures in the extreme Martian cold. Operating on the surface of a planet where night temperatures drop to –140 degrees Fahrenheit, the robots were warmed by continuously operating radioisotope heater units (RHUs) supplemented by intermittently operating electric heaters that turned off and on. Like the thermostat in a house—or the temperature regulation system in the human body—the rovers sensed and maintained their own internal temperatures.

At night, the rovers slept. This too solved a problem created by excessive cold. To function properly, instruments that served as appendages on the two rovers, such as the robotic arm and thermal emission spectrometer, needed to be made warm. This was accomplished with heaters, but the heaters needed electric power. During the day, solar arrays generated the electric power needed to produce heat. To run the heaters at night, the robots depended on internal batteries. Excessive use of the batteries, however, caused them to degrade. Sleep resolved this problem. As the sun began to set, the heaters shut down, and the appendages became quite cold. Inside a specially insulated box, the constantly running RHUs warmed the spacecraft core, but not its instruments or arms. Once the sun rose high enough to power the solar arrays, the rovers woke up, warmed their appendages, and began their daily routine. "Deep sleep," noted one spacecraft scientist, "is going to buy us back a huge amount of capability to drive farther, take more pictures, use the arm more."[42]

In addition to the dangers posed by extreme cold, robotic spacecraft need to be protected from excessive heat—even on a frigid planet like Mars. The robotic geologists *Spirit* and *Opportunity* possessed radiators that switched on when battery temperatures in the spacecrafts' core approached 68 degrees Fahrenheit, which could happen on an otherwise chilly Martian day if the robots worked too long. The rovers worked within a temperature range similar to that affecting humans. Heaters switched on when battery temperatures fell below –4 degrees Fahrenheit; radiators switched on at 68 degrees.

The MESSENGER spacecraft (the name stands for *ME*rcury *S*urface, *S*pace *EN*vironment, *GE*ochemistry and *R*anging), launched in 2004 and due to explore the planet Mercury starting in 2009, uses a simple device to keep cool. Orbiting the planet in a location at which the sun is eleven times brighter than on Earth, the robotic spacecraft carries a sunshade that works like an awning on a porch patio. It shields the spacecraft from the excessive heat of the sun while leaving instruments on the opposite side of the spacecraft free to point at their intended objects of study. Without the economical sunshade, spacecraft designers would have been forced to install expensive, high-temperature–resistant electronics.

The sun poses dangers to robotic spacecraft that are as significant as those encountered by humans. A robotic spacecraft that "looks" at the sun can be "blinded"—deprived of imaging capabilities—as easily as an astronaut who fails to use the supplied sun visor. The solar arrays on the Hubble Space Telescope were designed to lay flat toward the sun, but the telescope adheres to rules that prevent it from ever pointing its mirror in that direction. The telescope constrains its motion through a "50-degree Sun-avoidance zone." When the telescope finishes studying one object on the edge of the 50-degree zone and prepares to move toward another object on the other side, a computer program guides the instrument around an imaginary circle marking the exclusion zone until the telescope reaches the second target. In that way, the telescope avoids the dangers of looking too closely toward the sun.[43]

When serious difficulties occur, the telescope enters a number of safe modes, which its designers characterized as varying from the equivalent of "taking a 'nap'" to "being in a deep coma." In 1999, four of the gyroscopes that help point the instrument toward its targets failed. Telescope managers suspended scientific investigations and placed the instrument in a mild safe mode. They closed the telescope's aperture door to protect instruments against the possibility of an inadvertent drift toward the sun. The telescope maintained its attitude with the two remaining gyroscopes, enough to ensure stability but without the necessary force to permit scientific investigation. The telescope returned to work shortly after astronauts completed a serving mission at the end of the year.[44]

Radiation poses significant hazards for robotic spacecraft, just as it does for *Homo sapiens*. Machine parts are susceptible to radiation damage too. When NASA's *Galileo* spacecraft attempted to inspect Io, the Jovian moon orbiting just 422,000 kilometers above Jupiter, it encountered the intense energy collected within the gas giant's huge radiation belts. Io is a fascinating object. The close proximity of Jupiter generates tidal forces that create active volcanoes, the first such eruptions witnessed on an extraterrestrial landscape. Scientists wanted to inspect Io's volcanoes closely, but doing so threatened the survival of the robotic spacecraft. An earlier radiation burst during an encounter with the Jovian moon Callisto caused four faults within the spacecraft's computer and onboard tape recorder. As the *Galileo* spacecraft approached Io, another radiation burst triggered an error in the onboard computer, flipping the robotic spacecraft into a safe mode. Flight controllers worked to untangle the difficulty as the closest approach neared.[45]

Engineers utilize various shielding technologies to protect robotic spacecraft from excessive radiation and electromagnetic interference. The computers on the Mars rovers *Spirit* and *Opportunity* employed a series of resistors and capacitors that grounded radiation energy before it could creep into the RAD6000 microprocessors. "This has become a real workhorse for space missions," said a supplier of the devices. "We currently have about 150 of these in space today." Without such radiation hardening, X-rays, gamma rays, and other high-energy particles "could create short circuits, create fake bits or burn up electronics."[46]

The most effective protection, however, consists of directives written into spacecraft software that allow it to adjust quickly when onboard failures occur. The computer on the *Galileo* spacecraft, for example, was smart enough to collect data even as intense radiation triggered a safe mode and flight controllers on Earth fought to restore normal operations.[47]

Like people, robotic spacecraft under extreme conditions become lost and disoriented. As the Near Earth Asteroid Rendezvous (NEAR Shoemaker) spacecraft approached the asteroid Eros in 1998, an overly sensitive thrust monitor put the robotic vehicle into a safe mode. The spacecraft began to wobble and tumble. Unable to hold its solar panels in the correct position relative to the sun, the robotic spacecraft switched off its transmitter in an effort to conserve power. Far away on the planet Earth, one-third of the way around the solar system, flight controllers scrambled to unravel the problem, unable to communicate with the small spacecraft.

On its own, the spacecraft fought to maintain a proper position and conserve electric power. After twenty-four hours, as programmed, the robotic craft turned on its transmitter and broadcast a weak signal through its fan beam antenna in a rotating motion toward the center of the solar system. The signal reached Earth, where flight controllers observed the difficulty and set the spacecraft on a new course that allowed it to orbit Eros—albeit after a fourteen-month delay.[48]

The Lewis satellite was not as lucky. A small, technologically advanced spacecraft designed to observe Earth with sophisticated remote-sensing technologies, Lewis began to tumble on its fourth night of operation. Flight controllers normally monitor new spacecraft until confidence grows, but this spacecraft contained an advanced guidance system that allowed it to fly alone without continuous human control. The flight controllers had gone home. Through the night, the satellite fought to maintain a proper attitude relative to the sun and conserve power. By the time flight controllers returned to work, the satellite had depleted its batteries. Dead in space, it fell back to Earth one month later and burned up.

Robotic spacecraft certainly differ from human space travelers in the degrees of harshness that they can endure. These are largely matters of degree, however, not fundamental differences. Robotic spacecraft need protection from heat, cold, and radiation. They need rest and a source of energy, which they may deplete too quickly. They suffer from blindness and disorientation, and they can die. Increasingly, designers of robotic spacecraft treat these qualities through their organic analogies.

In one respect, the treatment of robotic spacecraft departs significantly from the human analogue. While humans and robots in space must be kept intact so that they can complete their work, humans must also be kept alive so that they can return home. With few exceptions, robots do not return. Like most of the rockets that dispatch them, they are expendable. The cost and difficulty of robotic flight would increase exponentially if the requirement of a return trip so uniformly imposed on human expeditions was forced on robotic ones. The human attitude toward robotics approximates the occasional treatment of animal subjects on early spaceflights. The Russian dog Laika rode into orbit on the second Earth satellite, *Sputnik 2*. After a few days in space, the dog was allowed to die as its life support systems wore down. Wernher von Braun proposed that the monkeys sent to ride on his "baby space station" be relieved of the trauma of uncontrolled reentry through the release of a "quick-acting lethal gas." Von Braun assured his audience that "the monkeys will die instantly and painlessly."[49]

The prospect of animal euthanasia proved too forbidding for a program that depended so substantially on public support. Scientists worked hard to return many animals safely to Earth. The Russian dogs Belka and Strelka became the first living creatures from Earth to survive orbital flight after they returned safely in 1960. The chimpanzees Ham and Enos and the monkeys Able, Baker, and Sam, all of whom helped prepare the United States for human spaceflight, all returned safely to Earth, although Able died during the post-flight surgical procedure required to remove its flight sensors. France became the first nation to return a cat safely from space with the suborbital flight of a black-and-white Parisian stray named Felix. Like their human counterparts, some animals destined for safe returns perished in space accidents, including the Russian dogs Lisichka, Pehelka, Mushka, and Bars.[50]

Perhaps robots will be treated in a similar fashion someday. In 1999, advocates of humane treatment founded the American Society for the Prevention of Cruelty to Robots (ASPCR). "Robots are people too," the society's website proclaims—"or at least they will be someday. And when that day comes, the ASPCR will have in place a set of guidelines designed to protect and uphold the rights of all intelligent artificial beings."[51]

With robots confronting many of the same types of challenges that humans encounter in space, the outcome of the human versus robotic flight debate has come to depend significantly on supporting technologies. People favoring human flight anticipated substantial advances in the technologies needed to support life in space and to protect it from hazards such as excessive radiation. Yet advances in life science research floundered as work on the International Space Station slowed. In contrast to the broken pace of technologies supporting human flight, the technologies advancing the ability of robots to survive and accomplish their missions steadily improved. In that sense, technological developments during the latter decades of the twentieth century again favored robotic flight.

TORTOISE BYPASSES HARE

Contemplating future developments in spaceflight, visionaries such as Wernher von Braun and Willy Ley anticipated that humans would continue to outperform robots for many centuries. Eyes, ears, brains, hands, and the power of speech invest humans with capabilities that permit exploration and discovery. The notion that human capabilities such as these would remain inexorably superior to their machine counterparts guided much of the early thinking about spaceflight. Expressed in many forms, the belief became part of the gospel of spaceflight. Simply put, the advocates of a human presence continued to defend the notion that humans could do in space many things that robots could not.

In fact, the ability of robots to see, hear, speak, think, and move around improved far more rapidly than most people had expected when the space age began. More exactly, great advances have occurred in the technologies allowing robotic spacecraft to sense and record their surroundings, especially in visual imagery; to transmit information back to Earth; to process information and solve problems; and to land on and move about other celestial bodies. Additionally, scientists and engineers building robotic spacecraft resolved the conundrum posed by the unyieldingly high cost of space transportation in a unique and clever way.

Imaging

The advances in spacecraft imaging—the "eyes" of robotic spacecraft—have been remarkable. In 1959, shortly after the creation of the civil space agency, NASA officials began work on Project Ranger, a mission aimed at gathering closeup pictures of the Moon. The state of spaceflight technology and its associated photographic technology was primitive. Officials at the Jet Propulsion Laboratory designed a spacecraft that could fly straight toward the Moon and record pictures of ever-increasing detail—until the spacecraft crashed into the lunar surface. Pictures had to be transmitted instantaneously, precluding use of automated cameras like those developed for spy satellites that advanced, developed, scanned, and collected information on exposed film. Reconnaissance satellites ejected film canisters and sent the exposed rolls home, an approach ruled out by each Ranger spacecraft's mission-ending demise.[52]

Engineers working on Project Ranger addressed this challenge in an ingenious manner. They installed television cameras on the spacecraft. Six vidicon television cameras with appropriate transmitting devices broadcast images over two independent channels. The cameras on *Ranger 7*, the first spacecraft to complete its mission successfully, began transmitting 17 minutes prior to impact. Back on Earth, broadcast signals were received, recorded, and displayed on a standard cathode-ray tube. A 35-millimeter camera took pictures of the images appearing on the television screen. It was a primitive system, but it produced images of better resolution than those available from the best telescopes on Earth.[53]

A dozen years later, NASA scientists placed two Viking landers on the surface of Mars. Each lander carried two cameras; each camera was 22 inches tall and 10 inches wide at the base. The cameras used a relatively new technology that captured light rays on a set of light-sensitive diodes. The diodes recorded the intensity of the light. Transmitted back to Earth, the information allowed scientists to construct pictures in much the same fashion as publishers create black-and-white newspaper images using dots of varying shades.

This was a more advanced imaging technology than the one used on Project Ranger, but it still possessed significant limitations. Cameras recorded light only one dot—or pixel—at a time. The light fell on twelve diodes, which were set to measure different focal lengths and colors. Those diodes, however, only recorded a single beam of light. To take a complete picture, the cameras, with their assemblages of motors and moving mirrors, scanned single vertical lines 512 pixels tall, recording each dot individually. The camera then rotated slightly and recorded information from another vertical line. Full-scale panoramas of the landing sites took thirty minutes to assemble. Scientists complained that any living creature darting across the path of the camera would appear as a disturbance in only a few pixels unless it conveniently decided to stand still and pose.[54]

Scientists had improved on this technology considerably by the time the Pathfinder lander arrived on Mars twenty years later. The Pathfinder camera sat in a small cylinder just four inches wide. The cylinder contained a CCD detector, a technology invented at Bell Labs in 1970 by George Smith and Willard Boyle using what is sometimes characterized as "electronic film"; the imaging device on the Pathfinder lander recorded the light intensity simultaneously falling on a pair of detectors each 256 pixels wide by 256 pixels tall. For panoramic views, the imaging system stitched together a series of smaller pictures. Motors, lenses, and filters allowed the camera to rotate, take pictures at different focal lengths, and record images in breathtaking color or black-and-white. Designers of the Hubble Space Telescope also utilized CCDs.[55]

The science pancams, or panoramic cameras, attached to the Mars rovers *Spirit* and *Opportunity* were smaller still.[56] Yet they possessed much larger capabilities, measured by the number of pixels they contained. The teams that created the cameras for *Spirit* and *Opportunity* explained that the devices "basically mimic the resolution of the human eye." Spacecraft designers placed the cameras twelve inches apart on what the designers characterized as the rover's "head," supported by its "neck" or mast. In a fashion similar to human eyes, the two cameras captured full-color stereoscopic images from a height duplicating the view that would be seen by a person standing on the Martian soil. The heads on which the cameras rested had more mobility than human ones. The heads rotated 180 degrees up and down and a full 360 degrees around. Each rover carried a total of nine cameras, including the two "eyes."[57]

Like comic book heroes, robotic spacecraft "see" in many different wavelengths. NASA's four great observatories were designed to view the universe in visible light, X-rays, gamma rays, and the infrared portion of the spectrum. In one of the more dramatic extensions of visual technology, scientists developed multispectral scanners that could detect the composition of objects simply by observing the light they reflect. Different types of rocks, vegetation, and materials produce distinct reflective values at different parts of the electromagnetic spectrum. This technology has been used extensively in Earth resource satellites searching for specific types of vegetation or minerals and in military reconnaissance satellites that can detect whether objects like tanks in a field are real or fakes cut out of plywood. The *Sojourner* rover that accompanied the Pathfinder lander to Mars carried an instrument called an Alpha Proton X-Ray Spectrometer, which could determine the basic chemistry of rocks it approached by shooting alpha particles at them and observing the energy spectra of the returning particles, protons, and X-rays. After nearly fifty years of technological development, there is not much that a robotic spacecraft cannot see.[58]

Communication

Much of the early skepticism about robotic capabilities arose from factors presumed to limit radio communication across the vast distances imposed by space. Early spaceflight advocates puzzled over the manner by which expeditions both human and robotic might communicate with Earth. Robert Goddard suggested that the first rocket to reach the Moon, an automated one, might carry a load of flash powder whose ignition would signal the successful arrival of the craft. Contemplating the first human expedition, members of the British Interplanetary Society, a group that included Arthur C. Clarke, suggested the use of flashing lights to communicate between Earth and the Moon. These exploration advocates, accomplished space experts, could not envision the advances in telecommunications that would soon occur.[59]

No single factor has done more to advance the capabilities of robotic spacecraft than their ability to communicate over the expanses of outer space. Robotic spacecraft built by humans have traveled incredible distances, even toward the edges of the solar system, transmitting data back to Earth all the while. After nearly thirty years in space, the *Voyager 1* spacecraft approached the heliopause boundary, the transition zone where the sun's solar winds cease to flow and the interstellar medium begins. Along with its *Voyager 2* twin, the two spacecraft continued to communicate with Earth, over ten billion miles away.[60]

The ability of machines to communicate over incredible distances has been expedited by advances in three related technologies. First, robotic spacecraft need devices that can store information for later transmission. In the beginning, this was done with tape recorders. The two Voyager spacecraft, products of 1970s technology, each utilize an 8-track digital tape recorder. The tape recorders will remain on the Voyager spacecraft as they wander, perhaps for eternity, through the Milky Way. Initially, the Hubble Space Telescope, launched in 1990, also stored data with mechanical tape recorders. With their many moving parts, reel-to-reel tape recorders are among the most failure-prone devices to be found on traditional spacecraft. In 1999, astronauts on a Hubble servicing mission removed the mechanical tape recorders from the orbiting observatory and installed a digital solid state recorder (SSR). With no moving parts or fragile tape, the SSR is much more reliable and can store far more information than conventional tape recorders. The advent of SSRs, a commercial technology applied to space, advanced the reliability and capability of robotic spacecraft substantially.

Having stored information, robotic spacecraft need some method of transmitting it back to Earth. This is the second great technology. People contemplating the future of spaceflight in the first half of the twentieth century understood radio science. They knew that any change in electric current propagates electromagnetic waves. The waves in turn induce electric currents in any conductors they encounter. This scientific principle provides the basis for radio transmission and reception. The transmitting power and range of early radios, however, was severely limited, a fact well known to consumers who purchased the first receivers. Reflecting on these limitations, early spaceflight advocates concluded, not unreasonably, that people on Earth would encounter great difficulties communicating with spacecraft on distant missions. Spacecraft, especially robotic ones, possess far less radiating power and must communicate over distances far greater than radio stations on Earth. Spaceflight advocates naturally assumed that astronauts would need to ride on the spacecraft, keep records, and eventually share their findings with people at home.[61]

Spacecraft engineers resolved this restriction by employing microwave frequencies transmitted via reflectors that could concentrate electromagnetic waves into narrow beams aimed at Earth. Engineers selected frequencies containing few competing sources of radio noise, such as those emitted from natural sources in the Milky Way. Back on Earth, engineers constructed the third technology, larger but similar reflectors, known as Cassegrain antennas, that contain secondary reflecting surfaces that concentrate the radio waves better. Through a combination of Cassegrain antennas, cryogenically cooled amplifiers, highly sensitive receivers, and complex error-correcting computer programs scientists can, to cite one NASA publication, "detect, lock onto, and amplify a vanishingly small signal that reaches Earth from the spacecraft, and . . . extract data from the signal virtually without errors."[62] The ease with which scientists can communicate with distant spacecraft allowed a far higher degree of automation than originally supposed.

Shortly after the formation of the civil space agency, NASA engineers constructed a series of stations to receive and transmit signals. The stations are located near Canberra, Australia; close to Madrid, Spain; and in the Mojave Desert of southern California. Frequent and steady improvements to what is called the Deep Space Network has resulted in a system that can detect remarkably faint signals from objects on extraterrestrial voyages of discovery. The network received a severe test during the robotic expedition to Jupiter, when the umbrella-like main antenna of the Galileo spacecraft failed to open. Like most robotic vehicles, the spacecraft carried auxiliary low-gain antennas that did not have to be so tightly focused on Earth. Although the backup antenna worked, the signal was very weak. The main antenna, had it worked, would have transmitted information nearly a thousand times more rapidly than the low-gain antenna was designed to do. NASA officials resolved this problem by instructing the robotic spacecraft to store images and other data on recorders and read those back during the long apogee periods when the spacecraft's orbit carried it away from Jupiter and its moons. They also modified the spacecraft's communication software so as to compress the transmitted data better and improved the Deep Space Network's listening devices. These adjustments allowed humans on Earth to receive what they characterized as "Galileo's whisper of a signal from Jupiter."[63]

Without such sensitive communication devices, humans on Earth could never control distant spacecraft. Humans would need to board the spacecraft and ride along. This was the method utilized for terrestrial voyages of discovery and the one anticipated by early advocates of spaceflight. The ability of space scientists to engage in what is known broadly as the science of remote sensing vastly increased the capability of robotic spacecraft. Remote sensing refers to the ability of flight engineers to mount instruments of increasing sensitivity on robotic spacecraft and receive detailed information from distant locations. The sensitivity of robotic spacecraft to their surroundings well exceeds the sensory capabilities of humans, and the capacity of robots to store and transmit this information far exceeds early expectations.

Computers

The robots invented by Isaac Asimov for his various stories possessed "positronic brains" that allowed them to process vast amounts of information and solve problems without direct human intervention. Asimov, who introduced the concept in his earliest robot stories, admitted that he did not know how such devices would work. To Asimov, the devices represented "some unknown power source that was useful, versatile, speedy, and compact." In later years, reflecting on the stories begun in 1939, he realized that he had described the fully developed modern computer.[64]

Space travel, even with humans on board, would not be possible without computers. Human brains cannot effectively monitor the incredibly large number of activities that take place during critical phases of flight, especially liftoffs and landings. Engineers designed electronic components for the Apollo-Saturn launch vehicle that propelled humans to the Moon that could relay the equivalent of three hundred pages of typewritten data to flight control centers on the ground every second. Large computers checked the data against previously established parameters, noted anomalies, and flagged serious concerns for human intervention. The computers that processed this information were too large to fit inside the rocket or spacecraft. This led to a fundamental design philosophy guiding early space endeavors. "When we first went into space," explained a seasoned NASA engineer, "the fundamental design approach was to keep all the complication on the ground and put the simplest design vehicle in orbit."[65]

Such a design philosophy favored the presence of many humans on the ground and only a few in the flight vehicle. Humans in both places interpreted anomalies and overrode balky computers. When a rapidly firing radar altimeter tripped an alarm in a sluggish flight computer on the *Apollo 11* lunar lander, flight commander Neil Armstrong overrode the computer and landed the vehicle himself. Close to the lunar surface, the computer could not process altitude information as fast as it was being received. Armstrong's effort to land was encouraged by a 26-year-old flight controller in Houston, Texas, who recognized the nature of the difficulty and, like Armstrong, concluded that it did not pose an unacceptable risk to the crew.[66]

Placing the oddly shaped lunar module "Eagle" on the surface of the Moon required the joint effort of humans on Earth and humans in space. Landing robots without the opportunity for direct human control proved even more demanding. Prior to the arrival of humans in 1969, NASA engineers sent a series of robotic Surveyor spacecraft to test the surface of the Moon. As one of the engineers explained, "the new and most significant challenge for Surveyor was the landing." To touch down safely on the lunar surface, the robotic spacecraft employed four devices: a high-performance solid rocket motor that slowed the robot's descent, adjustable liquid-fueled vernier rockets that controlled the spacecraft's attitude and completed the touchdown, a landing radar system capable of measuring both vertical and lateral motion, and an onboard computer that calculated the position of the descending spacecraft and issued instructions. The system operated in a closed-loop mode, meaning that the onboard computer controlled the entire landing, beginning fifty-nine miles above the surface of the Moon. From the radar signal that marked the ignition point for the solid rocket burn to the shutdown of the vernier thrusters fourteen feet above the lunar surface, the spacecraft was on its own. Without modern lightweight computers, landing the *Surveyor* spacecraft would not have been possible.[67]

The technical requirements for landing *Surveyor* were very demanding, compounded by the necessity that the landing system and computer fit within a total spacecraft whose mass could not exceed 650 pounds, an amount that also had to include scientific instruments, communications equipment, batteries, and solar cells. Five of the seven Surveyor spacecraft completed their missions, although the vernier rockets on *Surveyor 3* did not shut down at the proper point, causing the 650-pound robot to skip twice across the lunar surface before stopping near the rim of a small crater.

The advent of smaller and more powerful computers allowed even more onboard capability, reversing the philosophy that had kept the most powerful components at home. The Mars rovers *Spirit* and *Opportunity*, for example, each utilized a sophisticated wind-shear detection device as part of their overall landing system. About thirty seconds before landing, downward-pointing cameras on each landing shell took three pictures of the landing zone approximately four seconds apart. Computers compared the pictures, detected evidence of horizontal motion, and ordered the firing of three transverse rockets so as to dampen the effects of surface winds. Since the landings took place at a time when radio signals from the spacecraft required approximately ten minutes to reach Earth, all of this had to be accomplished by the spacecraft without external human control.[68]

In spite of the growing capacity of onboard computers, the software on nearly all spacecraft to fly during the first half-century of exploration consisted of instructions written in advance to handle foreseeable operational events. When confronted with unforeseen circumstances, robotic spacecraft are programmed to flip into safe mode and await further instructions, which occurs with frustrating frequency. Spacecraft engineers would like to develop autonomous capabilities that could resolve small operational anomalies without the craft freezing up and waiting for human intervention.

In 1998 NASA officials launched an experimental spacecraft called *Deep Space 1* that tested a number of advanced computer techniques. The computer on the 831-pound spacecraft contained an automated navigation system that checked its course through space against the predicted positions of 250 asteroids and 250,000 stars. A software program called "Remote Agent," designed with the help of individuals at Carnegie Mellon University, allowed the computer to make flight decisions on the basis of constraints and general instructions. The resulting increase in onboard autonomy allowed the spacecraft to communicate using tonal signals, which could be captured by small ground antennae instead of large and expensive listening devices. NASA officials likened the philosophy guiding the Remote Agent program to the HAL-9000 computer from *2001: A Space Odyssey*.[69]

Increased computer capability has advanced both human and robotic flight. NASA's space shuttle orbiters could not land without the five computers that control reentry. Shuttle computers, developed during the 1970s, are far more sophisticated than the ones used to guide Apollo flights to the Moon.[70] Yet they still required an onboard human presence and, by comparison to the devices that pilot contemporary robotic spacecraft, are ancient technology. Advances in computer technology have enhanced the ability of flight engineers to detach spacecraft operations from ground control and may someday allow the sort of autonomous flight that early space visionaries believed only humans could provide.

Mobility and Endurance

The history of robotic spacecraft has been one of increasing mobility and endurance. The earliest spacecraft to use robotic techniques flew by their targets, not pausing to orbit or land but collecting data and capturing a few images as their targets rapidly appeared and disappeared from view. The first spacecraft to approach the Moon was *Pioneer 4*, which slipped by its destination in 1959. The first spacecraft to fly by another planet was *Mariner 2*, which inspected Venus in 1962. As flight techniques improved, automated spacecraft orbited celestial bodies and attempted to land on them. The Soviet vessel *Luna 10* became the first object to orbit the Moon in 1966, the same year that Soviet scientists achieved the first soft landing on the lunar surface. *Mariner 9*, a U.S. spacecraft, became the first to orbit another planet when it reached Mars in 1971; five years later the first of two Viking landers touched down on the surface of Mars. In 1997 the Mars Pathfinder lander dispatched the first semiautonomous vehicle to roam another planet, the *Sojourner* rover, following a tradition initiated by the first lunar rover *Lunakhod 1*, sent to the Moon in 1970.

Improved mobility and endurance owe much of their force to the technology of solid-state electronics. This technology preceded the space age, having been launched by the Bell Laboratories research that led to the development of the transistor in the 1940s. Scientists subsequently combined millions of transistors on tiny integrated circuits, creating the microprocessors that drive a wide range of commercial, military, and aerospace products. When people like Arthur C. Clarke first contemplated the means by which scientists could construct communication satellites, the supporting technologies were primitive. The integrated circuit, along with related technologies such as the solid-state power amplifier, were not well developed. As technology improved, engineers were able to install communication transfer stations on satellites with masses and electric power needs much smaller than anticipated and with no need for a human presence.

With their integrated circuitry, robotic spacecraft complete their work using very small amounts of electric power. The *Sojourner* rover that explored Mars in the summer of 1997 drew power from a small solar array on its topside and three tiny batteries, which could not be recharged. At its peak operating level, with the sun overhead, the solar array provided 16 watts of power. That is the equivalent of the electricity drawn by a small light bulb, such as an oven light. With this power, the rover did its work and communicated to Earth through the nearby lander. The rovers *Spirit* and *Opportunity*, which began to traverse Mars in 2004, each produced 140 watts of power at full capacity, enough to power a light bulb used to illuminate a typical living room. As a point of comparison, the International Space Station with its human crew will require over five hundred times that amount of power to complete minimal operations.

Solar power sources are limited by the availability of the sun and by their tendency to degrade with time. The twenty-two-pound *Sojourner* rover was still operating when the Pathfinder lander through which it communicated died. Brian Muirhead, the project manager, speculated that the onset of the Martian winter caused the lander's power system to fail.[71]

The technology for a longer-lasting power source has existed since the beginning of the space age: the radioisotope thermoelectric generator (RTG). This device relies upon a well-known principle of thermoelectricity in which two intertwined materials (metals or semiconductors) produce an electrical current when exposed to heat. The RTG, which resembles a cylinder, uses a radioactive material such as plutonium to produce the heat. The device has no moving parts, is nearly indestructible, and can operate for very long periods of time. RTGs on spacecraft can easily provide electric power for twenty years or more.

Space scientists would like to use RTGs to power planetary rovers. Rovers powered by RTGs could explore for years, not weeks, over hundreds of miles. RTGs are most commonly used to power spacecraft bound for the outer planets, although they also have been employed on military navigation and communication satellites and the two stationary Viking spacecraft that landed on Mars in 1976.

RTGs do not generate a great deal of electricity. The RTG used on the first Navy navigation satellite produced a dismal 2.7 watts. The Viking landers each utilized a pair of SNAP-19 RTGs that together generated hardly enough electricity to power an average light bulb (70 watts total). Scientists installed three RTGs on the Cassini mission to Saturn to power a spacecraft as large as a thirty-seat school bus. The Cassini units generated about 850 watts of electric power, a substantial amount by robotic standards. With their integrated circuitry, robotic spacecraft require very low levels of electric power.[72]

Invested with an adequate source of electric power, extraterrestrial rovers could explore for very long periods of time. The first remotely controlled vehicle to roam another sphere, *Lunakhod 1*, utilized solar panels to produce electricity and recharge its batteries. Flight controllers in the Soviet Union drove the eight-wheel, solar-powered rover about six miles over eleven days. Soviet scientists drove a subsequent lunar rover twenty-two miles. Like the Pathfinder lander, it died when its power system failed. During the long lunar night, *Lu-*

nakhod 2 slept, drawing a small amount of warmth (but not electric power) from a radioactive heater unit. Soviet scientists could not revive *Lunakhod 2* after the fifth lunar night. An RTG would have provided sufficient power to keep the rover warm.[73]

The lunar rovers did not have the capability to chart their own path. Flight controllers in the Soviet Union controlled the rovers much as a child might steer an electronic toy. This proved to be a clumsy and ineffective technology that severely limited rover mobility. The rovers *Spirit* and *Opportunity,* which landed on Mars in 2004, possessed a much more sophisticated roaming technology. At top speed, the rovers could cover about six hundred feet in an hour. The rovers were programmed to stop every ten seconds and spend twenty seconds assessing the terrain. When employing its autonomous navigation system, each rover charted a path using images from two stereographic hazard-identification cameras, a front-mounted device for moving forward and a rear-mounted device for backing up. The computer identified obstacles based on their shape, while a hazard avoidance program limited the degree to which the rovers could tilt on sloped soil. The rovers could also drive blind, with the navigation system shut down, following a path charted with the help of flight controllers and based on images provided by the stereoscopic navigation cameras mounted on the rovers' heads.

Space scientists are working to provide rovers with "reactive control" software to accelerate their progress. The traditional technique, "deliberative control," requires the rover or its controllers to painstakingly create a sequence of mathematically precise instructions that guide its path. The rover follows these instructions one step at a time, much like "a blindfolded pirate looking for buried treasure."[74] Reactive control would allow a rover to plot its own path using live observations as it moves through its terrain, following safe courses and avoiding features that look like obstacles recorded in the robot's computer brain. With reactive control software and a steady power source, extraterrestrial rovers would advance their mobility and endurance once more, becoming more like true robots in the process.

Transportation

For fifty years, the cost of moving robots and humans into space has remained intractably high. The cost of space transportation is generally expressed as the expense of moving a given mass of payload from the surface of the Earth into space. The current expense of transporting each pound of payload mass to low-

Earth orbit approaches $10,000. To move the same mass from the Earth into the solar system can easily exceed $25,000 per pound.[75] These figures, adjusted for inflation, were essentially the same ones that existed thirty years earlier. Rocketry at the beginning of the twenty-first century was neither cheaper nor much more reliable than when humans first walked on the Moon. Efforts to "take the astronomical costs out of astronautics," a longstanding objective of rocket scientists, have generally failed.[76]

The intractably high cost of space transportation contrasted sharply with developments in aviation. Approximately fifty years into the history of atmospheric flight, aircraft manufacturers began to install gas turbine engines in production-line airplanes. The development of the modern jet aircraft served to mark a number of technological advances that significantly reduced the cost of flight operations. Assisted by a steady growth in personal income, groups of people that could not afford to fly during the first fifty years of flight entered the aviation market, vastly expanding the use of jet aircraft as a means of transportation.

People who contemplated the future of spaceflight predicted a similar path for rocketry. In 1972, NASA executives set out to revolutionize the burdens of space transportation by developing the reusable Space Shuttle. They did so with the goal of reducing space access costs to a tenth of their original value. So the goal to low-Earth orbit, expressed in 2007 dollars, was $1,000 per pound (down from $10,000). Motivated by early optimism in the 1970s for the cost-saving capability of the planned Space Shuttle, NASA officials committed robotic spacecraft such as the Galileo mission to Jupiter and the Ulysses solar satellite to the shuttle flight manifest. Yet transportation expenses did not fall.

Through the first half-century of rocket-propelled flight, the cost of space transportation remained stubbornly high. For a peculiar reason, this development favored robotic flight. With the cost of space transportation relatively fixed, designers of robotic spacecraft could turn to the other side of the "cost per pound" equation. They built smaller spacecraft, precipitating an impressive decline in the mass necessary to accomplish a given activity. In some cases, they packed more capability on spacecraft with a relatively constant mass; in other cases, they produced lighter spacecraft. Both results were impressive.

A dramatic demonstration of this development occurred in NASA's "great observatory" program. As originally conceived, the program called for NASA to launch four large astronomical observatories—what became the Hubble Space Telescope, the Compton Gamma Ray Observatory, the Chandra X-Ray Observatory, and the Spitzer Space Telescope. These were big machines, ranging in mass from 12,930 (Chandra) to 34,643 pounds (Compton). As planned, the first three observatories were launched on NASA Space Shuttle, whose operational costs averaged about $400 million per flight during this period. By the time NASA planners reached the Spitzer Space Telescope, they had run out of money. Congress refused to fund another large observatory. Undaunted, scientists redesigned the telescope so that it weighed only 1,650 pounds. The lower weight allowed it to be launched on a much less expensive Delta rocket, whose 7000 series permitted launch costs in the $50 million range. Through technological advances affecting the spacecraft mass, NASA scientists were able to cut transportation costs, even though the cost per pound did not fall.[77]

As part of their "faster, better, cheaper" initiative, NASA officials encouraged scientists to propose missions that could be launched on rockets like the *Delta* 2. The *Delta* 2 is a small to intermediate-sized rocket capable of pushing 11,000 pounds to low-Earth orbit. (The Space Shuttle, by contrast, can carry 55,000 pounds.) Since most of that mass consists of the rocketry and propellant necessary to move the robotic payload away from Earth, qualifying spacecraft had to be very small. The spacecraft alone typically possesses a mass of no more than two thousand pounds. Delta rockets were used to launch Mars Pathfinder, the Mars exploration rovers *Spirit* and *Opportunity,* the NEAR Shoemaker spacecraft, and a series of other low-cost Discovery-class missions.

In this respect, robotic spacecraft benefited from the same advances in microtechnology that shrank the size and improved the performance of commercial products such as the hand-held telephone and personal computer. The microelectronic revolution reduced weight while increasing capability. Concurrently, the weight of human beings during this period did not shrink at all, permitting no such advances in the realm of human spaceflight.

Improved reliability in solid-state electronics further helped to reduce the cost of space transportation by allowing robotic spacecraft to make longer and more circuitous voyages. The transportation burden imposed on spacecraft moving beyond Earth orbit is typically measured by the changes in velocity (called delta-v) necessary to move a spacecraft from a parking orbit above the Earth to the spacecraft's final destination. A lower delta-v translates into lower mass and reduced transportation costs. The savings, however, are not free. A lower delta-v typically requires a longer transit time and a more intricate trajectory, which increase mission length and the commensurate burdens imposed on spacecraft reliability. In the tradeoff between transportation savings and longer transit time, advances in solid-state electronics favor the longer voyage.

The *Cassini* spacecraft that arrived at Saturn in 2004 followed a seven-year path that carried it twice by Venus and once by Earth and Jupiter. Each planetary flyby imparted additional momentum, conserving the amount of fuel required for trajectory changes. The *MESSENGER* spacecraft launched toward Mercury in 2004 did not speed straight toward the inner solar system. If it did, it would fly into the gravitational well created by the sun, imposing substantial braking requirements once it reached its destination. Instead, scientists created a lengthy trajectory designed to carry the spacecraft by Earth once, Venus twice, and Mercury three times before settling into its ultimate orbit seven years after launch. Robotic spacecraft like these could not conduct extended missions without the reliability and weight-loss advantages obtained through miniaturization and solid-state electronics technology.

THE RACE AFTER FIFTY YEARS

Much of the debate over human versus robotic spaceflight has taken place in public forums, where combatants use words and concepts to promote their cause. On Earth, the public debate is largely inconclusive, in part because the two sides do not argue on common ground.

At the same time, a more tangible race between robotic and human spaceflight has taken place in outer space. In this respect, the two approaches confront each other in a common and demanding realm. Through what is roughly the first fifty years of spaceflight, the outcome of this practical contest has been more conclusive than the public debate. Here, the tortoise-like robotic flight program has enjoyed a distinct advantage. Advances in technology have provided space scientists with the tools to accomplish more work for a given expenditure of resources using robotic spacecraft, a development not enjoyed by engineers pursuing human flight.

The NASA Mars exploration program provides a good example of how this occurred. In the mid-1970s, NASA officials sent two robotic spacecraft to the surface of Mars. The Viking mission utilized a pair of landers assisted by a pair of orbiting spacecraft that helped scientists select the landing sites. The mission cost slightly more than $1 billion at that time, a considerable sum in a NASA budget that averaged only $3 billion annually during those years.

Viking's substantial cost deterred efforts to launch another Mars lander for twenty years. During that period, the technology of robotic flight advanced considerably. In 1997, NASA officials placed a new spacecraft on the surface of Mars, called *Pathfinder*, with a companion micro-rover named *Sojourner*. The

Viking mission, had it been conducted using Pathfinder-era dollars, would have cost nearly $3.9 billion. The Pathfinder mission cost one-fourteenth of that amount—$265 million. Some of that difference can be ascribed to differences in capability. (The Viking mission included orbiting spacecraft and a biology package; the Pathfinder mission did not, but it carried a rover.) Most of the difference, however, was due to advances in technology and the methods for managing robotic flight.[78]

The Pathfinder team kept project spending to a remarkably low level, one that could not be sustained for long. After the failure of an even less expensive Mars Polar Lander in 1999, government officials allocated $820 million for the mission that successfully placed the *Spirit* and *Opportunity* rovers on the Martian surface in 2004. The twin rovers provide a solid counterpoint to the twin Viking landers that arrived in 1976. Equalizing for inflation, the twin rovers cost roughly one-fourth of the sums allocated for the two Viking landers. That is due almost entirely to advances in robotic technology and mission management.

NASA executives promoting the human flight program wanted to achieve similar cost savings. Had this occurred, it would have advanced this endeavor as well. Yet such cost savings did not appear. The much-desired goal of a 90 percent reduction in the cost of piloted vehicles bound for low-Earth orbit, set for NASA's reusable Space Shuttle, did not materialize. The original cost goal for NASA's space station, set in 1983 at slightly more than $8 billion, was not achieved.

The history of NASA's efforts to establish a large orbiting space station is particularly instructive for understanding the obstacles encountered by people seeking to advance human spaceflight. In 1973, NASA officials launched the Skylab facility, which they called an orbital workshop in deference to their desire to build a permanently occupied space station. In the original concept for this mission, NASA executives hoped to use the workshop as the building block for a much larger artificial-gravity station. One configuration envisioned four workshops assembled in a pinwheel configuration with a nuclear power–generating plant placed at the far end of a central spar.

In the dollars of its day, the Skylab program cost $2.6 billion. That is the total program cost, including the orbital workshop, launch vehicles, and flight operations.[79] The people who prepared the initial plans for the NASA space station that followed envisioned a permanently staffed facility with at least four times the capability of the Skylab orbital workshop, measured in terms of power, mass, and other characteristics. This was to be accomplished for a sum of money slightly more than the inflation-adjusted cost of the Skylab program.[80]

Such an achievement would have significantly advanced the capacity of humans to live and work in space. Yet it did not occur. The United States will spend over $30 billion designing and fabricating its share of the International Space Station. The government will also incur the additional costs of transporting components to orbit and operating the station with assembly under way. Estimates vary, but the burden of completing the International Space Station to the United States alone will easily exceed $50 billion.[81]

NASA officials could have built, launched, and operated a Skylab-type facility for about $12 billion (in 2007 dollars). Instead, they attempted to build a more advanced facility. Had they been able to do so for slightly more than the cost of the Skylab project, space station managers would have achieved an impressive goal. They would have produced a more advanced facility for a relatively small increase in price. That would have represented a significant advance in human flight capability, matching the technological gains flowing to the robotics program.

During the 1990s a sea change in the approach to robotic space science took place, with a striking reduction in the cost of the majority of individual space missions, as small, relatively inexpensive projects replaced more costly ones. One of the last of those large expensive space projects was the Cassini mission dispatched to Saturn in 1997, which cost $3.3 billion, a considerable sum by both robotic and human standards. Meanwhile, scientists and engineers developed a large number of less expensive robotic spacecraft, such as the NEAR Shoemaker mission that orbited and landed on Eros in 2001 (total mission cost $212 million) and the Deep Space 1 project that provided the first flight test of an ion propulsion engine two years earlier (total mission cost $156 million).[82] The emphasis on low-cost innovation allowed space scientists to launch small robotic craft with more frequency than a preoccupation with Cassini-class expeditions would have allowed. More projects allowed more experimentation, with commensurate gains in technology.[83]

During the 1960s, the driving force of a space race between the United States and the Soviet Union forced engineers working on human spaceflight to develop a number of important new technologies. These included hydrogen-powered rocket engines and the technologies associated with orbital rendezvous. With the conclusion of the space race and, subsequently, of the Cold War, international competition disappeared. The human flight program, like the hare pausing to rest, stalled.

By investing in a larger number of low-cost missions, space scientists created among themselves some of the same conditions that had earlier motivated human flight engineers. Competition appeared. Scientists proposing new robotic missions in areas such as the Discovery Program competed for tax-financed funding against space scientists proposing other missions. NASA field centers competed against each other and against nonprofit organizations such as the Applied Physics Laboratory in Laurel, Maryland, to conduct flight activities. The frequency with which new projects were approved, a result of their relatively low cost, encouraged a large number of proposals. Proponents who lost one competition applied for others. Competition bred innovation. Within the realm of human spaceflight, the subsequent paucity of new starts in the human space effort discouraged competition, both domestically and internationally. Not surprisingly, the human flight program to emerge was characterized by a relative lack of innovation, persistently high costs, missed deadlines, and unachievable objectives.

In the beginning, when no one had yet flown in space, the people who envisioned the future convinced themselves that human capabilities would advance rapidly. They bet on the hare. In reality, developments subsequent to the first phase of exploration benefited the tortoise-like robotics effort. The race is not over, but the original prognosis has been severely tested by actual events.

CHAPTER SIX

Interstellar Flight and the Human Future in Space

Since people first contemplated rocketry as a means of transportation, visionaries have anticipated the possibility of travel from the Earth to other solar systems. "A discussion such as this may seem academic in the extreme," wrote Robert Goddard, the first person to launch a liquid-fuel rocket, but "it nevertheless poses a problem which will some day face our race as the sun grows colder."[1]

Carl Sagan, an articulate astronomer who helped build popular support for space exploration in the late twentieth century, argued that no civilization could ever avoid space travel in the long run. The danger posed by asteroids, comets, and exploding stars inevitably would cause technological civilizations to migrate. "Every surviving civilization is obliged to become spacefaring—not because of exploratory or romantic zeal, but for the most practical reason imaginable: staying alive," Sagan wrote in a book published two years before his death. "In the long run, putting all our eggs in a single stellar basket, no matter how reliable the Solar System has been lately, may be too risky."[2]

Although constrained from officially committing the government to such an audacious objective, various NASA leaders have expressed their personal commitment to interstellar travel. George Mueller, overall director of human spaceflight during the Apollo era, announced after leaving NASA that "the future of Mankind lies in populating first, the solar system, from there developing the technology to visit the stars and to begin populating the universe as we now know it."[3] A White House aide to President George H. W. Bush managed to plant a similar sentiment into a major presidential speech addressing the future of space exploration. "We will travel to neighboring stars, to new worlds," announced President Bush. "It will not happen in my lifetime, and probably not during the lives of my children, but a dream to be realized by future generations must begin with this generation."[4]

Predictably, such sentiments have crept into fictional depictions of space travel. E. E. Smith presented *The Skylark of Space* in 1928, a galactic "space opera" in which living beings speed between solar systems at speeds far exceeding the transit of light. To cover the vast distances encompassed by their television series, the creators of *Star Trek* invented warp drive, propelling various versions of the starship *Enterprise* to destinations throughout the Milky Way. Rapid transit permitted the construction of galactic sagas such as those found in Isaac Asimov's Foundation novels and the *Star Wars* movie series.[5]

This chapter examines the manner in which interstellar investigation will likely occur, framed in the context of the "man-machine" debate. Underlying this subject is a common challenge. The distances are great and the energy sources required to propel any object to interstellar velocities prodigious. People contemplating interstellar travel attack the energy-distance predicament in a number of ways. Some envision multigenerational spacecraft, a huge structure housing thousands of people who live and die as the spaceship plods between the stars. Those who initiate the voyage do not live to see their descendants complete it. Robotic advocates prefer the efficiencies afforded by machine flight. The energy requirements for most forms of robotic flight, while less than those imposed by human transit, are still prohibitively large. Additionally, any robots dispatched on an interstellar voyage could not be controlled from Earth and would need the capability for self-repair and perhaps even self-replication, a technical challenge of profound proportions. Frustrated by the slow pace of conventional flight, whether human or automated, some people search for strange phenomena within the laws of physics that might allow objects to evade the conventional barriers imposed by space and time. In practical terms, the technical obstacles contained in all of these approaches are daunting and possibly insurmountable.

Compared to the techniques for moving objects around the local solar system, the prospects for any sort of interstellar flight by humans or their machines seem exceedingly remote. Nonetheless, the motivation to contemplate interstellar travel remains strong. In fact, interest is likely to grow, a consequence of ongoing investigations into the nature of the galaxy and emerging accounts of the cosmic neighborhood.

In combination with the growing interest in interstellar flight, the practical difficulties of actually doing it have caused some people to step back from the challenge and ask whether any sort of star travel initiated by human beings might be achievable. The results of such detachment, to say the least, are startling. Once one removes the requirement that the objects transported be human or under human control, fascinating possibilities emerge. If creatures from Earth venture into the galaxy, they may do so in ways that depart considerably from conventional views of rocket travel in fiction and reality.

VISIONS OF THE COSMOS

As space-related technologies have changed, so have human views of the cosmos. This has occurred in the past, and it will surely happen again. New types of space telescopes are the most likely instruments of future change. Utilizing the science of interferometry, these instruments possess the capability to capture images of extra-solar planets and identify through studies of their spectra those that possess Earth-like characteristics. The first detailed images of extra-solar planets promise to revise the ways that humans visualize their cosmic neighborhood and provide a substantial impetus for interstellar exploration.

Understanding the means by which humans have perceived the cosmos in the past suggests how new discoveries might alter human perspectives in the future. The first humans viewed the night sky with unaided eyes, sharpened to perceive the cosmos by the absence of the unnatural light that pollutes modern skies. Among the fixed points of light, they observed patterns that took the form of familiar objects and mythical beasts. The Greeks saw a large bear toward the north (Ursa Major), a bull (Taurus), and a centaur, half human and half horse (Centaurus). The Egyptians and Persians saw a lion (Leo), while the Chinese created groupings that included a dragon, a tiger, and a black tortoise. Ancient astronomers recorded the motion of wanderers, which the Greeks called *planetes,* moving through the fixed points of light. Aristotle suggested that these objects, including the sun, sat on spheres that surrounded the Earth, with the stars on the outermost sphere.[6]

The invention of the telescope revealed the planets to be objects like the Earth. On certain nights, when the planets of the solar system are appropriately arrayed, observers can detect the paths along which the planets move. A line drawn through the planets points toward the setting or rising sun, tracing the ecliptic plane. At the equator, the ecliptic plane is vertical across the night sky relative to the horizon and shifts toward the horizontal as the observer moves north or south.

The telescope created a new cosmology in which the Earth and planets moved around the sun. To help novices envision this arrangement, scientists constructed models that depicted the solar system. One such device, called an orrery, was a common feature in classrooms and museums prior to the space age. Developed by George Graham, an eighteenth-century watchmaker working under the patronage of the 4th Earl of Orrery, this mechanical model accurately portrayed the principal planets. By necessity, the planets were larger and more tightly packed than in nature, but their periods of revolution were accurately scaled. More elaborate versions provided the principal moons or a source of central illumination demonstrating the apparent rising and setting of the sun. In clockwork fashion, the mechanism helped people visualize the motion of the spheres.

Although they are still available in mechanical and electronic form, orreries tended to disappear from common use with the advent of spaceflight. Again, the dominant perspective changed. Images of the Earth witnessed from afar helped to motivate a new perspective. The last Apollo astronauts to return from the Moon captured one of the most impressive, a color photograph of the Earth and its oceans and clouds, with the African and Antarctic continents prominently displayed. As Norman Cousins observed shortly after the expeditions ceased, "The most significant achievement of that lunar voyage was not that man set foot on the Moon, but that he set eye on the Earth."[7] The image became an icon for a new consciousness of the Earth as a living planet on which the delicate balance of land, sea, atmosphere, and living creatures created conditions permitting the maintenance of life.

Further images from the solar system reinforced the uniqueness of the Earth. A succession of planetary spacecraft returned images of hostile and often violent spheres—active volcanoes on Io, methane mist on Titan, desolate channels on Mars, and clouds of sulfuric acid on Venus. Quite in contrast to earlier expectations, which anticipated favorable living conditions on a number of local bodies, the other planets and their moons turned out to be relatively inhospitable places, further reinforcing the image of the Earth as a fragile and special place.

The special properties of Earth are readily apparent when compared to the properties of scientific data from other worlds in the solar system. From a distance, the Earth bears the signatures of a living planet. Its oceans contain liquid water, its atmosphere abundant oxygen, ozone, water vapor, and trace amounts of carbon dioxide and methane. Any cosmic sphere that displays such characteristics probably harbors life. No other body within the local solar system, however, possesses these features. To locate other bodies that might harbor life, scientists must look to other stars. What they find could once again revolutionize human perspectives of our cosmic locale.

In 2003, a NASA spacecraft provided a suggestion of what the search for life-harboring planets might reveal. The Mars Global Surveyor paused in its usual task of mapping Mars, turned its high-resolution camera toward the Earth some 86 million miles away, and took a picture. Using techniques similar to those employed to tint old black-and-white movies, NASA scientists "colorized" the image. The picture shows the Earth half-lit. The photograph is fuzzy, but cloud cover is clearly visible, as is the color of the seas. The Moon sits a few Earth diameters away, in a similar phase. This is the first image of the Earth captured from another planet, and it offers a suggestion of how the first image of a planet with Earth-like conditions circling another star might appear.[8]

The technological key to capturing an image of an Earth-like planet lies in the science of interferometry. Scientists know that other planets orbit other stars. In ordinary telescopes, the reflected light from such planets disappears in the glare of the central star. By slightly delaying wavelengths from the central star as they strike separate telescopes, the troughs and crests can be made to cancel each other out. The dominant light from the central star disappears, allowing its planets to appear. Less technologically sophisticated detectors have been proposed, but interferometry seems the most promising.

To collect sufficient light to produce images of extra-solar planets similar to that of the Earth seen from Mars, scientists want to launch clusters of small, space-based telescopes and fly them in formations that are very far apart. A sufficient number of terrestrial planet finders, examining light waves from reflected light sources, could give humans their first images of Earth-like planets circling nearby stars.

Scientists began to locate extra-solar planets in the mid-1990s. They did so by indirect means, measuring very slight variations in the positions of certain stars caused by the gravitational pull of large planets revolving around them. The presence of planets can also be confirmed by the slight dimming of stars as such objects pass in front of them. By 2005, more than 150 planets had been detected.[9] The number has grown rapidly. Many of these are large planets, however, gas giants with masses that approach or exceed the planet Jupiter.

Rocky objects like Earth are smaller and therefore more difficult to locate. To possess liquid water, they must travel within their solar "Goldilocks" zone—the area around the central star where solar radiation is not too hot, not too cold. The retrieval of information on extra-solar planets continues aggressively. Someday humans will possess images and spectrographic studies of Earth-like objects around other stars. When that occurs, it may displace the Copernican view of the cosmos, with its focus on movement within the solar system, and inspire mental pictures of the galactic star clouds within which such planets lie.

The principal means for re-visualizing the cosmic neighborhood is likely to arise from the existing system of galactic coordinates. This is a star map similar in nature to the lines of latitude and longitude that transcribe the surface of the Earth. Galactic coordinates are easy to visualize. To an observer standing on the Earth's surface on a starry night, the path of the Milky Way traces the galactic plane or equator across the night sky. The galactic plane is a large circle that best fits the disk of our local galaxy. Galactic latitudes are used to visualize the position of stars above and below the central plane. As on Earth, the latitudes run from 0 degrees at the galactic equator to +90 degrees at a point directly above the galaxy (the north galactic pole) and to –90 degrees at the south galactic pole.

Galactic longitudes define the position of stars relative to the center of the Milky Way. The galactic center, with its massive black hole, sits behind a cloud of stars in the constellation Sagittarius, best observed by people situated in the northern hemisphere during the August night sky. Out of respect for the curiosity of the creatures that invented the galactic coordinates, a line drawn from Earth to the center of the galaxy defines zero longitude in the same manner that the Royal Observatory in Greenwich, England, defines the prime meridian on the surface of the planet. Galactic longitudes run around the galactic equator for a total of 360 degrees. A star on the outer reaches of the galaxy, away from the galactic center relative to Earth, sits at 180 degrees.

The galactic perspective from Earth's surface, as at night, is somewhat disorienting since the plane of the solar system does not match the plane of the Milky Way. The solar system is severely tipped, with the celestial equator (a projection of the Earth's equator into the solar system) sitting at a 62-degree inclination relative to the disk of the Milky Way. The galactic equator, nonetheless, is still visible in the dark night sky. The galactic disk stretches like a continuous band across the night sky—not just at the galactic center but also along the stellar arms that embrace the Earth on all sides.

With a little practice, a most astonishing perspective appears. Humans can visualize the location of nearby solar systems. About twelve light years from Earth lie Epsilon Indi and Tau Ceti. Both are sun-like stars, with 77 percent and 81 percent of the sun's mass, respectively. Both have objects circling them.

The position of these two stars within the galactic neighborhood can be identified by combining three bits of information—their longitude on the galactic equator, their latitude above or below the galactic plane, and their distance from Earth. Epsilon Indi sits below the galactic plane, in the direction of the center of the Milky Way galaxy, 11.8 light years away. Tau Ceti rests on the opposite side, toward the outer reaches of the galaxy, like Epsilon Indi below the galactic plane, 11.9 light years distant.[10] (For perspective, the Milky Way is about 100,000 light years across.)

With more practice (and help from a computerized star program), observers can visualize the galactic neighborhood as if they were standing at a point in space outside the local group of stars.[11] This provides a three-dimensional perspective of the local neighborhood that is more revealing than the two-dimensional dome-of-the-sky approach as seen from Earth. What appears at first to be a shapeless cloud of lights becomes more familiar as the viewer identifies the most interesting star systems. No classical figures materialize, neither bulls nor lions nor dragons to provide humans with earthly and mythical surrogates for the apparent arrangement of the stars. With time, however, the locations grow familiar.

In the past, new conceptions of the cosmos inspired new forms of exploration. Human characteristics like curiosity and greed made it occur. In the centuries immediately following the Copernican revolution, the Earth and local solar system was thought to contain destinations in sufficient number to satisfy those motives—inhabitable planets, extraterrestrial life, and opportunities for migration. (The location of the solar system in the galaxy was then not well known.) Closer inspection of the solar system revealed a landscape more desolate than originally imagined. The vision motivating exploration is now being extended to nearby stars. The nearest life-supporting systems may be a few light years distant or they may be very far away indeed. Most experts believe that such systems exist. It is just a matter of where.

As Earth-like systems are discovered, the same impulses that motivated exploration of the local solar system will extend to other stars. As surely as the Copernican perspective replaced the Aristotelian, interest in the local neighborhood will grow. The motivation is real. The key question concerns the method of exploration. Which approaches to exploration and potential migration will technology favor?

INTERSTELLAR TRAVEL

The obstacles to interstellar travel are exponentially more difficult than those confronting exploration within the local solar system. The prospects for interstellar travel depend considerably on the proximity of enticing targets, developments in propulsion technology, and the nature of the payloads being moved. As presently understood, none of these factors favor the transport of human beings. One need not conclude that travel from Earth to other solar systems is impossible. Rather, what can be observed from the present situation is the likelihood that the transport of humans in their present form remains one of the least efficient means of reaching objects in other solar systems capable of harboring complex life.

Proximity of Targets

Disks of material circling nearby stars—that is, solar systems—appear to be remarkably common. Since beginning the search in the twentieth century, astronomers have been able to confirm the existence of many planet-bearing stars. Systems like the one in which humans find themselves, however, may be quite rare. The Earth and its life forms reside in a metal-rich solar system dominated by a single yellow-orange dwarf star with gas giants of moderate size moving through outer orbits of relative circularity. If for some reason inhabitable planets exist only in systems like this, then the probability of finding Earth-like spheres nearby may be quite remote. The best candidates may exist far from Earth rather than nearby, substantially increasing the difficulties of direct exploration or migration.

The nearest star system is Alpha Centauri, approximately 4.3 light years away. It consists of three stars, two of them in a close binary relationship. The two main stars (Alpha Centauri A and B) are separated by an expanse that approximates the distance between the planet Uranus and the sun, also known as Sol. Venturing a bit further, three small stars appear. Their distances from Earth range from 6 to 8.3 light years. These are red dwarfs (Barnard's Star, Wolf 359, and Lalande 21185), all with masses considerably less than that of Sol. The habitable zone in such systems, defined as that area permitting surface water in liquid form, would be extraordinarily close to these weak parent stars. Any such planets within those zones would race around them, completing their orbits in a few days. The strong gravitational attraction of the star would place such objects in a tidal lock, with perpetual night and day.

Sirius 2, at 8.6 light years from Sol, is a binary system like Alpha Centauri. The main star, Sirius A, is young, hot, and bright. Twice as large as Sol, but twenty-one times brighter, it bathes its solar region with intense radiation. Not much hope for life-bearing conditions exists there.

The next two significant star systems are again red dwarfs, slightly less than ten light years away. The first is a binary red dwarf system (Luyten 726–8 A and UV Ceti), the second a single red dwarf star (Ross 154). UV Ceti and Ross 154 are a particular type of red dwarf known as flare stars, an additional obstacle to living beings. Although UV Ceti, otherwise known as Luyten's Flare Star, possesses only 10 percent of the mass of Sol, its brightness has been observed to increase by a factor of seventy-five when flares periodically erupt. Such blasts of radiation would prove lethal to any creatures unfortunate enough to find themselves on a planet close enough to the red dwarf to stay wet and warm. Red dwarfs are one of the most common star types in the galaxy. Stars like Sol are rare, at least in the local neighborhood.

More interesting candidates exist at ranges slightly more than ten light years away. Epsilon Eridani, at 10.5 light years, is an orange-red dwarf star with a mass about 85 percent of Sol. Epsilon Indi, at 11.8 light years, is also an orange-red dwarf star. The two stars sit below Earth on the galactic plane, on opposite sides of the longitudinal coordinates, thereby forming a neat triangle of roughly equal sides. Epsilon Indi possesses a mass equal to about 77 percent of Sol. Objects have been detected circling the two stars. Epsilon Eridani appears to have a Jupiter-like object revolving around it, further enhancing its candidacy as a life-bearing system. In the local solar system, Jupiter serves to clear the inner solar system of dust and debris that might otherwise pulverize life-growing spheres.[12] These two candidates, however, suffer from a troubling deficiency. The two stars each emit less than 30 percent of the light produced by the Earth's sun. Because of this luminosity problem, habitable planets would need to be much closer to these stars than the Earth or Venus are to Sol. A planet around Epsilon Indi, which has just 15 percent of the sun's luminosity, would have to be in what scientists characterize as a "torch" orbit—closer than the planet Mercury sits with reference to the sun.[13]

The suitability of Epsilon Indi may be further complicated by the presence of two brown dwarfs. The first brown dwarf sits at a distance quite far from Epsilon Indi, well beyond the equivalent orbits of the major planets in the Earth's solar system. The second brown dwarf circles the first. Brown dwarfs are substellar objects larger than gas giants like Jupiter but too small to fuse ordinary hydrogen and become true stars. Nonetheless, they do generate heat, a consequence of their ability to produce a fusion reaction from deuterium, an isotope of hydrogen. The existence of brown dwarfs in the equivalent of what is Sol's Oort cloud might disturb dormant blocks of ice and send them racing toward the inner solar system of Epsilon Indi, where they could do catastrophic damage to any small, rocky planets revolving there.

Tau Ceti, at 11.9 light years from Earth, is more intriguing. This star is a yellow-orange dwarf, like Sol, with about 80 percent of its mass and about 60 percent of its luminosity. Tau Ceti may be as much as ten billion years old, more than twice the age of the sun. Before that estimate conjures up images of an ancient and technologically advanced civilization, consider that the Tau Ceti system appears to be full of debris. One estimate places the mass of comets and asteroids at ten times the amount found in the Earth's solar system.[14] Any complex life forms that arose on a planet orbiting Tau Ceti might have been blasted into extinction before they could develop the technology to deflect comets and asteroids. Tau Ceti also lacks the abundance of heavy elements found on Sol.

To find stars that duplicate the conditions of the sun, one must travel far. Chara is a yellow-orange star that possesses 108 percent of the sun's mass, 1.2 times its luminosity, and may be just as rich in iron. The star 37 Geminorum is practically identical to the sun and, at 5.5 billion years, is about the same age. It seems free of excessively large companions. Chara and 37 Gem were listed as two of the most promising candidates for investigation in a recent survey of potentially inhabitable solar systems.[15] By migration standards, however, these systems are awfully distant. Chara is 27 light years from Earth—nearly seven times the distance of the nearest star system—while 37 Gem sits a seemingly impossible 56 light years away. That is not much by galactic standards—the Milky Way is 100,000 light years across. It is a substantial distance, nonetheless, relative to the frequency of stars.

Optimists hope that inhabitable planets may appear under conditions more varied than those presented by the Earth's solar system, anywhere that liquid water flows. This might include the moons of conveniently placed gas giants or the habitable zones of multiple star systems. No one knows how likely this is. The orbit of any planet within a multiple star system might prove too unstable to permit living conditions given the presence of the second star. The distance separating Alpha Centauri A and Alpha Centauri B, for example, varies considerably. At times, they are almost as close as Sol and Saturn; at their furthest point, the distance approximates that of Sol and Neptune. Recent calculations, however, suggest that the orbit of any planet circling Alpha Centauri A at about 1.25 astronomical units would be stable and permit liquid water. (An astronomical unit, AU, is equal to the distance from the Earth to the sun.)[16] Like Sol, Alpha Centauri A is a yellow-orange dwarf star with about 110 percent of Sol's mass and 155 percent of its luminosity. Perhaps an inhabitable object resides there, a mere 4.3 light years from the Earth. New discoveries in the field seem to occur faster than the public knowledge base can absorb them.

If inhabitable planets require conditions that exactly duplicate those favoring Earth, the possibilities decline. Consider the conditions that have allowed complex life to arise on the third planet in the solar system of Sol. The sun's orbit around the galaxy is remarkably circular; its rate of speed matches that of nearby spiral arms. Both factors prevent the Earth and Sol from entering the inner galaxy or more densely populated arms where the greater frequency of supernovae could strip away conditions favoring life. The heavier elements out of which Earth and its living creatures were formed occur in disproportionately large amounts in the local neighborhood. These elements are the products of older stars that have blown apart and seeded younger systems like the one in which the Earth resides.

Sol is a single star in a galaxy populated with multiple-star systems. It is a larger star than most in the local neighborhood, but not so large that it would burn out quickly. The concentration of heavy elements in Sol helps to prevent solar flares with their accompanying radiation. In a region where astronomers have found many extra-solar planets in elliptical orbits, the planetary paths around Sol are more circular in form. The orbits of Jupiter and Saturn help to stabilize the inner solar system. If Earth had a more perturbed orbit, it would endure greater extremes of hot and cold. At the worst, the orbital instability engendered by a different set of companions could cause an Earth-like planet to be ejected from the inner solar system. The size and placement of Jupiter, with its debris-cleaning capability, is just right for the development of life on Earth. Yet had Jupiter been much closer to the sun or much larger, it could have prevented the creation of the Earth in the same way it seems to have interfered with the formation of a planet in what is now the asteroid belt.

The Earth has a large moon, probably the result of a freak encounter with a Mars-sized object early in its development. This unusual characteristic for an inner planet creates Earth's tidal zones, within which a rich variety of life has appeared. It also keeps the Earth from tilting too far or spinning too fast. The crust of the Earth floats on a mantle. This phenomenon, known as plate tectonics, lifts the continents above the sea, recycles the chemicals necessary to maintain a life-enhancing atmosphere, and allows for the formation of a protective magnetic field.

Scientists do not know the extent to which factors such as these constrain the development of inhabitable spheres. To some, like Peter Ward and Donald Brownlee, the rare confluence of such factors suggests that the frequency of Earth-like planets possessing complex life forms within the galaxy's many stars may be astonishingly low.[17] Translated into exploration requirements, it may mean that the distances between such planets is forbiddingly large. Rather than a few light years—the distance to the nearest stars—the nearest Earth-like planets might be fifty to one hundred light years away.

The existence of complex or intelligent life on extra-solar planets, however, need not be a requirement for habitability. Observations of the Earth's own solar system have revealed no large plants or animals on Mars. Yet that has not significantly dampened enthusiasm for human visitation and the possible transformation of that planet into a place where humans might live.

Preliminary evidence suggests that the distance between planets capable of supporting complex life is great rather than small. The whole history of interplanetary exploration has been one of expanding horizons, in which the potential proximity of living planets has grown larger with time. First humans looked to Mars and Venus, later to nearby stars. This substantially increases the burden imposed on those contemplating the size of the payloads that might be transported to those objects and the propulsion techniques needed to move there.

Propulsion

The vast expanse between Earth and potentially inhabitable extra-solar planets and the very long time spans required to reach them at currently attainable speeds have driven proponents of interstellar flight to consider radically new methods of propulsion. Proposed methods are creative, exciting, and probably unworkable. Any serious discussion of inner galactic flight, nonetheless, must include a discussion of propulsion technology.

The velocities required to reach other stars rest well beyond the capability of existing propulsion methods. When NASA officials launched the two Voyager spacecraft on paths toward the outer planets and eventually out of the solar system, they employed conventional chemical fuels—Aerozine 50 and nitrogen tetroxide (a hypergolic combination), solid rocket boosters that burned powdered aluminum and ammonium perchlorate, and an upper stage powered by liquid hydrogen and oxygen. The two spacecraft attained a heliocentric velocity in excess of 27 kilometers per second (60,000 miles per hour). A heliocentric velocity measures the speed of an object relative to the center of the solar system.

Each spacecraft carried a twelve-inch gold-plated record containing a message from Earth along with instructions on how to play the recording for the benefit of any curious beings that might ever recover it. As the two Voyager spacecraft moved away from the inner solar system, they slowed down. The heliocentric velocity of *Voyager 1* declined to about 15 kilometers per second (33,000 miles per hour) before the gravitational tug of Jupiter caused it to accelerate again.[18]

Both spacecraft flew by Jupiter and Saturn; *Voyager 2* visited Uranus and Neptune. In 1993 *Voyager 1*, heading out of the solar system toward the galactic north, encountered the heliopause, the place at which the solar wind begins to lose its ability to deflect the interstellar medium. At that point, after sixteen years of travel, *Voyager 1* had moved away from the Earth a total of 51 AUs. Due to the acceleration boost imparted earlier by the Jupiter and Saturn flybys, *Voyager 1* at that time was traveling more than 17 kilometers per second (38,000 miles per hour).[19]

To leave the solar system entirely and commence an intergalactic voyage, a spacecraft needs to depart from the heliopause and move through the Oort cloud. The outer reaches of our solar system are marked by three features. The Edgeworth-Kuiper belt of comets, set along the solar system's planetary plane, reaches from the orbital path of Neptune to at least 50 AUs. The heliopause may extend outwards 90 to 120 AUs from the center of the solar system—no one yet knows.[20] The Oort cloud of small icy objects extends even further. It forms a sphere surrounding the solar system in all directions. Some estimates place it as far out as ten to twenty thousand AUs. The Oort cloud may contain large planets or even substellar objects such as a brown dwarf gravitationally bound to Sol.[21]

If estimates of its size are correct, the Oort cloud reaches less than one-tenth of the distance to the nearest star system, Alpha Centauri 3. That star system lies 4.2 light years, or 267,000 AUs, away. Employing the average rate of speed achieved by *Voyager 1* (51 AUs from Earth over 16 years), a spacecraft would require more than eighty thousand years to complete its journey to the Alpha Centauri star system. The distances are so incredibly great that scientists prefer to calculate them using parsecs, a concept derived from the galactic coordinates. One parsec equals 3.26 light years.

To complete an interstellar flight over a reasonable period of time necessitates vastly increased speeds. Advocates of interstellar travel typically aspire to velocities approximating one-tenth of the speed of light. This would be an incredible achievement, one not attainable with conventional means. Such a speed would allow a spacecraft to cover the distance to the Alpha Centauri system in forty-two years. If that seems like a slow transit, consider that a spacecraft traveling at one-tenth light speed would traverse the distance covered in the solar system by *Voyager 1* in just three and a half days, a transit that took that spacecraft sixteen years. The radical increase in velocity required by this objective would necessitate radically new methods of propulsion.

The optimism with which proponents of interstellar flight approach the propulsion challenge is often extreme. In the mid-1950s, shortly after the first detonation of the atom bomb, scientists at the Los Alamos scientific laboratory proposed the use of such devices to propel spacecraft. Specifically, they proposed the successive detonation of about fifty "small yield" fission bombs at one-second intervals at a distance of approximately fifty meters from a spacecraft. According to calculations performed by the scientists, the ship would reach a velocity of 220,000 miles per hour. The resulting effort became known as Project Orion.

The technique imposed a number of practical challenges. "The critical question," the proposing group modestly observed, concerned the ability of spacecraft designers "to draw on the real reserves of nuclear power liberated at bomb temperatures without smashing or melting the vehicle." Designers proposed a saucer-shaped plate accompanied by a "sufficiently powerful magnetic field" to absorb heat and shock. Another challenge arose from the potential for nuclear fallout over the initial launch site. Addressing this challenge, the proposing group allowed that "the whole scheme presupposes elevation of the entire structure beyond the earth's atmosphere by a chemical booster rocket."[22]

Although the velocities potentially attainable by nuclear-powered rocketships exceed those available through chemical propellants, they are not efficient for interstellar trips. The speed is insufficient to allow a trip to Alpha Centauri, only about four light years away, in less than a century. Such a spaceship would probably still be confined to the solar system. For interstellar flight, more powerful propulsion systems are needed. In the 1970s, enthusiasts at the British Interplanetary Society responded with the calculations necessary to construct a spaceship capable of traveling at approximately 12 percent of light speed. Their effort was known as Project Daedalus.

As with Project Orion, the promoters of the Daedalus project investigated the means by which explosive pulses of nuclear energy could propel their spacecraft. The engine in the proposed Daedalus spacecraft was much more efficient. It relied on fusion reactions rather than the fission bombs proposed for Project Orion. The designers of Project Daedalus proposed to ignite small pellets made up of deuterium and helium-3 inside a relatively confined open-mouth chamber.

Set off by electronic beams, the pellets would react like miniature thermonuclear bombs.

According to the calculations performed by the study group, the continuous firing of a thermonuclear engine would allow the spacecraft to attain 12.2 percent of light speed after 3.8 years of work. A trip to Barnard's Star, the primary destination, would take less than fifty years.[23]

This was an exciting prospect, but one containing significant technical challenges. A principal one concerns the gathering of helium-3. The substance is not readily available on Earth. It does exist on the Moon's surface and in the atmosphere of Jupiter. Project proponents hoped that widespread interest in the use of helium-3 fusion reactors to generate all of the Earth's electricity would motivate recovery efforts throughout the solar system. Apart from resolving the physical difficulty of scooping helium-3 out of sources like the Jovian atmosphere, proponents had to envision a system for transporting it to the spacecraft. Mastery of the solar system, proponents of Project Daedalus agreed, would be a prerequisite for interstellar flight. "With propulsion systems of the Daedalus type," the proponents optimistically proclaimed, "there is no reason why the entire Solar System could not be easily accessible."[24]

Enthusiasm regarding Project Daedalus encouraged advocates of interstellar travel to investigate even more effective means of propulsion. Rocket scientists measure propulsion efficiency using a concept called specific impulse, which refers to the change in momentum delivered by a given unit of fuel and is the rocket scientist's equivalent of miles-per-gallon. For every pound of fuel burned, the liquid hydrogen / oxygen engines on the NASA Space Shuttle deliver a pound of thrust for 450 seconds. Estimates vary, but in theory, the engines on the Daedalus vehicle would produce a specific impulse of 100,000 seconds. As impressive as that might seem, other propulsion technologies do even better. Matter-antimatter engines could produce a specific impulse ten times that great—as much as one million seconds.[25]

The potential thrust provided by antimatter engines drew a great deal of attention during the latter decades of the twentieth century. "It is the most intensive source of energy theoretically possible," gushed George Mueller, retired from his position as head of NASA's human spaceflight programs. "Antimatter, either alone or in hybrid combinations with other energy sources, should be expected to figure prominently in space propulsion for the twenty-first century." Forced to explain how the various versions of the Starship *Enterprise* achieved such incredible galactic speeds, the creators of *Star Trek* announced that their fictional spacecraft employed an antimatter drive.[26]

By the mid-1980s, scientists were routinely creating antimatter in large research facilities. Antiparticles carry the negative value of the additive quantum number of their normal counterpart. An atom of antimatter hydrogen, for example, consists of a positron with a positive charge orbiting an antiproton with a negative charge—the reverse of nature's normal arrangement in which a negatively charged electron orbits a positively charged proton. When the two atoms meet, they annihilate each other in a spectacular burst of energy.

The basic obstacle to the use of antimatter is the difficulty of obtaining sufficient quantities. Mueller estimated that a spacecraft powered by only 180 kg of anti-hydrogen would reach the Alpha Centauri system in just twenty-five years. In terms of mass, this is the equivalent of about sixty-five gallons of gasoline. Scientists do not create antimatter in kilograms, however. They generate it in nanograms (a billionth of a gram) through a process that requires far more electric power than the potential energy contained in the particles produced. "The production and control of antimatter on practical engineering scales is still well in the future," Mueller confessed. He suggested that antimatter factories in space using solar power eventually might produce perhaps one kilogram per month.[27]

The difficulties inherent in producing sufficient fuel prompted advocates of distant flight to consider interstellar ramjets. In the realm of aeronautics, a ramjet is a type of engine that uses the forward motion of the aircraft to compress the flowing air into a combustion chamber. In space, the interstellar ramjet collects hydrogen atoms from the interstellar medium and forces them into a nuclear fusion reactor. The concept was proposed in 1960 by the physicist Robert Bussard.[28]

The concept, while theoretically intriguing, faces a number of practical difficulties. To scoop up hydrogen in regions containing one atom per cubic centimeter, the Bussard Interstellar Ramjet would require a collector some seventy miles wide. Furthermore, the vehicle would need to attain a velocity equal to 6 percent light speed before the collector could scoop up sufficient hydrogen to fuel the engine, and the ordinary hydrogen atoms it collected would make poor fuel for a fusion reactor compared to heavier isotopes or substances like helium-3.

Supporters of the concept suggested a number of alterations to counteract these challenges. Instead of using hydrogen atoms to fuel a nuclear reactor, the hydrogen could be used to increase the amount of mass moving through the reactor and out of the exhaust nozzle. Under the right circumstances, this produces a great deal of thrust. The hydrogen could be mixed with a much smaller amount of antimatter or made to interact with a lithium-6 or boron-11 fusion reactor.[29] By ionizing the hydrogen atoms located in front of the accelerating spacecraft with a powerful laser, the atoms could be collected with a large magnetic field instead of a massive and heavy collector. Each of these suggestions, however, creates challenges of its own. The proposed magnetic field has the effect of repelling ionized hydrogen atoms rather than inviting them in, requiring the added complication of a pulsating magnetic field.

Ultimately, propulsion methods such as these may be exercises in imagination. Proponents proceed from the assumption that the pursuit will yield insights into technologies that can hardly be imagined now. Suspend the laws of physics for a moment and imagine that a group of settlers began a journey to Mars 150 years ago employing the transportation technology used to transit the Oregon Trail, an exercise suggested by science fiction writer Joe Haldeman. In other words, the settlers would walk, accompanied by oxen and a wagon train. Using such methods, they would average less than twenty miles each day. In about sixty years, their pace would be exceeded by people on Earth traveling in cars. In another twenty years, commercial airliners would appear. A few decades later, they would be passed by spacecraft traveling at interplanetary speeds. Still moving less than twenty miles per day, the settlers would take more than five thousand years to reach Mars. In a small fraction of that time, their pace would be eclipsed by more advanced technologies.

Advocates of interstellar flight believe—or rather imagine—that a similar progression of discoveries awaits humans in the cosmic realm. The desire to begin the journey will provide the motivation to complete it more effectively. "You start out on a centuries-long project," Haldeman suggested, "and before you've gone a tenth of the way, someone passes you with a faster, cheaper way to travel."[30]

Payload

Proceeding from known laws of physics, advocates of interstellar journeys imagine propulsion methods more advanced than the ones used for early solar system exploration. They envision atom bomb–driven spaceships, fusion reaction, antimatter drives, and interstellar ramjets. Other enthusiasts contemplate beam-powered starships, in which the energy required to drive a spacecraft is shot along its path from a source at the port of origin. Still others dream of magnetic sails.[31] None of these proposals provides a relatively advantageous method for moving adequately large payloads between the stars.

In the most visionary statements, spaceships carry human beings to distant worlds. Moving people remains the ultimate dream. Proponents of this achievement quickly encounter the reality that ships traveling through interstellar space do not move as fast as people with their limited life spans want to go. To compensate for slower speeds, advocates of interstellar flight propose spacecraft massive enough to provide adequate living conditions for voyages of hundreds or even thousands of years. The people who begin the voyage would not complete it. They would set in motion an enterprise finished by their descendants, many generations removed. Really big spacecraft, however, have incredibly large energy requirements, a reality that works against the attainment of very high speeds. In the unhelpful tradeoffs between mass and velocity, spacecraft large enough to carry humans to the stars will lack for energy to reach their destinations within a reasonable period of time, while those small enough to attain reasonable speeds will not have enough mass to keep humans alive.

The cultural inspiration for generational starships can be found in the biblical account of the Great Flood, in which Noah constructs an ark as a means of preserving species during the catastrophic event. The Russian rocket scientist Konstantin Tsiolkovsky speculated on the manner in which humans might repeat the process with rocket ships. Someday, he posited, humans will need an effective means of escaping a devastating global disaster, and he recommended a starship for that purpose. But while Noah sailed for less than one year, Tsiolkovsky's starships would have to travel for tens of thousands.[32]

The practical realities of maintaining multigenerational starships are so daunting that much of their contemplation remains squarely within the realm of science fiction. How many passengers would be needed to maintain a sufficiently diverse genetic pool? How would people govern themselves during a long interstellar voyage? Would new generations accept the purpose of an expedition they did not initiate? In one of the earliest treatments of this subject, novelist Robert Heinlein imagined conditions on a multigenerational spacecraft heading for the Alpha Centauri star group. The ship is enormous, shaped like a cylinder. Its rotation creates artificial gravity. The necessity of rotation also results in the placement of various decks against the inner wall of the external shell, a configuration that prevents the inhabitants from easily viewing the stars.

Heinlein's story takes place many generations after the ship's departure. Early in the voyage, a mutiny occurs, during which most of the inhabitants die. The survivors live in a primitive state. Simple, religious folk occupy the inner rim, while mutants dwell in the ship's core. The simple people are unaware that they reside inside a spaceship. When a few clever people discover this reality, their leaders resist its revelation on the grounds that it would threaten their otherwise stable society. Struggles ensue. Heinlein's starship is a nightmarish place, not a utopian dream. In the end, a few of the people escape from the starship and descend to an inhabitable sphere.[33]

The social processes necessary to preserve the original purpose of a voyage lasting thousands of years would be elaborate. They would not be unlike those existing on Earth, which itself is a multigenerational transit mechanism traveling through the galaxy at a velocity of approximately a half-million miles an hour. During the late twentieth century, the physicist Gerard O'Neill addressed the social and technical problems of maintaining people in space for extended periods of time. O'Neill proposed the construction of large space colonies. He believed that the social benefits of living in colonies with unlimited solar energy and an absence of earthly pests would inspire most people to leave Earth, reversing its overpopulation. "I believe we have now reached the point where we can, if we so choose, build new habitats far more comfortable, productive, and attractive than is most of Earth," O'Neill professed in 1974.[34]

O'Neill's space colonies were not meant to fly but, according to his proposals, would sit relatively still in gravitationally stable points around the solar system. The experience, nonetheless, would prepare humans for the challenges of maintaining space habitats for extended periods of time. After a few hundred years, he predicted, "I feel sure that . . . a modified version of a space community will have traveled to a nearby star."[35]

One of the most ambitious spacecraft ever conceived appeared in the late 1980s in *The Futurist* magazine. Not content to attempt an interstellar voyage, the author proposed a trip out of the Milky Way to the nearby Great Spiral galaxy. The ship would measure thousands of miles across and weigh one trillion tons on the date of its departure. Its creators would spend thousands of years preparing the spacecraft, constructing it in the outer reaches of the local solar system. Tens of millions of people would board the intergalactic ark, and its population would grow substantially during a multi-million-year voyage. Using antimatter engines, pilots would spend fifty thousand years accelerating the ship to a velocity of 40 percent light speed. Having attained that speed, the ship would cruise for a period of five million years, moving away from the galactic plane of the Milky Way. Millions of years after leaving the local solar system, the descendants of the original crew would restart the ship's engines and begin the long deceleration toward their future home.[36]

The interstellar spaceship is an astonishing vision, breathtaking in its ambition. It is also one with substantial energy requirements, a reality that drives the most serious advocates to consider alternatives other than human flight. In proposing his antimatter-driven expedition to Alpha Centauri 3, George Mueller calculated that 80 percent of the ship's mass would be devoted to fuel. No matter how Mueller adjusted the size of the spacecraft, the ratio of propellant to payload remained roughly four to one.[37] On the fictional interstellar spacecraft that astronaut Buzz Aldrin and novelist John Barnes created for *Encounter with Tiber,* the ratio of propellant to payload stood at thirteen to one.[38] Robert Burruss, the space engineer who proposed the intergalactic spacecraft for *The Futurist,* calculated that 99 percent of his ship's initial mass would consist of fuel. The first 90 percent would be consumed during the acceleration phase; the remaining 9 percent would be needed for deceleration. The ship's payload—its people and structure—would make up only 1 percent of the initial spaceship mass.

In 2002 NASA, in conjunction with the American Association for the Advancement of Science, sponsored a symposium on interstellar travel and multigenerational spaceships. Participants examined two sets of issues—the engineering challenges related to the design of interstellar spacecraft and the "anthropological, genetic, and linguistic" concerns of maintaining the crew. "We must get out of the solar system at some point," the symposium leader wrote in the resulting book. "The time to start thinking about it is now."[39] Participants agreed that the energy requirements were formidable. Said one: "Any sort of rocket, even an antimatter rocket, has marginal performance for interstellar missions."[40]

Confronted by formidable energy requirements, scientists and engineers contemplating expeditions through interstellar space inevitably find themselves reconsidering the size of their payloads. As a result of this process, two of the most famous proposals for interstellar travel do not contemplate any room for humans at all. George Mueller's plans for a twenty-five-year expedition to Alpha Centauri 3 powered by antimatter engines resulted in a payload that weighed only one ton. To dispatch the craft, Mueller proposed a two-stage rocket ship. The first stage, containing twenty tons of reaction mass and 180 kg of anti-hydrogen, would propel a five-ton second stage to two-tenths light speed. The five-ton second stage would need four tons of reaction mass and 36 kg of anti-hydrogen to slow down. Mueller utilized the reaction mass to improve the efficiency of the antimatter engines, which would use small amounts of antimatter "to heat and accelerate a much larger amount of propellant, such as liquid hydrogen, as a reaction fuel."[41] That leaves only one ton of equipment to do the exploring—hardly enough to maintain a human crew.

Members of the British Interplanetary Society, proposing a fifty-year expedition to Barnard's Star, contemplated a larger payload. Using deuterium and helium-3, they hoped to deliver five hundred tons. That is impressively large—roughly the total planned mass of the International Space Station. Yet to do so required 50,000 tons for fuel. Even that much fuel would not be enough to slow down the spacecraft. It would arrive at Barnard's Star at its cruise velocity of one-eighth light speed, traversing nearly one AU every hour. Following a fifty-year transit, it would cross the destination solar system in about one Earth day. Such a short period of investigation following such a lengthy voyage hardly seemed to demand a human crew. Instead, society members proposed the use of four telescopes and a small crew of robots to investigate the star and any objects orbiting it. Some of the robots, called wardens, would be programmed to maintain and repair the ship, while others would detach themselves from the main spacecraft upon its arrival and assist with the high-speed inspection of the star system.[42]

Sober analysis of these limitations has pushed serious advocates of interstellar flight to consider a number of unconventional alternatives. Some would abolish the rocketship altogether, dispatching a spacecraft that consists of nothing but payload and structure. The fuel source for transporting any such spaceship would remain near Earth, in the inner solar system, where the sun provides a plentiful source of energy. Propulsion engineers would direct energy from the solar system to the spacecraft, which would ride the resulting beam to its final destination. A number of energy sources have been suggested—laser beams, microwave photons, charged particles, and small pellets accelerated by an electromagnetic mass driver. The crew might carry a reflector that could be placed in front of the spacecraft and be used like a mirror to reverse the direction of the beam, slowing the spacecraft once it approaches its destination. Fully deployed, the reflector would permit a return voyage. By one calculation, a human crew in an 80,000-megagram (176-million-pound) spaceship could complete a round trip to Epsilon Eridani, 10.5 light years away, in just fifty-one years Earth time. Traveling at close to light speed, the crew would age only forty-six years.[43]

Most of the mass in such a spacecraft would be devoted to structures like its thousand-kilometer-wide collector. Only 4 percent of the spacecraft mass would be allocated to the crew and its equipment. Squeezed together for about fifty years in living quarters with a total mass of 3,000 Mg (seven million pounds), including exploration vehicles, the crew might grow restless, even homicidal. As a number of analysts have noted, the social and medical challenges of completing such voyages would be as daunting as the technological ones.

Very small payloads could be accelerated to interstellar speeds with relatively modest amounts of energy. One intriguing proposal contemplates the use of tiny objects riding microwave beams toward the stars. The calculations for such an undertaking suggest that a spacecraft weighing no more than 20 grams (0.7 ounces) could be accelerated to 20 percent light speed in just two weeks utilizing a microwave transmitting station emitting ten gigawatts of power. Barring revolutionary developments in nanotechnology, however, an object so small would not be able to do much work. Designs for the proposed object, called Forward Starwisp, provide for a small sail, a camera, and an antenna for beaming images of any extra-solar plants back to Earth.[44]

Assuming that good images of extra-solar planets could be captured by telescopes near Earth, little justification exists for sending small spacecraft on such a long voyage. Scientists would want to send spacecraft with better capability, and such spacecraft would be large. With added mass, of course, the original challenges of energy and distance resurface. In addition, confinement of such missions to robotic payloads is disheartening to people primarily interested in sending earthly life forms to the stars. Repeated calculation of the mass and velocity problem has led some people to conclude that interstellar flight may remain physically impossible.[45]

Searching for a way out of this conundrum, advocates of interstellar flight have cut deeper into their fundamental assumptions. Some of the resulting scenarios move well beyond the usual human versus robotic debate. One set of proposals seeks to break through the conventional limitations imposed by space, while another set examines potential solutions proffered by rethinking time.

PHYSICS AND SPACEFLIGHT

Serious proposals for moving people or robots through interstellar space aspire to fractions of light speed, leaving visions of post–light speed to the realm of science fiction. The fractions vary from about 10 to 40 percent, with the most common targets set toward the lower end. In short, humans contemplating interstellar travel using even advanced propulsion methods under the best of circumstances are likely confined to velocities well below the cosmological speed limit.

Some entities do travel through space at light speed and require comparatively little energy to do so. Radio waves travel that fast in the vacuum of outer space. So do television broadcasts. Like all portions of the electromagnetic spectrum, including visible light, they travel through free space at 186,282 miles per second—light speed. Persons who wish to engage in interstellar exploration at light speed can do so through the use of electromagnetic radiation. This simple realization has inspired a number of unconventional approaches, some already executed.

In 1974, astronomers Frank Drake and Carl Sagan aimed the Arecibo Radio Telescope in Puerto Rico at the globular star cluster Messier 13 and sent a message. Delivered in binary code at light speed, the message, when properly read, produces a diagram depicting a human being, the chemical makeup of life on Earth, and the position of the Earth in its solar system. Messier 13 contains over one hundred thousand stars, and Sagan estimated that the chances of hitting a technological civilization capable of receiving the signal were 50 percent. The star cluster is located in the halo of the Milky Way, beyond the galaxy's main disk, so the message will not arrive for 24,000 years.[46]

The effort encouraged advocates of interplanetary communication to seek government funding for equipment capable of detecting any radio messages dispatched our way. Like scientists on Earth, intelligent beings in other solar systems may have settled on the use of the electromagnetic spectrum as the most feasible means of establishing contact with other civilizations. Sagan and Drake optimistically predicted that "there are a million civilizations in our galaxy at or beyond the earth's present level of development." If randomly distributed, the nearest civilization would be about three hundred light years away.[47] Transporting messages through radio or television waves would prove far more efficient than placing objects on slow-moving spaceships. Members of other galactic civilizations, if they exist, might have decided to stay home and communicate through the electromagnetic spectrum.

Government officials provided modest funding for radio investigations during the 1970s and 1980s. NASA Administrator James Fletcher, in a 1975 address to the National Academy of Engineering, assured his audience that the galaxy "must be full of voices, calling from star to star in a myriad of tongues." Fletcher's assertion was rooted as much in faith as in science. As a lay minister in the Church of Jesus Christ of Latter-Day Saints, he subscribed to the religious doctrine that God had created a plurality of worlds occupied by intelligent beings. A NASA publication assured its readers that the processes that had produced life on Earth might have worked elsewhere in the galaxy to produce beings "who stare at the heavens and wonder about other occupants of their universe."[48]

Encouraged by the prospect of finding the proverbial needle in the cosmic haystack, NASA officials proposed an elaborate ground-based listening system to search for messages that might have been broadcast by distant civilizations. Known as Project Cyclops, the $20 billion effort would have linked some fifteen hundred antennas, each one hundred meters in diameter, to enormous computers capable of discerning signals in the cacophony of natural electromagnetic noise. When Congress refused to provide funds, NASA officials scaled back the proposal and gave it a new name. They proposed to spend $135 million on a "Microwave Observing Project," searching through that portion of the electromagnetic spectrum providing the most expeditious means for interstellar communication.[49]

Comparing the expenditure to the funds lavished on human spacecraft, one observer called the less expensive alternative "the biggest bargain in history." Accustomed to traditional visions of space travel, however, lawmakers had difficulty envisioning electromagnetic searches as a viable alternative to robotic and human flight. Opponents likened this "Search for Extra-Terrestrial Intelligence"—SETI—to the general but unverifiable belief in flying saucers and "little green men." Senator William Proxmire bestowed his Golden Fleece Award on the project, an award of dubious honor that he periodically delivered to public activities that misused the public treasury. Another legislator likened SETI spending to "buying fur coats for your cows while your children were freezing."[50] Congress briefly provided monies for the Microwave Observing Project, then terminated all public funding in 1993. Supporters of the effort at the SETI Institute in Mountain View, California, a nonprofit organization devoted to "scientific research and educational projects relevant to the origin, nature, prevalence, and distribution of life in the universe," continued to search the skies for intelligent signals but were obliged to do so with private and philanthropic funds.[51] Leaders of the institute hope to construct an Allen Telescope Array, a collection of 350 six-meter antennas capable of scanning as many as one million stars and the planets they might harbor for intelligent life.

The SETI effort received considerable attention in 1985 after Carl Sagan published his novel *Contact*, which was inspired by Project Cyclops.[52] In the novel, humans receive a message from the Vega star system. As promising as this development might seem, it does not immediately overcome the distances involved. Vega is twenty-five light years from Earth. Any two-way conversation between intelligent creatures on Earth and those on a planet orbiting Vega would require a half-century for the first exchange. In the novel, Sagan offers an ingenious alternative to a two-way conversation. The message electronically received on Earth provides plans for a machine that can circumvent the cosmic limitations imposed by space and time. Once constructed, the machine creates a wormhole.

Sagan showed a draft of his science fiction novel to physicist Kip Thorne. In the draft, the machine produces a black hole. Thorne, who was completing a book on black holes and hyperspace, suggested that Sagan instead employ a series of wormholes. The laws of quantum gravity, Thorne observed, allow for the natural existence of "exceedingly tiny wormholes." A wormhole is a short tunnel connecting two distant points within the universe outside of the four dimensions that humans conventionally experience. Natural wormholes disappear as soon as they arise, but Thorn speculated that a technologically advanced civilization might employ the laws of quantum gravity to hold a wormhole open long enough for something to pass through.[53]

A committed proponent of human spaceflight, Sagan could not resist the temptation to dispatch a human crew through the transit device. In his novel, engineers construct a device on Earth that, when activated, creates a pathway to an exit located in the vicinity of Vega. This cosmological tunnel provides access to additional passageways that lead throughout the galaxy. In the novel, five individuals travel in a dodecahedron-shaped chamber to Vega and beyond. Movie producers simplified the narrative to a single passenger, the central character Ellie, played by actress Jodie Foster. The journey, while it does not lead to the architects of the message, places Ellie closer to an understanding of the sense of immortality that pervades the universe and so motivates interstellar flight.[54]

In his book and a series of accompanying articles, Thorne has explored whether humans might be able to use a wormhole as a conduit for space travel of the conventional sort. Unlike a black hole, whose gravitational forces would stretch and annihilate any object or message that entered it, a wormhole provides some possibility of transit. To hold the wormhole open so that it does not snap shut around the object traveling through it, Thorne suggested that it be lined with an exotic material that possesses negative energy. The laws of quantum gravity are not well enough understood to calculate how this might be done, but as Thorne and Michael Morris observe, neither does current understanding "rule out traversable spacetime wormholes."[55] In this respect, fantastic tales in which children drop into rabbit holes or step through wardrobes and emerge in different worlds might unwittingly describe quantum reality and provide the cultural inspiration for future space travel.

The possibility of traversable wormholes solves two of the principal challenges associated with interstellar flight. An object traveling into a wormhole would not require much propulsive force. Relative to the wormhole, it could travel quite slowly, well below the significant fractions of light speed proposed for spacecraft traveling between stars. In the movie *Contact*, the designers of Ellie's spacecraft simply drop her into the artificially created tunnel, relying on the force of gravity for the initial acceleration. Inside the tunnel, she travels at a relatively modest velocity. Relative to conventional spacetime, however, she travels extraordinarily quickly, covering distances in no time at all that might take thousands of years to traverse otherwise. From the perspective of observers on Earth, in fact, her entire journey consumes far less than a second of time. Such warping of space and time thus alleviates the need for lengthy journeys or multigenerational spacecraft.

Outside of science fiction, the wormhole does not provide a reasonable technology for interstellar travel.[56] Rather, it represents a collective yearning for a method yet unknown that avoids the implacable constraints of space and time. The practical challenges of creating an actual wormhole would be quite formidable. One NASA publication suggests that the creation of a wormhole would require all of the force contained in a neutron star, charged with "incredible voltage" and spun to near–light speeds. Even if that much energy could be produced, the designers of this space-time machine would face an additional complication. Someone or something would need to repeat the process at the other end. A wormhole has two openings, and both must be lined with negative energy to allow transit.[57] Science fiction writers typically resolve this challenge by placing an alien civilization at the opposite end. Without such help, which occurs much more easily in fiction than in reality, someone from Earth would need to dispatch a spacecraft to the ultimate destination, resurrecting the original difficulties that prompted the search for alternative methods of travel in the first place.

AN ALTERNATIVE PERSPECTIVE

In their present state, nearly all of the imagined options for interstellar flight confront substantial barriers. The energy requirements for conventional spacecraft, even those powered by exotic drives, are impossibly large. Plans for the construction of multigenerational spacecraft, an approach favored by advocates of human flight, do little to resolve the propulsion challenge. To conserve energy, the most feasible proposals dispatch robots rather than humans, some very small. This approach suffers from its own deficiencies. Not only does it require the construction of machines with extraordinary capabilities, it ignores the primary motivation for interstellar travel—the dispersion of humanity. Electromagnetic signals provide an interesting method for listening, but the approach is not a practical method for carrying on an interstellar conversation. The possibility of wormholes that evade conventional space and time seems fascinating, but no one knows whether humans could construct them or use them to deliver objects across the wide reaches of space. Regarding the latter approach, one disheartened commentator admits: "In terms of real physics, they are just one step removed from sprinkling the top of your head with fairy dust and wishing you were somewhere else."[58]

Humans beings may be confined to their earthly home for as long as the species remains in its present form. A few may travel to nearby objects like the Moon and Mars, but the interstellar medium may prove impassable. Given the general interest in space travel and the desire for human immortality, this is not a popular view. Yet it is periodically issued by those determined to throw some realism at the overall yearning for inner-galactic travel.

Public interest in galactic exploration, however, is not likely to disappear. Humans will likely develop telescopes that reveal conditions on extra-solar planets. A commensurate revision of the manner in which humans visualize the cosmic neighborhood should follow. Perhaps the people at Mountain View will pin down an electromagnetic broadcast. Together or separately, these developments are likely to increase interest in interstellar exploration in the same manner that earlier revelations about the solar system excited the first ventures into space.

The desire for interstellar exploration is on a collision course with the physical barriers confronting it. Humans are a clever lot, however, capable of solving puzzles that at first seem impenetrable. To date, most of the solutions have focused on space—how to move humans or their machines across the vast distances between Earth and other potentially livable spheres. Conventional space provides only one set of dimensions in the universal equation, however. As Albert Einstein noted, another dimension is time. The contemplation of time suggests other avenues as well.

"Our conception of time itself is now turning out to be very incoherent and superficial," stated Olaf Stapledon, one of the first people to address its effects on evolution and space travel.[59] Born into the same generation as rocket pioneers Hermann Oberth and Robert Goddard, Stapledon was a philosopher who lectured at the University of Liverpool in Great Britain until his success at writing science fiction novels allowed him to depart his teaching career. He produced his first works of fiction at age forty-four, the most famous being *Last and First Men* (1930) and *Star Maker* (1937). At the time, his literary standing in the English-speaking world was clouded by his attraction to left-wing causes and his opposition to the Second World War.[60] He nonetheless had a great influence on future science fiction writers such as Arthur C. Clarke, who invited Stapledon to address the British Interplanetary Society in 1948, two years before the latter's death.

Stapledon questioned his qualifications to address the gathering: "I feel much as a man might feel who merely because he once wrote a children's story about a magic carpet, has undertaken to discourse to a society of aeroplane designers about the future of aviation." In his novels, Stapledon addressed the alterations that might occur among species living on various planets over exceptionally long spans of time. Briefly stated, the creatures change, in both form and mind. In *Last and First Men,* Stapledon examined the manner in which humans might evolve over a period of two billion years. Many changes occur. Some of the alterations produce creatures preoccupied with physical prowess, while other alterations produce pure minds. *Star Maker* takes place over five billion years. In it, the narrator visits other worlds and communicates with species he encounters as he attempts to fathom the purpose of the universal creator or "Star Maker." Unconstrained by the narrow perspectives imposed by conventional time, the narrator comprehends the transformations that the creatures occupying these worlds undergo.

Confronted with the request to address members of the British Interplanetary Society, Stapledon's philosophical training led him to pose a profound question—exactly what part of humanity might one want to move into the solar system and beyond? "If one undertakes to discuss what man ought to do with the planets, one must first say what one thinks man ought to do with himself." Would humans go as conquerors, exploiting the planets? Would they go in order to extend their control over the environment, as a way of proving that they once existed? Or would they dispatch creatures adapted to live on other worlds, physically not human in form, but sharing with humanity its spiritual desire for "love, kindness, fellowship, [and] genuine community between persons?" Social values such as these, Stapledon said, had great "survival value" and might be viewed as more persistently human than the physical form.[61]

Looking to the stars, Stapledon doubted that human beings would ever colonize the galaxy, at least in their current physical form. The "spatial immensities," he worried, were too large. Yet that did not eliminate the possibility "that a human race far more advanced than ours" might acquire the capability to engage in some sort of interstellar communication. Returning to one of his favorite fictional themes, Stapledon suggested that creatures on widely separated worlds might, through some sort of transformation in their physical form, acquire the capability to converse with each other in unforeseen ways. Perhaps it would be a "highly developed technique of telepathy." He admitted no knowledge of how this might actually occur. The possibility, nonetheless, "should not be ignored."[62]

To reach for the stars, Stapledon believed, humans had to change. Transformation of physical beings over vast periods of time and the emergence of mental processes that extend beyond the individual mind characterize Stapledonian thinking. The alternative, he believed, was self-destruction. Although he did not use the word *transhuman,* the term later used to characterize such creatures as might emerge, the concept was clearly implied in his work.

Given enough time, Stapledon suggested, humans could transform themselves. The physical barriers to space travel would not change, but the ability of some post-human entity to interact over vast cosmic distances could. Even if human beings failed to reach other stars, their humanity might arrive through what Stapledon characterized as a higher state of spirituality folded into a "cosmical community of worlds."[63] This was an enormously powerful message. It influenced people like Arthur Clarke, who employed humanistic transformation as a central theme in his prolific writing career. Indirectly, at least, Stapledon's message reached people who, some ten to twenty years later, began to contemplate the manner in which alterations in the human form might permit radically new approaches to space exploration.

CHAPTER SEVEN

Homo sapiens, Transhumanism, and the Postbiological Universe

In the conventional human and robotic approaches to flight, the time assigned for the completion of space missions—relatively speaking—is short. The longest human expedition to the Moon (Apollo 15) took thirteen days. Individual astronauts may remain on the International Space Station for as much as six months. According to the most cautious plans, humans traveling to Mars would be gone twenty-nine months, hardly longer than sailors participating in the great terrestrial expeditions that traversed the Earth's seas. The robotic spacecraft dispatched to Mars as part of the Viking program returned data for seven years. Only with the *Voyager* spacecraft, which took twelve years to investigate the four gas giants, did the length of time needed to complete an expedition exceed ten years. Compared to the average life spans of human beings—or to the overall length of their professional careers—these are not exceptional time spans.[1]

The effects of time on methods of space exploration are analogous in some ways to Einstein's special theory of relativity. Einstein suggested that the passage of time was not fixed but varied according to the motion of the observer. Scientists have confirmed this phenomenon with atomic clocks launched into space. Traveling faster than their counterparts on Earth, the clocks keep slower time. Locally, the differences are tiny. Within the solar system, the effects of time relativity among bodies moving at different speeds are so slight as to be imperceptible without highly sensitive instruments. For investigations of the galaxy and universe, however, the effects become more pronounced.[2]

In a similar fashion, the incorporation of time relativity into visions of space travel seem unnecessary as long as explorers confine their activities to the solar system. Time effects became strikingly more important when exploration advocates began to think in galactic terms. Should any creatures respond to the electronic message aimed toward the globular star cluster M13 by astronomers Frank Drake and Carl Sagan in 1974, their response would not be heard for another fifty thousand years. The spacecraft *Pioneer 10,* which carried a plaque announcing the location of the civilization that sent it, would reach the vicinity of the red star Aldebaran in about two million years. Who knows what creatures might then occupy Earth?

How might the consideration of the eons of time needed for space-exploring civilizations, including the human one, affect the effort? Will the human form evolve during this lengthy period, and how might it be altered because of its spacefaring experience? The results, to say the least, are strange.

SO WHERE ARE THEY?

Scientists and other people who contemplate space travel have noticed a curious feature in the Drake formula. About the time that Frank Drake began to listen for interstellar messages in 1960, he prepared an equation useful for estimating the number of extraterrestrial civilizations that might respond to a cosmic greeting card. The formula begins with the number of stars in the galaxy and gradually reduces that figure until it approximates the number of worlds harboring civilizations with technology capable of communicating with one another at any single point in time. The formula contains variables for the number of stars in the galaxy, the fraction with planets, the proportion of planets with conditions suitable for life, and the fraction on which life develops, becomes intelligent, develops technology, and decides to communicate across interstellar distances. The last element in the equation, called *L,* generated much attention. It represents the average life expectancy of a technological civilization whose members decide to communicate, expressed as a fraction of the span of life on a planet.[3] In short, it represents time.

Enrico Fermi, the great physicist, noticed a curious feature, the Fermi paradox. If the values describing *L* are small—say, a few hundred years over a half billion—then technological civilizations have a fairly short period of time within which to communicate from world to world. The effect of a small *L* on the overall formula is profound. The predicted number of civilizations capable of communicating with each other in the Milky Way galaxy at any one point in time rapidly diminishes as the variable *L* shrinks in size. When the variable gets really small, it produces an outcome in the overall formula that approaches one. That is, the number of civilizations capable of communicating with each other at any one time is limited to whatever civilization is contemplating the activity at that moment—after which it dies out, to be replaced by another one at some point in the future. Simply put, a small *L* means that there is no glut of advanced civilizations at all. A small *L* could well mean that humans are currently alone. Humans may be capable of dispatching electronic messages into the galaxy, but the probability of receiving a signal from another extraterrestrial civilization within the same window of opportunity is vanishingly remote. Other technological civilizations may have existed in the past, but due to their short life spans, they are long gone. As if to confirm this observation, humans have been searching for extraterrestrial signals since 1960—if there really are many civilizations out there looking for us, it is fair to ask why we have not heard anything yet.[4]

Conversely, suppose that the value of *L* is large. If a planet lasts ten billion years and permits civilizations to persist for one billion years or more, the number of communicative civilizations existing at any one time is potentially huge. Given the age of the universe and the history of stars, the conditions suitable for the development of the first communicative civilizations could have appeared more than eight billion years ago. A large *L* would conceivably produce millions of civilizations capable of communicating with each other, both in the past and now.

Therein lies the paradox. Assuming that *L* is very large, the galactic airwaves should be teeming with extraterrestrial broadcasts. Insofar as interstellar communication encourages space travel, the extraterrestrials should already be here.[5] Since the airwaves seem silent and aliens do not walk openly among us, according to the best evidence it follows on the basis of the Drake formula that the longevity of technological civilizations must be very small. This proposition is terribly disappointing to those who view space travel as the means for extending human civilization into the cosmos, thereby ensuring the survival of the species. It suggests that humans at this time may possess the only intelligent communicative civilization in the galaxy and—worse still—are likely to revert to some pre-technological state, after which the opportunity for cosmic dispersion will disappear.[6]

The logic of this conclusion may be flawed. The parameter *L*, while incorporating time, contains no factor for the potential physical changes that complex creatures might undergo in conjunction with their interest in science, technology, and space travel. The parameter assumes that having attained a communicative civilization, the members of that civilization would remain in their original physical state for thousands and even millions of years.

What reasons do we have to accept this assumption? Science fiction writers whose stories span vast time periods often incorporate transformational change, even though the Drake formula does not. Some writers foresee the arrival of mutated life forms, often as the result of terrible wars. In *The Time Machine,* H. G. Wells describes a future world full of vicious Morlocks and peace-loving Eloi, and as the story unfolds its hero deciphers a complex story of evolution. In *Planet of the Apes,* Pierre Boulle anticipates the rise of intelligent chimpanzees coupled with a corresponding decline of *Homo sapiens* as the dominant species on the planet. In Robert Heinlein's classic science fiction novel *Orphans of the Sky,* conflict on board an intergenerational starship bound for the Alpha Centauri system divides the occupants into mutants, who reside in the ship's core, and simple farmers who occupy the outer shell. Such transformations often place the participants in pre-technological states from which they are incapable of organizing complex activities such as space travel.[7]

In other works of fiction, creatures progress to a more spiritual, post-technological condition unencumbered by their bodies. This optimistic vision guides much of Arthur C. Clarke's work. In *Childhood's End,* alien beings arriving in giant spacecraft oversee the total transformation of the human race. The extraterrestrials that created the gigantic starship in Clarke's *Rendezvous with Rama* have long since evolved into a nonbiological state. Most significantly, this vision dominates the central narrative in Clarke's classic screenplay and novel *2001: A Space Odyssey.* In that story, astronaut David Bowman encounters a cosmic passageway that leads him to a place where unseen alien forces transform him into a "star child" with supernatural powers. The movie concludes with the newborn creature orbiting Earth in an embryonic sac. Clarke wanted his audience to accept that image literally, as a real consequence of space exploration, in which the first human undergoes a metamorphosis into a new and higher form.

Unfamiliar with the transformational theme and puzzled by the surrealistic scenes preceding it, most viewers left this cinematic presentation confused rather than informed. Clarke attempted to clarify the misunderstanding with a series of novels: *2010*, *2061*, and *3001*. His thesis: transformed beings no longer need earthly conditions to survive. They can exist on alien worlds or even within machines.[8]

In some stories, the transformations pulsate. Olaf Stapledon described the alteration of humanity through eighteen different stages over two billion years in *Last and First Men*. Humans and their descendants change in physical form, producing variations as diverse as winged creatures and giant brains. In a manner reminiscent of Edward Gibbon's *Decline and Fall of the Roman Empire,* undistinguished civilizations occasionally follow greater ones. Nuclear conflagration, astronomical catastrophes, and extraterrestrial invasions prompt decline and migration to new planetary homes. Isaac Asimov built his *Foundation* series around a similar "decline and fall" theme. The central character, Hari Seldon, seeks to spare the galactic civilization from a thirty-thousand-year period of darkness between empires, but to do so he must combat the Mule, a mutant with extraordinary powers who, as the product of an unpredictable biological irregularity, defies the predictability of Seldon's mathematically precise theory of psychohistory. Again, transformations rule.[9]

Human transformation of some sort is a likely consequence of long-duration space travel and colonization. When English settlers first established a colony at Roanoke near the outer banks of present-day North Carolina, they proceeded to have children. The first of them, Virginia Dare, was born on August 18, 1587. Her physical condition did not differ from that of her parents or of the people in England whose citizenship she shared. Should humans colonize other worlds, similar births will inevitably follow. Presumably those individuals will retain citizenship in the nations from which their parents came, but will they still be *Homo sapiens*? The space environment has profound effects on living organisms, ranging from alterations in subcellular processes to changes in the structure and function of whole organ systems. The altered gravity of a different world, not to mention the different radiation effects, could have profound physiological and biological consequences. Would this result in another evolutionary stage for humans? This is a question of profound significance, both physically and metaphysically. Human transformation and transition is virtually ignored by those who advocate extraterrestrial migration.[10]

An astronomer and historian of science who has written extensively about human interest in extraterrestrial life, Steven J. Dick wrote a 2003 article that anticipated the effects of long-term transformational change on the nature of technological civilizations. He employed Stapledonian thinking to offer an explanation for the silence of the heavens that so confounds SETI researchers. "Long-term Stapledonian thinking is a necessity if we are to understand the nature of intelligence in the universe today," Dick wrote.[11] The model for detecting extraterrestrial signals assumes that aliens, once having invented radio and television, will continue to broadcast in the electromagnetic spectrum for as long as their civilization survives. Assuming that at least some technological civilizations survive and mature, they would need to broadcast the same signals for thousands and even millions of years. That is a lot of reruns. Stapledonian thinking, Dick observes, introduces the possibility that the beings inhabiting such civilizations would change during the periods involved. They might change biologically, lose their interest in broadcasting, or even alter their overall form.

Dick speculates further. How might a technological civilization change once it acquired capabilities such as broadcasting and space travel? Dipping into the realm of artificial intelligence and science fiction, Dick suggests that it might become "postbiological." Its members might grow tired of occupying fragile, gooey bodies. They might develop mental capabilities that extend well beyond the limits of biological intelligence. They might become pure intelligence. Referring to such creatures, Dick suggests that the progress of civilization will produce conditions "in which the *majority* of intelligent life has evolved beyond flesh and blood intelligence, in proportion to its longevity."[12] The universe might be teeming with life as Drake and Sagan suggest, but in a form unrecognizable to humans in their biological forms.

Dick's thesis provides an alternative to more conventional and pessimistic views of interstellar communication. The pessimistic view holds that humans are essentially alone. In their book *Rare Earth,* scientists Peter Ward and Donald Brownlee suggest that complex life forms like those found on Earth require local conditions that occur only rarely in physics. Simple life forms could appear under a variety of circumstances, but complex life might well be extraordinarily atypical.[13] This would account for the apparent absence of extraterrestrial signals. According to Ward and Brownlee, humans may have no one else with which to communicate. As an alternative to this point of view, technological civilizations might arise, persist for eons, and evolve into non–technologically dependent forms. This too would explain the silence humans hear, albeit in a more complicated manner. Changes in form and technology might steer the occupants of other worlds away from the electromagnetic spectrum. Perhaps such species prefer to communicate over vast distances at rates of speed that seem sluggish to humans, with their short life spans. "It is entirely possible," Dick speculates, "that the differences between our minds and theirs is so great that communication is impossible."[14]

Given the estimated age of the universe, plenty of time exists for such changes to occur. The best evidence suggests that the universe is between 11.2 and 20 billion years old; as noted above, the conditions permitting the development of complex life could have arisen in other star systems as much as eight billion years ago.[15] As a point of comparison, biologists believe that complex land-dwelling life forms did not appear on Earth until 450 million years ago. In astronomical terms, plenty of time exists for the natural biological transformation of any species that acquires technology and survives. Perhaps, however, biological creatures do not have to wait for the slow process of evolution to occur. Perhaps such creatures engineer that transformation themselves, through technology.

In that respect, Dick offers a fascinating observation. Cultural evolution, which includes technology, occurs at a much faster pace than natural biological change. Using their powers of invention, a technologically advanced species might accelerate the process of natural change that would otherwise occur slowly and randomly. Dick comments, "It is clear that *biological* evolution, by definition, over the course of millions of years would produce nothing but more advanced biology. . . . But even at our low current value of L on Earth, biological evolution by natural selection is already being overtaken by cultural evolution, which is proceeding at a much faster pace than biological evolution."[16]

Dick does not attempt to estimate the exact length of time needed for intelligent biological beings to morph into postbiological forms. Nature has provided billions of years with the proper ingredients for the development of complex biologies, suggesting that the available time spans are sufficient, should the means exist. "Lacking knowledge of advanced biological or postbiological motivations," observes Dick, "we are unable to predict the cultural evolution even of our own species in the near future, much less those billions of years older than ours." Transformation might require millions of years; it could be much more or much less. Accepting the inevitable nature of such change, any species that acquires an interest in interstellar investigation, joined with a new appreciation for the effects of time and change, soon will acquire what Dick calls a "sweeping reconsideration" of the manner in which they visualize the universe around them.[17] Whatever one thinks of Dick's thesis, it does advance the necessity of contemplating the effects of change and time on any spacefaring civilization, including our own.

SPACEFLIGHT AND THE POST-HUMAN FUTURE

A restless wind blew across the American landscape in the fall of 1960. The calming era of Dwight D. Eisenhower soon would end, to be replaced by a new political synthesis that ushered in the great social transformations of the 1960s. Black and white "Freedom Riders" went to the South to take direct nonviolent action to end the American apartheid; a new generation of intellectuals put forward radical political, social, and economic ideas; young people just coming of age questioned the status quo at every turn.[18] This political and social ferment found expression in the sciences as well, as new ideas in physics, biology, and other disciplines transformed human knowledge. In an era in which anything seemed possible, scientists seriously proposed Manhattan Project–style research efforts to subdue all manner of challenges present for humankind.[19]

Scientists Manfred E. Clynes and Nathan S. Kline epitomized this new sense of expectation in what has become a seminal article titled "Cyborgs and Space." It appeared in the September 1960 issue of *Astronautics*. Reporting research funded by NASA, this article set forth an agenda that challenged the dominant paradigm of human exploration laid out by Wernher von Braun and institutionalized in NASA's 1959 long-range plan. Rather than accepting the standard NASA approach to spaceflight—recreating Earth's environment beyond this planet—Clynes and Kline suggested that it made more sense to change humans through biological and technological alteration so as to make them better able to survive the existing conditions of space.

As Clynes and Kline remarked: "Altering man's bodily functions to meet the requirements of extraterrestrial environments would be more logical than providing an earthly environment for him in space." To fly into space—at the time of their article, no one had yet—humans would need to take with them all of the elements of earthly existence, supplemented by devices to protect them from hazards like radiation that did not penetrate the Earth's protective magnetic shield and atmosphere. This seemed like an unnecessarily complicated approach. "Artificial atmospheres encapsulated in some sort of enclosure constitute only temporizing, and dangerous temporizing at that, since we place ourselves in the same position as a fish taking a small quantity of water along with him to live on land. The bubble all too easily bursts." To overcome this problem, they suggested that any human committed to space travel should attempt "partial adaptation to space conditions, instead of insisting on carrying his whole environment along with him."

To describe the resulting creation, they offered a new word. "We propose the term 'Cyborg,'" Clynes and Kline stated, producing a concept whose use became widespread in the realms of both scientific investigation and science fiction.[20] This concept had appeared occasionally before 1960, but new scientific knowledge coupled with an expectation of continuing great discoveries made their article seem not so far-fetched. A cyborg is a living organism with a mixture of organic and electromechanical parts.[21] The authors were more technical. "The Cyborg deliberately incorporates exogenous components extending the self-regulatory control function of the organism," wrote Clynes and Kline.

The term is a portmanteau of *cyb*ernetics, the science of control processes in living creatures and machines, and *org*anism. They added, "In the past evolution brought about the altering of bodily functions to suit different environments. Starting as of now, it will be possible to achieve this to some degree *without alteration of heredity* by suitable biochemical, physiological, and electronic modifications."[22]

Clynes and Kline contemplated the bodily alterations that would allow people to function more freely in space. "What are the changes necessary to allow man to live adequately in the space environment?" they asked. Their proposals typically addressed self-regulatory functions of a biological sort, processes that take place more or less automatically. The emphasis on self-regulatory functions would leave humans better able "to explore, to create, to think, and to feel."[23] Such functions had the benefit of also being more congruent with the then-fashionable emphasis on cybernetic control mechanisms that featured feedback loops and automated control.

The two authors suggested means to modify sleep patterns, both to alter circadian rhythms and to induce hibernation. They proposed the use of an inverse fuel cell, implanted in place of the lung, that would make conventional breathing unnecessary. They suggested modification of the skin as an adaptive response to extremes of temperature and increased exposure to radiation. In one of their more interesting applications, they offered a means to internalize the environmental control and life support systems that process waste on human spacecraft. The procedure involved sterilization of the gastrointestinal tract and modification of certain exterior orifices in combination with the installation of internal feeding and waste-processing devices.

In nearly every case, their proposals employed some sort of technological device containing a biofeedback mechanism that could automatically sense and respond to various local conditions. Such devices drew their inspiration from the expectations associated with early cybernetic products such as the artificial pacemaker for controlling heart rhythms, which was first used in humans the same year that the article appeared. The first pacemakers were fairly simple devices, sensing and correcting the rhythm of the heart. Later versions became more sophisticated, responding to factors such as the level of physical activity of the user. Clynes and Kline believed that advances in technology might permit similar applications in space. "Scientific advances of the future," they wrote, "may thus be utilized to permit man's existence in environments which differ radically from those provided by nature as we know it."[24] The new cyborg, they concluded, would free humanity to explore space.

Clynes and Kline recognized that cyborgs already walked among us. To be sure, their numbers have multiplied since they first published on the subject in 1960. The broadest definition of *cyborg* encompasses any human with machine parts, which could possibly extend the concept to any person who wears eyeglasses or dons clothing designed to compensate for harsh conditions. It might even extend to communicative devices like the cellular telephone or the Internet. The narrower definition of *cyborg* requires that the mechanism be implanted in the body and possess some sort of a feedback mechanism that adjusts its operation to its surroundings. Even a definition as conservative as this casts a broad net. People with pacemakers, hearing aids, and insulin pumps meet that definition; all of those devices utilize computer chips to adjust their performance. A wide range of prosthetic aids that respond to nerve impulses grant their users mechanical advantages they would otherwise not possess. In a modest sort of way, we may already be in the process of becoming cybernetic organisms.[25]

NASA officials seemed quite comfortable with the merger of humans and machines when Clynes and Kline first proposed the concept. The space agency funded their research and continued to support investigations of human modifications for spaceflight. In 1963 space agency officials received a study titled "Engineering Man for Space: The Cyborg Study" that they had commissioned from a company called United Aircraft in Farmington, Vermont. The study emphasized how artificial organs, pharmacology, and psychology might enhance prospects for long-duration spaceflight. It explored a multitude of biological concerns that humans would face and outlined a research agenda to enhance the human/machine symbiotic relationship. A centerpiece of the study involved pathologies for controlling human homeostatic processes in space through artificial organs, drugs, hypothermia, sensory deprivation, and mineral dynamics, among other methods. "The need for this work," the authors advanced, "arises because man is basically a biological organism designed to operate within the parameters defined by the earth environment. Despite a remarkable degree of over design, there are many areas in which man's capabilities fall short of requirements posed by [space] missions."[26] While the authors recommended an appropriate research agenda, they also reluctantly acknowledged that the biotechnology required to enhance long-duration spaceflight was beyond current capabilities.

Public discussion of this matter continued after the 1963 study, but little serious NASA activity followed. In October 1964, *Life* magazine carried a story on engineering humans for space travel. The author, Albert Rosenfeld, suggested that the ideal astronaut would be someone who had "the oxygen requirements of a Himalayan Sherpa, the heat resistance of a walker-on-coals, who needs less food than a hermit, who has the strength of a [boxing champion like] Sonny Liston, and runs the mile in 3 minutes flat while solving problems in tensor analysis in his head."[27] In 1973, the television series *The Six Million Dollar Man* began its six-year run featuring Lee Majors as Col. Steve Austin.[28] The crash shown in the opening sequence of the series was real; it showed a 1967 accident that befell NASA test pilot Bruce Peterson and provided the inspiration for the idea that doctors could rebuild the body of a prospective space traveler who had survived a terrible crash. In the actual crash, which was quite spectacular, Peterson lost one eye; as introduced at the outset of each episode, his fictional counterpart received many bionic (or cybernetic) parts that gave him superhuman qualities:

> Steve Austin, astronaut. A man barely alive. Gentlemen, we can rebuild him. We have the technology. We have the capability to build the world's first bionic man. Steve Austin will be that man. Better than he was before. Better, stronger, faster.[29]

NASA officials chose not to invest more funds in cyborg studies and distanced themselves from the concept in the late 1960s, although some at the space agency still believed that such alterations would eventually be necessary for long-duration space missions. The realization that bioengineering humans for spaceflight was well beyond the technological reach of the 1960s even as the immediacy of the Apollo Moon landing took precedence over exotic concerns may have prompted NASA officials to end this research. Something deeper, however, might have been at work. Social scientist Chris Hables Gray offered this observation. Whereas the hero in *The Six Million Dollar Man* was still human, "a cyborg is potentially a post-human; a human modified beyond being human. I think this is why, without any conscious decision I can find record of, NASA refused the term cyborg so completely. Because at every level of the organization, from the astronauts who were being cyborged to the bureaucrats at the top who always had to worry about the public perception of the program, the idea of the cyborg was very threatening."[30]

A MOVEMENT BEYOND

As NASA abandoned the idea of a posthuman design for spaceflight, others embraced it. The concept moved from the realm of spaceflight, where it had begun, to more exotic terrain. Disparate advocates adopted a common concern with the philosophical implications of moving toward a posthuman condition in which medical science and machine technology would eliminate aging and vastly enhance the intellectual, physical, and psychological characteristics of human beings.[31] This outcome was viewed as a nearly inevitable consequence of changes emanating from the Age of Enlightenment in which the intellectual elite of Western civilization embraced the scientific method, accepted the value of progress, and elevated the natural rights of individual beings. Transhumanism, as the movement became known, deals largely with the transitory stage in which *Homo sapiens* move from the human condition to something beyond.[32]

Like the space exploration movement in America, transhumanism has strong roots in science fiction. One of its earliest proponents, Robert Ettinger, received his inspiration from a short story in a 1931 issue of *Amazing Stories*. In it, a human being is left to float in outer space for millions of years. Suspension in the frigid vacuum of space preserves the individual's body; upon its recovery, advanced beings supply it with a mechanical body and revive its organic brain. This story informed Ettinger's writing of his own science fiction story in which he argued that death is a relative concept defined by the state of medical knowledge existing at any one time. In 1964, Ettinger advanced these ideas in his nonfiction book *The Prospect of Immortality,* which went through many editions and became an influential tract in the transhumanism movement.[33]

In a later work, *The Age of Spiritual Machines,* Ray Kurzweil describes a similar progression. Humans again achieve a form of near-immortality, in this case through enormous advances in computer technology and artificial intelligence. The book is constructed around a set of conversations with a young woman, who over a period of one hundred years comes to understand the underlying principles of transitional change. She eventually sheds all of her biological parts and becomes an entity within a machine. The book introduced the concept of singularity, the point at which the pace of technological change becomes so rapid that old methods of anticipating future events can no longer be employed. Technological singularity is a key concept in the transhumanistic philosophy and marks the boundary between human civilization as we know it and the posthuman condition.[34]

Like writers of fiction, the embracers of transhumanism are often more interested in the consequences of immortality and personal enhancement than the actual method of its achievement. They are united by a common belief that such change is nearly inevitable given the commitment of modern civilizations to scientific investigation and the accelerating rate of technological change. Such anticipation has led many to embrace unproven remedies and strange procedures, causing the movement to diverge from mainstream scientific investigation. In addition to his otherwise sober insights, Ettinger advocated the cryonic storage of individuals until such time as technology would allow appropriate revival and enhancement. His *Prospect of Immortality* made him something of a media celebrity on behalf of this unique idea. In a similar fashion, the Iranian-American philosopher and futurist F. M. Esfandiary called for humanity to embrace a posthuman future that promised near-immortality, the elimination of disease, and biotechnological enhancement. He changed his name to FM2030 to celebrate his conviction that "in 2030 we will be ageless and everyone will have an excellent chance to live forever."[35] FM2030 died in 2000 from pancreatic cancer and undertook cryonic suspension at the Alcor Life Extension Foundation in Scottsdale, Arizona. Ettinger has made plans to be preserved in essentially the same manner at his own facility, the Cryonics Institute in Michigan.

This ferment of a posthuman future in which humans and machines might merge into what would be a fundamentally new life form sparked a debate about what it means to be human, to have a body, and to be superior to other forms of life. Successful efforts to decipher the human genome, to clone animals, and to expand computing capacity have convinced believers that a posthuman future is near. Of course, these expectations could go the way of others enthralled by the promise of cancer cures and space elevators. A feeling arose among those attentive to this discussion that if "we" (generally meant to be Western intellectuals) do not develop this technology, someone else will—an argument familiar from the early days of atomic weaponry and space travel. Once the genie is out of the bottle, the argument goes, it is very hard to convince the apparition to get back in.

Many intellectuals rose to challenge the more optimistic beliefs of the transhumanistic philosophers. To its critics, the transhuman concept represented technology gone wild. It was totalitarian in impulse, unpredictable in its consequences, and ultimately—by its very own definition—destructive of human civilization. Early on, Jacques Ellul, a leading theologian and philosopher in the 1960s, raised concerns about the implications of mechanical consciousness. He believed that technology would soon make humanity subservient.[36] The possessors of such technology would not contentedly allow their lives to be governed by the ignorant masses who do not possess it. With their superior qualities, they would want to rule rather than be ruled. Likewise, John Kenneth Galbraith, the influential American economist, warned that the ultimate pursuit of science and technology in a democratic society would result in decisions that "cannot be otherwise than totalitarian."[37] Galbraith extended these fears in a later publication: "If we continue to believe that the goals of the industrial system—the expansion of output, the companion increase in consumption, technological advance, the public images that sustain it—are coordinate with life, then all of our lives will be in the service of these goals. What is consistent with these ends we shall have or be allowed; all else will be off limits."[38] British scientist and philosopher Jacob Bronowski expressed much the same fears: "A world run by the specialists for the ignorant is, and will be, a slave world."[39]

Even within the U.S. space program, seemingly untouched by the transhumanism debate, officials acknowledged the growing concern over technology. NASA Administrator James C. Fletcher observed that "this country seems to be on an antitechnology kick." The civil space program was founded on a commitment to the peaceful uses of technology for the progress of humankind. "Now we are overreacting," Fletcher warned.[40] The first generation of spaceflight engineers believed that they were using technology for a noble purpose. "The solid ground of common national purpose," warned historian Sylvia K. Kraemer, by the time of the Moon landings "had already begun to shift ominously under their feet."[41] No wonder NASA officials backed away from a consideration of the cyborg as space traveler.

One of the most distinctive criticisms of twenty-first-century technologies appeared in the April 2000 issue of *Wired*, a magazine devoted to the examination of new trends in computing and technology. The author was Bill Joy, billionaire cofounder of Sun Microsystems, a handsomely successful computer company located in Silicon Valley. Joy raised concerns about an effort related to but not exactly part of the transhumanistic movement—the creation of artificial intelligence, that is, computers that are smarter than human beings. Transhumanists see such computers as devices for augmenting—in some cases, replacing—the organic human brain, giving human thought a postbiological quality. Yet artificial intelligence computers might just as well appear independently as a result of other pursuits, such as the effort to promote national security. Such computers would be something separate from (and potentially in competition with) human beings. Joy was concerned about this latter possibility.

Joy issued a stern warning about the pursuit of artificial intelligence. "Biological species almost never survive encounters with superior competitors," Joy warned.[42] Should the artificial intelligence movement succeed, machines would learn how to make their own decisions, in which case the remaining humans would have no control over their machines. The machines might decide to keep humans around, more or less in the form of domestic animals, but the decision would be up to the machines. Possibly their creators would continue to exercise some control, but they would be a small elite, and the vast mass of humanity would be reduced to nonparticipants.

Machine and biological technologies offer the promise of great advances, Joy admitted. Yet they could also replace humanity:

> The vision of near immortality that Kurzweil sees in his robot dreams drives us forward; genetic engineering may soon provide treatments, if not outright cures, for most diseases; and nanotechnology and nanomedicine can address yet more ills. Together they could significantly extend our average life span and improve the quality of our lives. Yet, with each of these technologies, a sequence of small, individually sensible advances leads to an accumulation of great power and, concomitantly, great danger.[43]

For the most part, this debate has taken place away from the public consciousness. In no small part this resulted from the self-serving nature of modern technological society, where new gadgets are unquestioningly embraced for their usefulness. Occasionally the protests of zany techno-terrorists such as the Unabomber, Theodore Kaczynski, make their way into the public discourse. Fringe transhumanists occasionally win time on television talk shows to discuss cryogenics. Yet few among the general public seem to care. Most Americans happily accept their place as the recipients of techno-scientific wonders that daily make their way into the marketplace; they do not worry a great deal about potentially destructive consequences.

While the public seemed to ignore this debate, the technological and intellectual elite absorbed it. The explosive development of microelectronics, computer capacity, and biotechnology enlisted new believers in a transhumanistic future. A handful of scholars, public intellectuals, policy analysts, and artificial intelligence software designers emerged to suggest that a future based on what some flippantly termed "robo sapiens" was close at hand. Much of the intellectual community views this alteration as inevitable.[44] Their concern in many ways mirrors mid-twentieth-century discussions in intellectual circles about the implications of cracking the atom and constructing the Bomb.

Andy Clark, director of the Cognitive Science Program at Indiana University and author of many studies of the subject, insists that humans have become inextricably linked to machines. Humans cannot survive without them. The future will find humans more closely tied to the technologies they have created until they reach a posthuman state. Rather than invoking fear that we will become nonhuman, Clark celebrates this possibility and the wondrous potentialities it offers. The transitional period will be risky, but generally Clark is optimistic. He asks: "If it is our basic human nature to annex, exploit, and incorporate nonbiological stuff deep into our mental profiles—then the question is not whether we go that route, but in what ways we actively sculpt and shape it. By seeing ourselves as we truly are, we increase the chances that our future biotechnological unions will be good ones."[45]

This community, not driven in any way by concerns about spaceflight, believes that at a fundamental level *Homo sapiens* as a species is endangered in ways never before conceptualized. The next stage in the evolution of life on Earth, they suggest, will be toward a human/machine symbiosis. "If the name of the game is processing information," social scientist N. Katherine Hayles opines, "it is only a matter of time until intelligent machines replace us as our evolutionary heirs. Whether we decide to fight them or join them by becoming computers ourselves, the days of the human race are numbered." Far from viewing this development with trepidation, Hayles concludes, "Although some current versions of the post-human point toward the anti-human and the apocalyptic, we can craft others that will be conducive to the long-range survival of humans and of the other life-forms, biological and artificial, with whom we share the planet and ourselves."[46]

This posthuman or transhuman movement has many enabling technologies—genetics, cryonics, pharmacology, nanotechnology, and biotechnology, among others—along with numerously disparate ideologies and fractious groups. Some concepts seem completely reasonable to virtually all participants, while other beliefs are adopted by fierce clusters of supporters. Both libertarian and progressive elements appear, invoking utopian visions of progress and ultimate perfection. Some are motivated by capitalism and the desire to attain vast wealth, while others are entranced by the possibility of immortality. The movement is as diverse as it is strange.

Regardless of persuasion, believers in posthumanism assert that humanity is on the verge of a great evolutionary leap that portends both promise and peril. They believe that the manner in which human society responds to this transformation will shape, fundamentally and unalterably, the future for at least a millennium. James Hughes teaches biomedical policy and ethics at Trinity College but is better known as the founder and executive director of the World Transhumanist Association. He visualizes a future in which babies are born free of genetic defects, most of the diseases that now drastically reduce our population are eliminated, nonhuman sentient beings achieve legal rights, and the successors to humanity achieve near-immortality. The mapping of the human genome, according to Hughes, represents just the beginning of a future in which human life can be radically improved. Hughes acknowledges that these possibilities harbor questions and fears, as anything new and different always has. Dubbing the more severe critics as "bio-Luddites," Hughes suggests that their ascendancy will only ensure that the nations they represent, first and foremost the United States but also other nations, will not participate in the next fundamental transformation of human history. The biotech revolution, he believes, has the potential to alter life on the planet more significantly than did the Industrial Revolution so vigorously embraced by the United States.[47]

Hughes insists that the battle lines in this debate have already been drawn, with skirmishes over stem cell research, pharmaceuticals, cloning, and related innovations foreshadowing more intense intellectual battles to come. Hughes asks, what do we do with cloned individuals? Do duplicate persons have individual rights, like identical twins, or are they the same person, say, with only one vote between them? What will prospective parents do once they have the capability, through mastery of the human genome, to ensure that birth defects are eliminated in their fetuses? Will they be allowed to select genes for greater intelligence or other potential advantages? These are only some of the challenges that Hughes identifies.

To Hughes, the bio-Luddites conjure up objections as a means of opposing any progress. Hughes takes a different approach. He argues that it is impossible to turn back these innovations. Rather than trying, he says, members of democratic societies should seek to regulate and control them. Through his faith in democratic institutions, he believes that self-governing people can reach decisions that will preserve human freedom and make possible a hopeful future.

While transhumanists such as Hughes are concerned mostly with biological alterations, others such as Bill Joy and Ray Kurzweil focus on machines. These are not robots in the conventional sense, whose programming is based on the science of cybernetics and control. Rather, they foresee the development of new entities—perhaps *machine* is not the best term to describe them—that combine advanced human and computer capabilities. A set of interrelated "GRIN" technologies (genetics, robotics, information, and nanotechnology), such advocates believe, are leading to a posthuman world. In his 2005 book *Radical Evolution, Washington Post* writer Joel Garreau suggests that all such scenarios are energized by "the Curve," an ever-accelerating growth rate in technological capability that is based on "Moore's Law," which observes that the capacity of an integrated circuit doubles every eighteen months.[48] At some point, fairly soon, a singularity will be reached in which machine intelligence surpasses human intelligence at its best. Kurzweil goes further, arguing that it will become possible to "jack in" to the mainframe, à la *The Matrix* feature film, and download the memory of a human being. The result, he suggests, would be a new silicone-based life form. In his 2005 book, *The Singularity Is Near,* Kurzweil examined in detail how this evolutionary process leading to the union of human and machine might take place. Kurzweil's analysis is full of optimism and long on hyperbole. Many have commented on his naïveté, asserting that his failure to see problems that might arise in the posthuman future is both confounded and confounding. For Kurzweil, nonetheless, the technology will create a Panglossian "best of all possible worlds."[49]

As anyone who observes the transhuman discussion realizes within minutes of encountering it, techno-utopian ideas are open to a wide array of ethical considerations. Opposition has ensured a rollicking debate. Opponents question the morality of altering the human body through genetics, chemicals, or technology. They conjure images of Nazi eugenics and the selective breeding of human beings. They emphasize the freakish nature of scientists involved in these studies, invoking analogies ranging from Mary Shelley's *Frankenstein* to Hubertus Strughold's Nazi experiments on concentration camp prisoners to José Manuel Rodríguez Delgado's bizarre experiments with brain implants undertaken in the 1960s and 1970s.[50] "It's not nice to fool with Mother Nature" is a phrase more than once invoked by critics of these transhuman concepts. So is the term *mad scientist,* with the implication that those involved in the research are overzealous in their efforts and that not everything that is scientifically feasible is also desirable. Joy takes the criticisms to their ultimate conclusion. "Unless we take strong action," Joy warns, "we are liable to find ourselves with a whole new category of massively destructive technologies all able to be put into action by widely available commercial devices." He concludes, "If we use technology to create robotic intelligences that are superior to ours, they might come to view us as expendable."[51]

At the same time, the possibilities are staggering. Significant advances have already occurred in areas such as neural implants. In 2005, the U.S. Food and Drug Administration approved "brain pacemakers" for use as a last-ditch treatment for depression. Applications could increase as experience accumulates. Treatment of ALS, also known as Lou Gehrig's disease, through the creation of human-machine interfaces has certainly made life easier for individuals such as the remarkable Stephen Hawking, a brilliant mind trapped in a limited body. Yet the practice is new and the implications unclear. Skeptic Adam Keiper's observations regarding this new technology can in many ways be applied to the whole transhumanist movement.

> In time, the technology will improve, and implant-based brain-machine interfaces will be worthwhile for a great many patients. But as things stand today, they make sense for almost no one. They involve significant risk. They are expensive, thanks to the surgery, equipment, and manpower required. They can be exhausting to use, they generally require a lot of training, and they aren't very accurate. Only a locked-in patient would benefit sufficiently, and even in some locked-in cases it wouldn't make sense. For all the *technical* research that has been done, there has been very little *psychological* research, and we still know very little about the wishes and aspirations of severely paralyzed patients.[52]

Everyone agrees that the human-biotechnology-machine interface will change with time. How much it will change, and what results will become manifest, is open to debate.

ALTERNATIVES TO CONVENTIONAL FLIGHT

Essentially none of the transhumanistic activity involves NASA, astronauts, or the space programs of various government agencies. Almost none of the research, on which billions of dollars a year are being spent by private and government entities, emanates from agencies with a direct interest in human spaceflight. At the same time, the activity portends significant implications for space in a fashion similar to the application of other critical technologies invented outside of the space science field. The rockets that propelled the first American astronauts into orbit, to cite one historic example, were not invented by space scientists. They were prepared in their original incarnation by engineers seeking to use ballistic missiles to deliver nuclear weapons between points on the Earth.

The dominant spaceflight paradigm, in all of its elements, has been predicated on the assumption that people will carry their own life support systems into space. Orbiting space stations and excursion vehicles are pressurized to create appropriate atmospheres. Plans for lunar and Martian colonies anticipate that their occupants will extract basic commodities like oxygen and water from local resources. The effort to maintain Earth-like conditions conducive to the maintenance of human life reaches its zenith in proposals to terraform or change the atmospheres of planets like Mars. Even the redoubtable Carl Sagan, who thought much about intergalactic travel, emphasized the necessity of changing other planets to become like Earth rather than changing humans to fit their new surroundings. Sagan embraced space migration as a means to preserve the human race; changing the race to permit space travel gained little consideration.[53]

The transhumanism debate suggests another possibility. In pursuing a variety of goals, humans may change the creatures that ultimately do the exploring. As applied to space travel, this resurrects the question raised by Olaf Stapledon more than sixty years ago. To repeat the issue as posed in the previous chapter, exactly what part of humanity do the partisans of spaceflight wish to transport? As Stapledon predicted, before humans decide what to do on other planets, they must decide what to do with themselves.[54]

To the advocates of the traditional paradigm, the answer is clear. Humans will go in their existing physical form, as *Homo sapiens,* surrounded by expediting technologies. This viewpoint is consistent with the vision of humans as the end product of evolution, a species so godlike that its preservation and protection from environmental change requires interplanetary dispersion.

Advocates of the transhumanist perspective, to say the least, do not accept this thesis. To them, humans are fragile creatures with inadequate life spans, barely suited to thrive for extended periods of time on Earth. Although they do not apply their line of reasoning to interplanetary dispersion, it is clear that any humans so assessed would be wholly unsuited for this enterprise as well. "Transhumans regard our bodies as sadly inadequate, limited by our physiognomy, which restricts our brain power, our strength, and, worst of all, our life span," says Brian Alexander. "Transcendence will not be found in the murky afterlife of usual religions, but in technological and biological improvement. 'I'd like backup copies of myself,' one told me at a transhumanist conference, 'something more durable than a carbon-based system.'"[55] Extensive discussions have taken place in recent years on the relationship between artificial computer intelligence, biotechnology, and human development. Its relevance to spaceflight seems obvious. In fulfilling the spacefaring dream, the intelligent life to leave Earth and colonize the galaxy may not be entirely human in form. No one knows whether this might occur, but its consideration broadens the possibilities.

The conventional view of spaceflight holds that people in a bubble using technology to preserve the human form will venture into hostile surroundings for ever-increasing lengths of time. On Mars, that would require humans to surround themselves with supportive technologies that compensate for a thin atmosphere, weak gravity, relative cold, dry weather, high winds, and increased radiation exposure. Supportive technology might take the form of flexible space suits, oxygen tanks, exoskeletons, pressurized ground vehicles, and secure shelters. The first humans on Mars would subsist under conditions similar to but much harsher than scientists spending the Antarctic winter at the South Pole.

For interstellar travel, the requirements would be even more severe. The most conventional proposals for interstellar travel by human beings presume that their spacecraft travel at exceptional speeds or, barring that possibility, that humans construct multigenerational spacecraft that travel for exceptional periods of time. Both alternatives dispatch humans in their current form. Science fiction writers achieve faster-than-light speeds effortlessly, but real people would probably confront insurmountable barriers in the attempt. The various fictional versions of spacecraft powered by matter/antimatter "warp" drives require the suspension of at least five laws of physics. Antimatter exists, but faster-than-light travel appears at present to be a physical impossibility. In *Contact,* Carl Sagan evaded such constraints by dispatching a crew—still in human form—through a wormhole created on the basis of information supplied by a SETI message. Might breakthroughs in physics allow the creation of wormholes that evade the conventional limits of space and time? Perhaps, but at present scientists doubt that they could be constructed in such a fashion as to provide tunnels sufficiently secure for limitless travel.[56]

The multigenerational spaceship is similarly flawed. Serious proposals exist for these types of ships, but thus far no one has demonstrated the ability to create a self-contained "terrarium" that could support humans on Earth, much less one moving through the near-vacuum of space. In the late 1980s, in an experiment funded at $150 million by Texas oil magnate Edward Bass, humans at Biosphere 2 in Arizona's Santa Catalina Mountains sought to test technologies that might be useful for sustaining life on the Moon or Mars. Unfortunately, their terrarium leaked. The experiment ended badly; no further human habitation took place after 1994. The goal of keeping people alive in an enclosed, self-contained environment whisking through interstellar space may be beyond human capabilities for many years.[57]

The idea that humans can move to other worlds and retain their form is certainly the most conventional of notions, but one that deserves the highest levels of skepticism. Frankly, we find it to be unrealistic. Although it fits well into the popular vision of spaceflight as created through novels, films, and television dramas, the goal of humans alive in technological bubbles creates challenges that are exceptionally difficult. For this reason alone, we believe that spaceflight by humans will ultimately prove futile. It might work as far as Mars, but probably not much beyond.

Even if human settlers could reach other worlds and construct Earth-like habitats, local conditions would encourage modification of the human form. Astrobiologists do not fully understand the relationship between gravity and the development of life, but it is unlikely that gravity plays no role at all. Children born to successive generations of settlers on Mars, for example, a one-third Earth gravity world, might change in unpredictable ways. Their bones may grow long and thin. In that sense, they might become transhuman creatures even if they did not intend to do so. Humans have adapted superbly for conditions as they exist on the surface of Earth, and there is no reason to suspect that they would not do so again on another world.

The transhumanist perspective accepts the inevitability of physical change. Its advocates, moreover, want such change to occur in a purposeful way, and they want it to happen soon. Humans reengineered to withstand the challenges of long-duration space travel or existence on other spheres represent one alternative to the conventional vision of humans who would otherwise be required to carry the old world with them.

Scientists have already investigated a few possible applications. The most modest applications address the burdens associated with long-duration space travel. One of the most common involves a form of hibernation known as suspended animation. Animals that hibernate use a form of controlled hypothermia to lower their body temperatures, reduce their heart and lung rates, and conserve energy. Scientists have induced this process in mammals that do not naturally hibernate, such as dogs and mice, though with some ill effects.[58] Could humans be taught to hibernate? Long a staple of science fiction, suspended animation has been suggested as a means of "freezing" astronauts set on tediously long journeys. Their caretakers, possibly machines, would wake the travelers at the conclusion of the voyage.

Induced human hibernation is a transhumanistic intervention. The technique has been effectively portrayed in films such as *2001: A Space Odyssey* and *Alien*. NASA officials investigated "depressed metabolism" during the first two decades of spaceflight but abandoned such studies on the grounds that it was technically infeasible. Whether such a technique will ever prove useful for space travel is problematic, but it does illustrate one method by which alterations in human physiology might enhance the prospects for long-duration flight.[59]

Somewhat closer to the philosophy of transhumanism is the possibility that humans will acquire vastly increased life spans. This would eliminate a number of obstacles otherwise encountered on multigenerational spacecraft. The same people that began a long voyage might be on hand to complete it. The crew would not need to worry about the challenges of maintaining a sufficiently diverse genetic pool for healthy reproduction, a substantial challenge for multigenerational expeditions. A long-lived crew, moreover, might not experience the flight as being exceptionally long given their extended longevity, eliminating the need to induce hibernation or other sleep-like states as a method of reducing tedium.

Induced physiological change is an interesting possibility, one in which space scientists have shown some interest. The bulk of the work in this area, however, is being done by people with little interest in spaceflight. They want to live longer and more fruitful lives on Earth. Advocates of the traditional spaceflight vision want to leave Earth, largely as a means of preserving the current human species and its civilization. They are not anxious to replace humans with some posthuman form. The reluctance of space scientists and their advocates to acknowledge the more radical elements of transhumanism suggests a disconnect that may be hard to bridge.

A more promising possibility may arise in the postbiological realm. Compared to biological modification, this is a more exotic alternative. From a spacefaring perspective, however, it is more familiar. Space scientists have plenty of experience creating machines bound to do the work of exploration. Machine capabilities keep improving with time. The approach receives substantial attention from people interested in robotics and substantial investment from the defense and computer industries. Humans have already shown that rovers can function effectively on Mars, and intelligent robots would be a feasible strategy for interstellar exploration. Robots set on long voyages do not have the same environmental requirements as human beings, reducing the need for an elaborate protective bubble.

People who embrace the transhuman philosophy or engage in GRIN research suggest that humans are on the verge of creating a postbiological life form. The implications for existence in space are as remote today from common experience as were the early images of space travel to the people who first encountered them. Nonetheless, they are not wholly impossible. In one of the more common proposals, a computer achieves artificial intelligence. It becomes smarter than a human being. While it would have metallic parts, it would not be robotic in the conventional sense. It might have characteristics that make it quite well suited to space travel, utilizing, for example, energy sources quite different from those supporting biological entities. Traditional robots set on spacefaring expeditions often carry radioactive generators to provide warmth and energy that would be destructive to animal cells. Such creations might have strikingly different and more robust centers of intelligence—the human brain loses consciousness after ten seconds if deprived of oxygen in the bloodstream and suffers death in five to thirty minutes. Silicon-based memory chips like those found in a common computer can be shut down and restarted without losing their memory or capacity to work.

The postbiological approach nonetheless quickly gets strange and possesses its own challenges. As discussed earlier, machines face survival issues in space that are metaphorically as challenging as those faced by people. For example, a compact disk has a life span that roughly approximates that of a human being. Its only comparative virtue is the ease with which its data may be migrated to the next storage container.[60] Machines wear down, and any postbiological entity sent on an interstellar voyage would need to possess some method of self-repair or possibly reproduction.

The need for postbiological reproduction suggests a fascinating alternative, one considered by rocket pioneers. If mechanical entities need to reproduce on a long voyage, why send them at all? As an alternative strategy, humans or their successors could dispatch a spacecraft that contains the instructions to produce some sort of intelligence once it reaches its destination. This strategy falls into a larger category of proposals that envision the dispersion of earthly life forms without the use of crewed spacecraft at all. American rocketry pioneer Robert H. Goddard discussed in 1918 the possibility of seeding other planets with simple "protoplasm" that would evolve in its own way thereafter. In his view, an earthly exodus of this sort would eventually take place to escape our dying sun. In time, after drifting onto other worlds, this protoplasm might reconstruct itself into complex beings. As biologists have shown, however, there is no guarantee that such reconstruction would produce earthly life forms. Even on Earth, biology proceeds in many unpredictable ways. On another planet, with different conditions, the introduction of rudimentary biological forms from the Earth would almost certainly not result in the eventual production of human beings. The process is wholly unpredictable, and humans would likely not be around to observe its conclusion. In part to resolve this challenge, Goddard additionally recommended that the protoplasm be accompanied by the accumulated knowledge of human civilization in a very compact and durable form, with the aim of supplying any creatures that might arise with instructions for becoming more like human beings.[61]

In their *Interstellar Travel and Multi-Generational Space Ships,* Yoji Kondo and his fellow editors include a unique proposal from Joe Haldeman for dispatching self-replicating machines. In the beginning, the machines would arrive on new worlds for the purpose of making them ready for human habitation. Following this logic a bit further, Haldeman makes an additional observation. By the time that the machines finish their work, "it may be that the self-replicating machines would be as human as we are. So why send old-fashioned flesh humans at all?" Individual personalities might be loaded into machines for the interstellar voyage. Once there, Haldeman adds, "it would be silly to download the personalities into squishy mortal human bodies."[62] The personalities would endure much longer if they remained in their postbiological form. The eventual consequence might be a postbiological space traveler silicon-based as opposed to carbon-based in form.

To summarize, the conventional visions of transporting humans and/or robots to other worlds is being augmented by transhumanistic approaches not yet fully applied to cosmic flight. In the conventional visions, humans and/or robots venture out in their traditional forms, although probably not much beyond this solar system. In the moderate transhumanistic alternative, selective modification for characteristics such as longer life spans allows humans to adapt to the rigor of space travel better. In the moderate postbiological alternative, intelligent nonbiological entities do the work. In the most radical alternative, the seeds of life (mechanical or organic) are dispatched to distant worlds, possibly with instructions for their reconstitution into a desired form.

The latter alternatives, moving as they do away from the necessity of maintaining organic life under Earth-like conditions, raise a fascinating issue. Exactly where might such potential life forms go? If they do not require Earth-like conditions to survive, they could select as their destinations non-Earth-like worlds. Many spheres, including those available within the local solar system, would prove acceptable. Eventually, such creations may not adopt a material form at all. They might prefer to avoid solid bodies with their commensurate hazards and exist in a free state anywhere. This line of reasoning, as can be seen, is completely speculative and goes pretty far.

Goddard's proposal, written secretly for fear that it would engender ridicule, contains a significant insight. He not only dealt with the mechanics of dispersion but also added a cultural component as well. His "protoplasm" dispatched life; his encyclopedia contained culture. Goddard's proposal provides for both the transmission of earthly life and the dispersion of human culture. As we shall see, the cultural component turns out to be a critical one in any discussion of transhumanistic alternatives.

CULTURAL IMPLICATIONS

The conventional vision of space travel foresees humans and robots working together to explore and settle space. Humans and robots venture out in their traditional forms. The breadth of options available for space travel now includes new alternatives. If they do not deserve serious consideration now, they will soon. The more radical alternatives inevitably force a reassessment of the ultimate purpose of spaceflight along Stapledonian lines. What exactly would humans propose putting into space?

The conventional issue, one that has preoccupied advocates of spaceflight in previous years, poses a confrontational choice between unnecessarily limited alternatives. Will space be explored by humans or robots? This was an important question at the beginning of the space age and will continue to have relevance for many years to come. If one considers possible modifications to the entities involved, interesting possibilities emerge. One branch of the transhuman movement favors the modification of human beings. It foresees the modification of *Homo sapiens* into biological entities with posthuman capabilities. If modified for space travel, such transformed creatures would still have their origins in the human genome. In that sense, it is appropriate to categorize this creation as a posthuman being, even a cyborg.

Another branch of the transhuman movement favors nonbiological entities. Its advocates foresee the development of a machine intelligence that could potentially incorporate elements of human cognition. As a human mind inside a nonbiological entity, it would have its roots in robotics and hence could be categorized as postbiological.

At a certain point, however, the human versus robot question ceases to be relevant. From the biological perspective, posthumans in biological form would be a new type of being. They might be unwilling or unable to reproduce with *Homo sapiens*. Like cyborgs, they might contain mechanical parts, although such transformations could be achieved strictly through biological alterations. From the mechanical perspective, any postbiological creation would clearly not be a human being. Yet it additionally would not be a machine. Machines in the traditional sense imply human control; the likelihood that postbiological entities would be controllable seems problematic. Isaac Asimov believed that highly advanced machines could be programmed to obey their human creators. Perhaps such subservience could be hardwired into their subconscious in the same way that curiosity seems to motivate us.

We are skeptical of Asimov's thesis. That which renders a postbiological entity capable encourages its independence. Any entity with the knowledge of a two-year-old child might show the same willingness to test authority. Anyone who has raised children understands how this occurs. With an adolescent mind, who knows what sort of resistance might prevail. A postbiological entity that has developed its own consciousness and ability for independent action (that is, free will) would cease to be a robot available for menial work in the service of human beings. Neither robotic nor human, such a creation would have characteristics of both but ultimately be something new.

Should any transformations occur—and be applied to space travel—they might still carry elements of humanity to other worlds. Transformed beings need not abandon the purpose of their creators. *Homo sapiens* are human in a number of ways. The term has a physiological component. Humans have an erect body and a large brain. The brain is capable of language, reason, emotion, and self-consciousness. Taking for a moment one of the more radical transformations, the development of artificial intelligence, it is clear that such creations would not need an erect body frame. They would not need to possess an organic brain, but their center of intelligence would have the capacity for reason, language, and, almost certainly, self-consciousness. As to the persistence of emotion, we leave that to the speculation of science fiction writers. Generally speaking, writers have suggested that advanced postbiological entities will abandon many of the more primitive emotions associated with human beings. Emotions are often presented in science fiction circles as a character defect. In psychological terms, postbiological creations might possess some of the characteristics associated with humanity.

Humans or *Homo sapiens* also have a social side. They tend to live in groups. They are curious and adventurous, and they communicate with one another. They are capable of making tools and have an ability to organize themselves in such a fashion as to manipulate their environment. They keep and transmit histories. It is likely that many of the social characteristics of human beings would be preserved in a posthuman species. It is hard to imagine a posthuman species maintaining its high level of development without systems of social cooperation like these.

Most important, humans possess culture. They have a sense of aesthetic appreciation that leads them to produce music and art. They explore, migrate, and report their findings. They engage in space travel. Would any posthuman entity share these characteristics as well?

Human imagination, consciousness, and powers of reason encourage space travel. As a consequence of their self-consciousness, humans are aware of their mortality. To deal with this unsettling aspect of life, humans in nearly every culture have developed beliefs such as the immortality of the soul, the expectation of an afterlife, reincarnation, or resurrection. As Joseph Campbell observed during his earthly travels, preoccupation with immortality is one of the most widely distributed cultural aspects of human life and defines in many ways what it means to be human.[63]

Migration and space travel address this need. By traveling to other continents and eventually other worlds, humans assure themselves that they will continue to thrive. Exploration and migration have great survival value for the societies that engage in it. It also helps to relieve the sense of anxiety created by conscious knowledge of one's mortality. As science writer John Noble Wilford commented, "Travel to other worlds is human destiny."[64] Perhaps it is hardwired into the human psyche. As astronaut Michael Collins has opined, "Call it genes, character, culture, spirit, ethos; by whatever name, it is within us to look up into the night sky and be curious, within us to commit our bodies to following our eyes."[65]

While any posthuman entity would certainly be motivated to maintain its own existence, there is no guarantee that it would be interested in space travel. To the extent that space travel provides a means to pursue immortality, a posthuman entity might accomplish that in other ways. People like Robert Goddard and Carl Sagan who have written about the need to disperse humanity have done so because of their knowledge that the Earth will not forever harbor conditions suitable for the maintenance of human life. If a posthuman entity does not need those conditions to survive, it has lost much of the motivation to move. Nor is there any reason to think that a posthuman entity would possess any affinity for the particular vision of space exploration created by humans in the twentieth century and promulgated through novels, magazines, television programs, theme parks, and films. It might have very different interests.

Humans attain a great advantage in dispatching astronauts and robots to carry out the work of space exploration. Astronauts and obedient machines fulfill the purpose of the humans that sent them. They pursue the vision. A truly posthuman creation may not desire to pursue the interests of human beings, whom it may very well view as inferior and misguided. It may have its own culture.

By moving toward the creation of entities better suited for space travel, humans may wind up displacing the vision that motivates the goal. The entities we create to fulfill our dreams may have visions of their own. While one can speculate about the possibilities, it is really impossible in advance to know. Beyond the point of singularity, as Kurzweil observes, our capacity to predict breaks down.

To answer the Stapledonian question posed above, some aspects of humanity would persist. Some would disappear. We think that the original vision of space exploration would be one of the most likely elements to go. The original vision was conventional and highly compatible with cultural norms existing in Western nations at a specific point of time, the middle years of the twentieth century. That vision has elicited billions of dollars in public funds and private spending. Yet the framework of technology and culture that supports it may in the long run be less realistic than the other, less conventional alternatives that have surfaced in recent years.

Ultimately, discussions such as these are intellectual exercises designed to broaden one's thinking about the options involved. Space exploration with astronauts and robots is not the only way to visualize this process. Carrying out the vision of space exploration may require changes so fundamental that the results may displace the cultural motivation that launched the activity in the first place. Exactly what will happen cannot be known, but the possibilities are certainly worthy of consideration.

Contemplation of the transhumanistic and postbiological thesis raises a final, provocative question. If other civilizations on other worlds attempted to journey outward from their homes, did they change in a similar manner? While no physical evidence yet exists to suggest that intelligent life arose elsewhere in the universe, many people believe that it did.[66] The challenges of moving to new environments different than those experienced on their own worlds may have changed those creatures as well. If so, they may have changed so fundamentally that they might fail to recognize the Earth as an abode of what they define as intelligent.

CHAPTER EIGHT

An Alternative Paradigm?

The civil space effort, especially in the United States, drew its inspiration from a set of proposals appearing in the mid–twentieth century that can be generally characterized as the von Braun paradigm. Those proposals, with virtually no exceptions, preoccupied themselves with human spaceflight. They appeared in a number of forums—books by Willy Ley and Arthur C. Clarke, a series of articles in *Collier's* magazine, illustrations by Chesley Bonestell, and a three-part *Disneyland* television series.[1] Wernher von Braun appeared prominently in the *Collier's* and in the *Disneyland* series and provided the technical instructions for most of Bonestell's spacecraft paintings. The plans emerging from that effort focused on the development of large rockets and spacecraft, the construction of Earth-orbiting space stations, rocket trips to the Moon, the establishment of lunar bases, and human expeditions to Mars.[2]

Officials at the newly created National Aeronautics and Space Administration adopted the von Braun paradigm as the centerpiece of their first long-range plan in 1959, modifying it somewhat to include robotic flight. The NASA approach can be described as one in which humans and robotic spacecraft explore space together, with the emphasis on human objectives. While half of the objectives in the NASA long-range plan proposed robotic or automated flights, most of those robotic missions served as precursors to human exploration. The automated portion of the plan foresaw the development of orbiting satellites and observatories, robotic flights to the Moon and planets, and remotely controlled tests of spacecraft that would later carry human beings. The overall program, nonetheless, steadfastly endorsed von Braun's principal objectives—spacecraft

TABLE 8.1.
Von Braun Paradigm

- Development of multi-stage rockets capable of reaching space.
- First launching of orbital satellites and "baby space station," the latter carrying animals into space.
- Selection and training of astronauts.
- First human flights into space and return to Earth through air and/or rocket braking.
- Development of large, winged, multi-stage, reusable spacecraft capable of carrying humans and equipment into space, making space access routine.
- Assembly of large, permanently occupied, rotating space station in Earth orbit.
- Launch of large automated space observatory, tended by humans from nearby space station.
- First attempt at a flight of humans around the Moon.
- Assembly and refueling of multiple spacecraft in Earth orbit leading to the first human landings on the Moon; exploration of the Moon and establishment of lunar bases.
- Using same techniques, first human expedition to Mars.

SOURCES: Chesley Bonestell and Willy Ley, *The Conquest of Space* (1949); Arthur C. Clarke, *The Exploration of Space* (1951); *Collier's* magazine, eight-part series on space, 1952–54; Ward Kimball, *Disneyland* television series on space flight, 1955–57. See also Symposia on Space Flight, Hayden Planetarium, 1951–54.

capable of carrying humans and their equipment into space, an Earth-orbiting space station, and human expeditions to the Moon. While the official long-range plan concluded with humans landing on the Moon, members of a special committee assisting the planning effort insisted that "the ultimate objective of space exploration is manned travel to and from other planets," reaffirming von Braun's ultimate goal. To the list of objectives on the long-range plan, committee members added the words "Mars-Venus landing," with the clear understanding that humans would be on board.[3]

In the decades that followed, the actual course of spaceflight unfolded in a manner not anticipated by the promoters of the von Braun paradigm. In general, events favored a situation in which robotic and human approaches competed for dominance, with human flight objectives prevailing as far as the Moon and robotic technologies expanding throughout the solar system. In formulating his proposals, von Braun preoccupied himself with the inner solar system, the region where his human-centered vision had its best opportunity to prevail. Had he emphasized wider vistas, a different conception might have emerged. Yet he did not, and proposals for human space travel to the Moon and Mars dominated planning efforts for the U.S. civil space effort. In the same manner that von Braun restricted his vision to the Earth-Moon-Mars region, advocates of robotic exploration confined their requirements to efforts within the solar system. This helped to maintain the traditional definition of robotics, in which humans in control stations on Earth oversee the robots traveling through space. Flight

TABLE 8.2.
NASA Long Range Plan, 1959

1960:
- First launching of a Meteorological Satellite and Passive Reflector Communications Satellite.
- First launching of a Scout vehicle, Thor-Delta vehicle, and Altas-Agena-B vehicle (by the Department of Defense).
- First suborbital flight of an astronaut.

1961:
- First launching of a lunar impact vehicle.

1961–1962:
- Attainment of manned spaceflight, Project Mercury.

1962:
- First launching to the vicinity of Venus and/or Mars.

1963:
- First launching of two stage Saturn vehicle.

1963–1964:
- First launching of unmanned vehicle for controlled landing on the moon.
- First launching Orbiting Astronomical and Radio Astronomy Observatory.

1964:
- First launching of unmanned lunar circumnavigation and return to earth vehicle.
- First reconnaissance of Mars and/or Venus by an unmanned vehicle.

1965–1967:
- First launching in a program leading to manned circumlunar flight and to permanent near-earth space station.

Beyond 1970:
- Manned flight to the moon.

SOURCE: NASA Office of Program Planning and Evaluation, "Long Range Plan of the National Aeronautics and Space Administration," December 16, 1959.

activities beyond the solar system received little attention from official government planners, allowing the two approaches (human and robotic) to maintain their traditional form. Any contemplation of activities beyond the solar system would have seriously undercut the two approaches and forced new thinking. The discussion that follows amplifies these observations, with final thoughts and suggestions in conclusion.

Observation #1: The von Braun paradigm assumed that most of what stood to be gained through space travel could be accomplished within the inner solar system. Although promulgators of the paradigm were aware of challenges beyond, the leading advocates of human flight focused their promotional activities on the challenges of getting people to the Moon and Mars.[4] Most important, supporters of this effort believed that local conditions on nearby destinations, especially Mars, would be sufficiently enticing to engage public attention for a very long time. Von Braun believed that the task of exploring Mars would consume human civilization for hundreds of years.[5] In the process, colonization would probably occur.

One of the primary beliefs supporting the von Braun paradigm was the expectation that Mars harbored life—or, absent living creatures, at least possessed conditions that would permit humans to plant it there. No planet has been more consistently held up as a possible repository of life than Mars. Percival Lowell's observations of that planet between 1894 and 1908 led him to argue that Mars had once been a watery planet and that intelligent beings had constructed the topographical features he saw as canals. The vision of intelligent life on Mars, whether native or transplanted, stayed in the popular imagination for a long time, and it was only with scientific data returned from robotic spacecraft sent to explore the planet that this began to change.[6]

The April 30, 1954, issue of *Collier's* magazine in which von Braun presented his Martian expedition featured a cover prepared by Chesley Bonestell. On the cover, a flotilla of spacecraft approached the theretofore unseen planet. As painted, the surface of the planet was crisscrossed with canals. In this instance Lowell's observations symbolically represented the expectation that conditions necessary for habitation, particularly liquid water, existed on Mars.

Writing in the influential *Conquest of Space* (1950), Willy Ley amplified beliefs that some astronomers on Earth had seen canals but also noted that others had not. Combining the instances where canals had been seen with what appeared to be seasonal color changes on the Martian surface, Ley concluded that Mars harbored life or could be made to do so. As Ley wrote, "We are justified in believing in life on Mars—hardy plant life. The color changes which we can see are explained most logically and most simply by assuming vegetation. From a terrestrial landscape—Tibet—we can get a good idea of what to expect. . . . Of terrestrial plants, lichen might survive transplanting to Mars and one may imagine that some of the desert flora of Tibet could be adapted."[7] Bonestell provided visual support by including a painting that showed water flowing in a straight line down a canal-like course emerging from a snow field on the polar ice cap. On either side of the water course, a greenish hue altered the normally red soil, suggesting the presence of vegetation.

The expectation that Mars presented conditions suitable for habitability encouraged promoters of the von Braun paradigm to anticipate colonization. Mars became the "new world" for people wanting to move to distant lands. Thomas Paine, chair of the National Commission on Space and a past NASA administrator, chose a Robert McCall painting of a Martian colony for the cover of his 1986 commission report, *Pioneering the Space Frontier*.[8] Astronomer Carl Sagan, like von Braun a principal spokesperson for human space travel, helped to popularize colonization by explaining how humans might alter the atmosphere and surface temperature of Mars in such a fashion as to make the planet fit for human habitation, a process known as terraforming.[9] For years one of McCall's paintings hung on the wall outside the NASA administrator's door. The rendition, titled "Terraforming Mars," showed the occupants of an extensive colony pumping greenhouse gases into a steadily warming planetary atmosphere.

Writers encouraged early interest in the habitability of Mars by assuring their audiences that other planets in the inner solar system might be capable of harboring life as well. A 1962 publication from NASA's Jet Propulsion Laboratory assured readers that life might emerge on Venus as the sun depleted its fuel and cooled. While it was now Earth's turn to harbor life, this popular concept suggested that Mars had once been habitable and that life on Venus was just beginning to appear. Beneath the planet's clouds, the theory offered, could be found a warm, watery world and the possibility of aquatic and amphibious life. "It was reasoned that if the oceans of Venus still exist, then the Venusian clouds may be composed of water droplets," noted JPL researchers. "If Venus were covered by water, it was suggested that it might be inhabited by Venusian equivalents of Earth's Cambrian period of 500 million years ago, and the same steamy atmosphere could be a possibility."[10] In 1961, Sagan published an article in *Science* magazine suggesting a means by which Venus might be terraformed, a proposal flawed by misestimates of the density of the local atmosphere.[11]

Some even suggested that more extreme forms of life might exist on the Moon, sheltered in protected areas afforded by crater walls. Belief in the possibility of lunar life persisted through the 1950s up until real exploration began. One such idea about lunar life was reported in a 1959 book, ironically titled the *Strange World of the Moon*:

> Yet seasonal changes there certainly were: some markings darken, others become paler, expand or contract, with a variation of hue, in the course of a lunar day; nor are these changes symmetrical as between evening and morning. The temperatures of the topsoil are in step with the phase. Most of the seasonal changes, on the other hand, lag behind the Sun two or three days, keeping pace with the subsoil temperatures. This shows that they do not depend on superficial alterations but have their seat some way below the ground. Such indeed would be the behaviour of vegetation sending long taproots down into the gas marsh.[12]

Many others investigating the Moon during the early years of space exploration willingly considered the possibilities of a Moon not fully dead and alien. In 1955 astronomer Patrick Moore suggested that vegetation might be present in the crater Aristarchus, where changing bands of color signaled the possibility of life precariously surviving near gaseous eruptions from underground.[13] The anticipated habitability of these spheres helped to generate substantial interest in proposals for large-scale research stations and eventual colonization.

Ironically, the anticipation of human occupancy was crushed by robotic spacecraft sent to inspect future potential homes. Robotic spacecraft revealed a far more hostile and uninviting solar system than visionaries had anticipated. When *Mariner 2* flew by Venus on December 14, 1962, at a distance of 21,641 miles, it dashed hopes for a tropical, watery planet filled with aquatic and amphibious creatures. Instruments confirmed a planet inhospitable to life with surface temperatures of some 900 degrees Fahrenheit.[14] Early NASA planners who wanted to include Venus as a coequal destination with Mars quickly abandoned that goal. Mars received a similar reception when *Mariner 4* passed that planet in July 1965 and found a cratered, lunar-like surface. Images from the spacecraft depicted a planet without artificial structures, canals, or any features that intelligent beings might create. *U.S. News and World Report* announced that "Mars is dead."[15] Even President Lyndon Johnson pronounced that "life as we know it with its humanity is more unique than many have thought."[16]

This was followed eleven years later by the inconclusive effort of NASA's Viking spacecraft to locate any living organisms or plant life on the planet's surface, an event that according to NASA scientist Gerald A. Soffen set back interest in Martian exploration for two decades.[17] Only then did many begin to perceive how much the expectation of extraterrestrial life on Mars had been oversold. At the time of the Viking landing, planetary scientist and JPL director Bruce Murray warned his colleagues of the consequences of ballyhooing the mission as a definite means of ascertaining the existence of life on Mars. The public expected to find it, as did many of the scientists. What would happen if those hopes went unfulfilled? Murray argued that "the extraordinarily hostile environment" revealed by previous NASA flybys "made life there so unlikely that public expectations should not be raised." Carl Sagan, who fully expected to find something, accused Murray of pessimism. Murray accused Sagan of excessive optimism. The two openly jousted over how to present the Viking mission to the general public. Murray, along with other politically sensitive scientists and public intellectuals, argued that any failure to detect life after optimistic forecasts and billions spent on spacecraft would spark public discontent that could lead to reduced public funding for future investigations.[18]

Alas, the Viking landers discovered no confirmable life on Mars. In the manner typical of true believers, however, advocates of the von Braun paradigm did not abandon their dream—they merely reinterpreted it.[19] If Mars did not harbor life now, maybe it did so when the solar system was young. Perhaps water had carved the canyons and channels visible from orbiting satellites, creating conditions conducive to the development of life. Perhaps life forms persisted in subterranean sanctuaries. Mars continued to occupy a prominent position in the exploration vision, and von Braun's paradigm did not vanish simply because the planet failed to confirm expectations. In 1989 and again in 2004, U.S. presidents called for exploration efforts that would lead humans to land on Mars. A collection of space groups amplified such proposals, including the Mars Society and the Planetary Society, cofounded by Carl Sagan. Exploration of Mars maintained its exalted status within NASA's strategic plan, and the presidential pronouncements regarding Mars owed much of their impetus to the persistent attractiveness of the von Braun dream.[20]

Simple reality, nonetheless, has repeatedly contradicted these impressions. Mars and Venus revealed themselves to be far more forbidding places than anticipated when the space age began. The Earth, by comparison, appears to be a far more unique and precious sphere.[21] In spite of efforts to maintain the original vision, the inner solar system does not enjoy the same exalted status that it did when the space age began. To the extent that exploration efforts shift away from the Moon-Venus-Mars area, they move away from the vision of space exploration most conducive to human preeminence. New challenges appear, but they are ones in which the traditional interest in human flight has a smaller place.

Exploration of the Moon and Mars, as von Braun predicted, may occupy the attention of human beings for many years. The impetus is not as powerful as it was fifty years ago, but it is still strong. Someday it will pass in favor of other concerns. This may happen soon, or it may happen far in the future, but it will occur. When that happens, the von Braun paradigm, attached as it is to the inner solar system, is likely to pass with it.

Observation #2: The von Braun paradigm proved more difficult to accomplish than expected, and the social movements providing its base of public support declined in importance as time passed. The vision of human spaceflight through the inner solar system, while attractive in its own right, drew much of its strength from the manner in which its proponents associated it with other social movements and events of great prominence. By association, human space travel drew strength from the history of terrestrial exploration and colonization, the American frontier, utopianism, and the Cold War.[22] With the exception of persistent utopianism, the social movements associated with the von Braun paradigm dwindled in appeal as the era of space travel matured.

Advocates of human spaceflight promised that the endeavor would extend the five-hundred-year-old tradition of terrestrial exploration and colonization by which Europeans and Americans "conquered" the surface of the Earth and its seas. From the navigation of the oceans to the inspection of the poles, exploration and settlement had played a principal role in the spread of European and subsequently American influence across the globe. Now it would happen again. Additionally, proponents promised that the spacefaring vision would continue the virtues ascribed to the American frontier, which had reached its terrestrial limits scarcely a half-century before the space age had begun. As such, advocates of space exploration associated their movement with two of the most popular narrative forms of the mid–twentieth century, the cinematic Western and the adventure tale.

Additionally, the space race offered a means to compete with the Soviet Union in the Cold War without resorting to the use of weapons of mass destruction. Like knights in personal combat settling tribal disputes off the field of battle, the superpowers expressed their technological prowess through peaceful feats of space exploration. Lyndon Johnson expressed a common sentiment when he announced, shortly after the launch of the first earth satellite, that "control of space means control of the world."[23] The early space race served as a surrogate for war, in which the United States and the Soviet Union challenged each other to a demonstration of technological virtuosity. The desire to win international support for the American or Soviet "way of life" became the raison d'être for extraterrestrial completion, and it served that purpose far better than anyone could have imagined when the race began.[24]

Finally, the vision of cosmic migration carried a strong utopian appeal—the notion that a better life and new beginnings could be found in distant places. Utopian sentiments played a strong role in the migration of Europeans and other people to the New World. Space exploration advocates extended these sentiments to other worlds.

The heroic models of terrestrial exploration, the dime store Western, and the Cold War do not occupy the same preeminent positions in society as they did when the space age began. The golden age of terrestrial exploration, maintained in its final stages through expeditions to the North and South Poles, ended thirty years before the first astronauts and cosmonauts appeared. Interest in scientific investigation of the cosmos remains strong, but public preoccupation with the heroic approach in which hardy individuals defeat forces of nature through their own talents did not persist. It was to this latter image that advocates of human spaceflight clung.[25]

Public interest in "frontiering" diminished as the twentieth century entered its last quarter. The American frontier is generally thought to have closed in the late nineteenth century, so persons who had participated in it were still alive when the first rockets appeared. By the twenty-first century, those who had experienced America's frontiering past were gone. With them, frontier myths, art forms, shootouts, and the romance of cowboy life lost much of their appeal.

With regard to the Cold War, it no longer exists. Advocates of human flight would like to think that superpower competition still requires government officials to lavish large sums on space. Certainly, in the early twenty-first century, China has its space ambitions. But the current conflict between Western civilization and Islamic fundamentalism will probably not be decided by appeals to feats of space exploration—certainly not in the same manner that people could assess the relative strengths of the U.S. and Soviet systems by examining their space programs.

As discussed previously, utopianism has been a central feature of American life since the first Europeans arrived. Beginning with religious settlements, dreamers have viewed colonization as a means of leaving behind the corruptible world and creating a more perfect society. This impulse still enchants advocates of space settlement and colonization. Trips to the Moon in the 1960s and 1970s only raised these hopes further, offering possibilities for humanity to move outward and start anew on other, more pristine worlds. The first human spaceflights suggested that Americans had both the capability and the wherewithal to accomplish truly astounding goals, provided that they possessed the will. Such expectations are invariably flawed, for prospective colonists invariably transplant the seeds of imperfection that they seek to escape.[26]

In each case, shifting public interests have undercut the social and cultural foundations of the von Braun paradigm.[27] This is not to say that the movements supporting it have entirely disappeared. Relatively speaking, however, they are much weaker than when von Braun and his colleagues first advanced their vision of human spaceflight.

Technological developments, or more precisely the lack of them, have further diminished the paradigm's appeal. The technological advances that were expected to accelerate the course of human space travel—advances that have become commonplace in the electronics, computer, and aviation industries—have simply not occurred. Gains in productivity come hard to labor-intensive activities, where the amount of work that people can do is severely limited by their physiological capabilities. Human space travel is labor-intensive, both in the ground preparation and the actual transport of human beings. Pound for pound, newer robots do more and consume less energy than people. The size of people, by contrast, has not decreased. In general, productivity improvements are easier to attain in capital-intensive activities than in ones dependent on human beings. The explanation suffers somewhat through its application to aviation, where great achievements in the cost of transport per passenger mile have occurred through advances in technology and the ability of corporate managers to cram increasingly larger numbers of people into airframes.[28] Nonetheless, efforts to repeat the aviation experience in space have generally failed.

The history of NASA's efforts to complete an Earth-orbiting space station and build a low-cost reusable Space Shuttle illustrates these trends. When Wernher von Braun pitched his idea for a large rotating space station in 1952 to the editors at *Collier's* magazine, he estimated that the structure would cost $2 billion. Members of NASA's Space Station Task Force offered an equivalent estimate in 1983, when they predicted that a fully operational space station would cost about $9 billion, although a fairer estimate would have been in the $12 billion range.[29] When adjusted for inflation, the two estimates are roughly the same. The projected estimates produced the equivalent of a $22 billion space station in 2007 dollars. The actual cost of the International Space Station (ISS)—for design, construction, and initial operations—will exceed $60 billion.[30]

Much of the cost growth was produced by factors not anticipated in the original estimates. These included the high cost of shuttle transport and the effects of assembly delays and redesign. Even compensating for these factors, the actual station remains a pale shadow of von Braun's dream. The von Braun space station was designed to hold an eighty-person crew; the ISS will perhaps house a half-dozen someday. Among its many functions, von Braun's space station—or, more precisely, the area around it—served as an assembly and refueling location for missions bound for the Moon and Mars.[31] So did the facility contained in NASA's original plans, what was known as "Space Station Freedom." Cost growth on the actual station forced government officials to remove that task. Bringing in Russia and placing the ISS in a high-inclination orbit diminished its usefulness as a "base camp" for expeditions to the Moon and Mars. If and when it is completed, the ISS will serve as little more than a microgravity research laboratory. Such an orientation may contribute indirectly to the understanding of human survival in space, but it omits a substantial element from the overall vision of human flight.[32]

Von Braun predicted that a substantial number of flights of a winged, reusable spacecraft would be needed to assemble the space station and deliver the equipment needed for trips to the Moon and beyond. The editors at *Collier's* magazine prominently displayed an illustration of such a spacecraft on the cover of the issue promoting von Braun's space station plan.[33] In 1972, NASA officials presented the cost estimates for the spacecraft. The total life-cycle costs for a reusable Space Shuttle, they said, would total $16 billion, including capital investment, operations, and payload preparation. For that sum, the nation could purchase 580 flights, more than enough to complete an international space station and accomplish other elements of the von Braun dream.[34] Politicians approved the estimate, and the NASA Space Shuttle was born.

The cost per flight of that 580-mission manifest, adjusted to the value of 2007 dollars, amounts to about $150 million per flight.[35] At that level, much of what the von Braun paradigm entailed might have been accomplished. Reality stood in the path of expectations, however, deflating hopes for sharply reduced transportation costs. Through 2002, NASA's Space Shuttle flew 110 times at an average mission cost (in 2007 dollars) of $1.4 billion per flight, nearly ten times the original estimate.[36] This was a terrible disappointment and a substantial diversion from the von Braun dream.

One could blame the ensuing gaps between vision and reality on a number of factors. Weak management, a lack of cost discipline, political interference, and technological challenges all proved more daunting than anticipated. The vision of frequent and inexpensive flight was based on airline models that called for the allocation of substantial investments and a large work force over a growing number of flights. When the number of flights dwindled, per mission costs soared, which in combination with two accidents encouraged government officials to reduce the number of flights even further.

In sum, technical difficulties intersected with a weakened social and cultural foundation to steal much of the momentum accorded the original von Braun dream. Social movements and events did not favor the human spaceflight movement to the degree that they did at the beginning of the endeavor. Contrary to expectations, what was once strong grew weaker in form.

Observation #3: While advances in human spaceflight lagged behind expectations, advances in robotic flight exceeded them. The obstacles thought to restrict robotic flight when spaceflight commenced melted away as experience accrued. Like developments in the microelectronics industry generally, advances occurred in technology, cost-effectiveness, and capability. The changes transpired far more rapidly than people standing at the cusp of the space age anticipated. From the perspective of the mid–twentieth century, advances in robotics have been revolutionary.

As recounted earlier, humans vastly improved the ability of robotic spacecraft to observe and record their surroundings (the science of remote sensing), transmit information (communication), process data and solve problems (computing), and move about their objects of inquiry (mobility). These advances paralleled—and in some cases helped to inspire—general developments in telecommunications, microelectronics, and computing. The rate of change has been rapid and exponential and even under the best of circumstances relatively hard to imagine by someone standing at the midpoint of the twentieth century in a world without communication satellites, cellular telephones, digital cameras, personal computers, or high-definition color television.

Concurrent with the growth of technology, the cost of building and operating robotic spacecraft declined. Much of this was due to smaller, more reliable components and improved methods of project management. In inflation-adjusted terms, the rovers *Spirit* and *Opportunity* explored the Martian terrain in the early twenty-first century for about one-fifth the expense of placing the Viking landers on Martian soil thirty years earlier. Project workers on the Mars Pathfinder mission demonstrated that it could be done for even less. That expedition was carried out at one-fourteenth of the inflation adjusted cost of the Viking mission, a level of cost reduction that on subsequent missions could not be sustained.

As a consequence of these changes, the capability of robotic spacecraft steadily advanced. New robots did more than old ones—for less money. The history of the Voyager mission illustrates the overall trend. As originally proposed, the Voyager mission was to be a "Grand Tour" of the outer solar system, taking advantage of a rare alignment of the outer planets that would not reoccur for 179 years. Burdened with the expense of funding the newly approved NASA Space Shuttle, presidential budget analysts would not approve the Grand Tour. Instead, they allocated funds sufficient for a spacecraft that would last five years and visit Jupiter and Saturn. The cost savings were substantial—the approved mission cost just one-third of the estimated Grand Tour. Using available funds, project workers at NASA's Jet Propulsion Laboratory built sufficient capability into the two spacecraft so that they could operate for decades rather than years. Consequently, *Voyager 2* visited Uranus and Neptune in addition to Jupiter and

Saturn, and both spacecraft continued to return data as they traveled into the outer reaches of the solar system, where they began to encounter the interstellar wind.[37]

With their newfound capabilities, robots covered the solar system. Just a half-century after the launching of the first Earth-orbiting satellite, robots had visited all of the major planets, with a spacecraft en route to Pluto; traveled into the heliosheath; landed on Venus, Mars, Titan, the Moon, and the asteroid Eros; mapped Venus and Mars; descended into the atmosphere of Jupiter; and returned samples from comet Wild 2 and the Moon. Humans sent large, automated telescopes investigating different parts of the electromagnetic spectrum into orbits around the Earth and trailing behind it. A variety of satellites operating without onboard human control investigated the cosmos and provided vital services in areas such as communication, weather prediction, navigation, and the monitoring of Earth resources.

In one of the more surprising developments of the early space age, no nation stationed soldiers in space. People forecasting space activities at the commencement of cosmic flight pictured military observation posts, communication centers, navigation stations, orbiting bomb platforms, and missile bases on the Moon—all staffed by members of the various armed forces. When NASA leaders first proposed the construction of an Earth-orbiting space station in the early 1980s, they listed military activities as part of its mission. Advocates of the National Aerospace Plane, proposed during the same period, lauded its use as a very rapid troop transport. Anticipating a substantial role for space-based systems in future military conflicts, officials in the U.S. Department of Defense established their own astronaut corps as the space age began.[38] With the exception of a small amount of military research conducted on the U.S. Space Shuttle and various Russian space stations, no military personnel ever flew in space. The U.S. disbanded its Air Force astronaut corps before any of its members flew (some transferred to the civil space program). Contrary to early expectations, the work of military reconnaissance and communication was completed entirely with robotic spacecraft. Military personnel were not needed to staff orbital observation posts, operate space-based communication stations, or change the film in reconnaissance satellites. Although space assets play a critical role in modern warfare, the work is carried out by robotic means. This was quite unexpected and a substantial endorsement of robotic capabilities.

In a related development, all early commercial space activities relied on automated or robotic devices. Today, business firms operate a wide range of space-based devices. Space is a multi-billion-dollar industry, larger worldwide than the combined expenditures of all government space activities, military and civil. Business firms operate communication satellites and Earth resource satellites. They build the equipment that utilizes signals from global positioning satellites (the satellites are maintained by the U.S. military). They build the rockets that launch the equipment.[39] In spite of forecasts touting space tourism and the advantages of letting humans repair satellites in space, the global multi-billion-dollar space industry remained robotic throughout its formative years.

Judged in terms of overall capability, humans still possess skills that robots do not. They can think better and, once on the surface of extraterrestrial bodies, move faster than machines. Although robotic capability advanced significantly in the decades following the first Earth satellites, and robots moved closer to a state of parity, humans are still superior to robots—at least at the start of the twenty-first century. That statement carries a significant qualification. Humans are generally superior to robots where both are capable of doing the same work. In lunar traverses and satellite repair, humans can outperform robots. Where humans cannot or choose not to go, however, robots under human control are capable of operating at very high levels of efficiency. In its purist terms, the NASA vision of "humans and robots together" applies only to those areas where humans might travel. The vision of robots acting as precursors to human exploration or in a coequal status applies to the area bounded by the Earth, Moon, and Mars. In other areas of the solar system, the dominant policy is clearly robots working alone under the guidance of humans back home.

In the region where the von Braun paradigm rules, a substantial case can be made for a "mixed" space program in which both humans and robots work together. Motivated by the vision proclaimed by von Braun and other space pioneers, this is the policy that government officials in the United States and the Soviet Union originally pursued. The overall exploration program after fifty years of spaceflight, however, has moved well past that realm. It physically extends to the whole solar system, with the capability to investigate conditions beyond. The enlargement of space exploration was due both to advances in space technology—especially robotics—and to the fact that nearby planets turned out to be much less interesting places than originally envisioned. In many respects, the von Braun paradigm was a victim of its own success. In its early years it spurred substantial public interest and extensive government expenditures, which, given the increasing ease with which robots moved about the solar system, widened the arena of investigation well beyond the regions where humans and robots might work effectively together.

Observation #4: In spite of substantial advances, the robotic alternative suffers from its own weaknesses, both technical and social. In assessing the capabilities of robots in space, it is important to consider the exact nature of these devices. As classically defined, robots are machines, operating under human control, that carry out tasks too menial or dangerous for humans to perform. Notwithstanding insights offered by science fiction writers, robots are not autonomous life forms, they are not a separate species, and they are not sentient beings. They are not capable of carrying out complex tasks that go beyond the intentions of their creators, which is to say that they can only accomplish what they are programmed to do. In the words of Isaac Asimov, they are simply machines.[40]

The work of machines in space is highly choreographed. (The same, incidentally, can be said for human beings.) Just as astronauts need to be elaborately trained, machines need to be extensively programmed. While increased autonomy might improve the range of exploration activities, increased improvisation would probably doom them. Outer space is simply too hazardous to permit either robots or astronauts to exercise the levels of discretion permitted ship captains and other adventurers during terrestrial voyages of discovery.

The absence of autonomy raises the fascinating (but somewhat theoretical) question of who exactly is exploring space. Every space mission conducted thus far has been tightly planned and highly supervised by flight controllers and project managers on the ground. As NASA space science project manager Oran Nicks rightly observed, humans have participated in every space mission ever flown, even when robots have done the actual work.[41] For that matter, so have machines. No human has ever flown in space without the help of a very complicated machine. In all cases, humans have controlled those machines, either from ground stations on Earth or through a combination of ground stations and astronaut control.

At the beginning of the space age, many of those people who stayed on the ground treated astronauts more like robots than human beings. The astronauts were viewed as "test subjects," not meant to pilot their spacecraft but rather to gauge the effects of space travel on a relatively inert human frame. Astronauts fought with flight engineers over issues such as these. In the United States, where the vision of human spaceflight was particularly strong, the astronauts tended to prevail.[42] Their supervisors nonetheless did not allow astronauts the high levels of flight autonomy achieved in fictional depictions of space travel. Buck Rogers and Captain James Kirk of the Starship *Enterprise* exercised far more discretion in operating their fictional spacecraft than early space travelers ever achieved.

In its early stages, all spaceflight proceeded from a philosophy of control. To put it crudely, robots and astronauts were treated like dogs on a leash. The philosophy drew its power from industrial-age models of machine and worker subservience and the organizing principles of large-scale project management. This proved to be a perfectly adequate approach for the early phases of space exploration. It was sufficient to land humans on the Moon and dispatch robots to the outer planets. The approach breaks down, however, in the face of really complicated missions. A human colony on Mars or a robotic mission to search for extraterrestrial life under the icy surface of Europa could not be treated this way. It certainly disintegrates when one contemplates voyages of discovery beyond the local solar system. Those missions cannot be controlled by operators sitting behind consoles at flight stations on Earth.

With a little contemplation, one can imagine mechanical or electronic entities completing space missions in distant places far removed from even indirect human control. These would not be robotic machines in the conventional sense, certainly not as the concept has been traditionally employed. Being on their own, their capabilities would vastly exceed those of present-day robots.

The machine-age model that has dominated robotic activities in space suffers from its association with traditional solar system missions. It is a concept deeply rooted in human control. For advanced flight activities, especially those that go beyond the local solar system, human control is not feasible—certainly not in the traditional sense. Entities might be dispatched with a general set of instructions and the capacity to carry them out. Control in that sense would be indirect, with the mission unfolding in unanticipated ways. (If all potential scenarios could be anticipated, direct control could be reimposed.) Traditional machine-age models are not much help in understanding how indirect control might work. With its emphasis upon preestablished rules and detailed anticipation of events, the traditional model draws attention away from the principal challenges faced by humans seeking to create postbiological entities or artificial intelligence machines.

Nor are the social issues surrounding robotics much help in this regard. Proceeding as well from an industrial-era perspective, the traditional social literature on robotics is too preoccupied with issues of jurisprudence and conformity to commands. Like discontented workers, would clever machines rebel? Would they seek legal rights and privileges? Could humans implant rules that would prevent such entities from harming their creators? To anyone familiar with the history of servitude and labor-management relations, these are salient issues. Yet control is not likely to be a central issue among entities that by the very terms of their creation are beyond direct human control. Issues of coexistence would be a much more pressing concern.[43]

From the perspective of people entranced by the concept of human flight, no very smart machine or postbiological entity will ever achieve the ultimate purpose of space travel. The robotic approach in all of its forms does not address the deep-seated desire to preserve the human experience through cosmic dispersion. Given the requirements imposed by cosmic travel, however, the quest for human immortality generates more confusion than illumination. This quest presumes that machines and humans will maintain their present form. In the long run, that is not likely to occur. As noted in the previous chapter, the requirements of space exploration may encourage human and machine characteristics to merge in strange and unexpected ways, raising a whole new set of social concerns.

To summarize once more, robotic capabilities in space have advanced much more rapidly than most experts believed they would. The capacity for human spaceflight has advanced less rapidly than anticipated and appears, under the best of circumstances, to be confined to the Earth-Moon-Mars area. The inner solar system no longer dominates the exploration enterprise to the degree that it did when the space age began. Attention is shifting beyond the solar system to the galactic neighborhood, a development likely to be accelerated by robotic telescope technology. The classic robotic model is adequate for addressing neither the requirements imposed by very advanced expeditions into the solar system nor the challenges of interstellar flight. If space travel proceeds into those realms, something new is likely to emerge.

Observation #5: Both human and robotic approaches in their classical form suffer substantial shortcomings. Confronted with the requirements of long-duration spaceflight and migration, the characteristics associated with the two approaches tend to merge. The discussion that follows is speculative, tentative, and a little bit strange. Nonetheless, it is one that inevitably presents itself when the relative advantages of human and robotic spaceflight are contemplated within the context of long time spans and immense distances. Under those conditions, the exploration enterprise tends to alter the creatures that undertake it.

The requirements of long-term exploration, especially in the galactic realm, severely test both human and robotic approaches to spaceflight. Robotic spacecraft are not well suited for extended travel into regions far removed from human control. Humans are not well designed for extended journeys beyond Earth and its protective environment. Both tend to wear out over time, a process typically accelerated by the harsh conditions prevailing in outer space.

If one were free to design an object capable of extended space travel, it would have many of the features associated with machines. It would resist radiation, be capable of operating under extremes of hot and cold, not require a breathable atmosphere, possess simple energy needs, and be able to sit still in a confined space for an extended period of time. It would need a long life span and a simple method of self-repair. Given the energy required to move objects to interstellar speeds, it would be as small as possible. Certain organic substances meet those requirements. E-coli bacteria, for example, would make good space travelers. They have survived unprotected on spacecraft instruments set on the surface of the Moon, a realization that has prompted some persons to propose them as likely candidates for interstellar migration.[44] Single-cell microorganisms, however, lack in intelligence what they gain in survivability. They may be robust, but they are awfully dumb. Ideally, the specially designed travelers would possess the capability to solve complex problems, interpret their surroundings, and redirect their flight. They would, in brief, need to possess the equivalent of human brains.

For a machine to fulfill these requirements, it would need to be more like a human being. At a minimum, the machine would need to possess cognitive capabilities. It would need the capacity to think and to solve problems. It would be required to make decisions on its own without human help. It would need to respond to local conditions and unforeseen events. Ultimately, it would need some method of repairing or replacing itself as its parts wore down. In short, it would need to be as much like a human being as like a traditional machine. In fact, it would cease to be a machine, at least as that term, with its industrial era connotations, is employed.

Should humans ever dispatch members of their own species to nearby solar systems, either on voyages of discovery or for purposes of migration, such travelers would benefit from certain physiological alterations. Extended life spans would help, as would characteristics better adapted to the interstellar medium. The ideal space travelers would be as hardy as bacteria and as smart as computer geeks. Once they arrived, they would adapt to local conditions, possibly changing in physical form. Physiological change is a far more likely consequence—and possible condition—of interstellar travel than the vision of humans forever immortal in their present frame. Such travelers would cease to be human in the classical form. They would, as the term is currently employed, become transhuman.

The same proposition, incidentally, can be offered regarding any humans attempting to settle Mars. Within a few generations, such travelers would adapt physiologically to conditions such as the gravity effects of a less massive world. No one knows exactly how this would occur, but it is possible that humans born on a one-third gravity world would grow tall and thin, with spindly bones. The interaction of gravity and radiation with skeletal DNA might produce colonists unable to return to Earth without the assistance of mechanical aids.

By what process might these requirements be fulfilled? Insofar as humans and their successors are concerned, the process could take place in the traditional way. To use a word that engenders much concern, the species could evolve. Were human civilization to persist for millions of years, the physiological changes needed to engage in interstellar flight might occur as a result of natural evolution. Earthly life forms have changed considerably since the Cambrian explosion some 500 million years ago and are likely to do so for many millions of years more. Should intelligent life forms continue to occupy the Earth for an equal amount of time, they too will change. It is inevitable. Nature cannot guarantee, however, that such alterations will favor interplanetary migration.

Yet the change need not take that long. And it need not occur randomly. As an alternative, alterations might occur as the result of advances in technology. To use a phrase from a widely discredited field, they might occur as a result of intelligent design. The intelligence in this case would not be supernatural but would take the form of people altering human and machine forms. The pace of change in technologically relevant fields is so great that some people, albeit clearly on the optimistic side, believe that fundamental alterations in robotic and human capabilities could occur within the next one hundred years.

Very few of these changes are promoted by people interested in space exploration. Most are occurring in fields like military robotics, artificial intelligence, and biotechnology. With the help of civilian scientists and engineers, military officers are attempting to produce robots of increasing sophistication for battlefield use. The goal is to produce machines capable of autonomous battlefield operations within ten to twenty-five years. One wonders whether such autonomous entities will be willing to die for their countries.[45] Scientists and entrepreneurs are working to produce computers whose capabilities for speech and problem-solving at least mimic those of human beings. Their goal is to build computers as intelligent as their designers. Medical researchers are working to advance human longevity as well as produce more resilient body parts. Their ultimate goal is substantially increased life spans. Much of this work is taking place at the microbiotic level, where scientists are tinkering with stem cells and DNA.

The amount of money directed toward medical care and research is staggeringly large, well exceeding the amounts spent on space travel.[46]

The degree to which such discoveries get applied to space depends substantially on the overall attractiveness of the spacefaring vision. Space exploration advocates command a minuscule share of any nation's gross domestic product and do not typically possess sufficient resources to propagate fundamental change in their supporting technologies. In the past, they have benefited greatly from investments in fields such as military rocketry that are tangentially related to space exploration. Perhaps this will happen again. If the overall vision is strong and up to date, it could motivate the appropriate change. At the moment, the original von Braun paradigm, we have suggested, is neither. It is rapidly fading in social significance and technological relevance. If a new paradigm emerges to supplement or replace it, that might spark technological innovation in ways not currently envisioned by space advocates.

FUTURE PRIORITIES

To summarize, the von Braun paradigm was directed toward the inner solar system. The new paradigm focuses on the galactic neighborhood. The von Braun paradigm visualized a very small role for robots; the new paradigm accepts what has already occurred—a substantial role for automated spacecraft. Von Braun foresaw humans in their present form traveling to nearby planets in the tradition of explorers like Christopher Columbus and those who followed; the new paradigm at least acknowledges the possibility that those who travel far in space may be intelligent entities of a nonbiological kind or humans reengineered for space travel.

The old paradigm received substantial impetus from the Cold War, the desire for national prestige, the great wealth of the United States, heavy investments in rocketry, and a public made receptive to space travel by widely disseminated images depicting how it might occur. The Cold War has disappeared, the United States ceased making substantial investments in rocketry, and the popular image of what humans might expect to find on Venus and Mars has been dashed by the reality of conditions on those spheres. National wealth remains a prerequisite to space travel (poor nations do not fly in space), but public officials in rich nations have shown little enthusiasm for budgetary munificence of the extraterrestrial kind. Public support for trips to Mars remains where it was forty years ago. About 40 percent of the American public favors it; about 55 percent are opposed.[47] If Mars had surface water or native vegetation, if it was not so awfully

TABLE 8.3.
A New Space Paradigm

- Development of multi-stage rockets capable of reaching space.
- First launching of orbital satellites.
- Flights of humans into space to determine capabilities and explore near-Earth environment.
- Exploration of the Moon, Mars, and other bodies in the solar system using robotic spacecraft and some human expeditions.
- Development of military and commercial applications in space, especially in low-Earth and geosynchronous orbits.
- First attempts to contact extraterrestrial civilizations through electromagnetic means.
- Construction of space telescopes, including terrestrial planet finders.
- First images and spectrographic studies of extra-solar planets.
- Development of robotic spacecraft with ever-increasing levels of autonomy for the purpose of solar system exploration.
- Identification of nearest extra-solar planet with Earth-like conditions, should any exist.
- Development of new propulsion systems.
- Advances in technology and management that significantly reduce the cost of spaceflight.
- First successful attempts to create machines with fully autonomous flight capability or to modify human physiology for long duration space travel.
- First attempts to explore extra-solar planets.

windy and cold, it would be a more inviting destination. As an earth-like planet somehow gone dry, Mars is certainly worthy of study. It is not likely to receive as much attention as space advocates would prefer.

Advocates of the old paradigm held up an unsubstantiated image of Mars (and occasionally Venus) as justification for space travel. It was an image built on a foundation of vivid imagination; it persisted through much of the twentieth century. The image provided the seed stock from which the early public interest in space travel grew. No image of similar power has risen to replace it. Without a replacement of adequate appeal, public interest and government support for space travel will continue to languish.

Priority One

A number of possible alternatives exist. None is likely to exert more power than the possible discovery of a new planet that has the characteristics of the old depiction of Mars—one that harbors conditions favorable for the maintenance of life. The old Mars (and to a lesser extent the Moon and Venus) inspired public interest precisely because they combined the anticipation of human migration with places whose actual conditions could, with a little technology, become known. The search for extrasolar planets promises to repeat this history. Like the previous history, habitable places may exist or the search may prove similarly disappointing. Regardless of the outcome, the implications for human civilization are profound.[48]

Space scientists would like to search for extrasolar planets in the following manner. They have already confirmed the existence of planets around nearby stars by measuring slight shifts in their radial velocity—alterations in the motion of parent stars produced by the gravitational effects of unseen planetary companions. Space telescopes are being produced that would extend the search by observing shifts in the position of stars relative to adjacent stars and by measuring the dimming effects produced when planets pass in front of their stars. NASA's Space Interferometry Mission (SIM) is designed to accomplish the former, while its Kepler Mission will use the latter technique. The United States, France, and the European Space Agency are all engaged in efforts to detect extrasolar planets using methods such as these.[49]

To observe such planets directly, astronomers would like to construct even more elaborate space telescopes. In contemplating the priorities that might motivate a new paradigm of spaceflight, this is number one. NASA scientists have proposed a project called Terrestrial Planet Finder. It would combine a cluster of telescopes flying in formation that would detect the reflected light of relatively cool Earth-like planets. The cluster would employ interferometry techniques to collect light waves in the infrared spectrum and employ a coronagraph to block out the visible light of the central star. By delaying the light signals received by different instruments in the telescope cluster, astronomers could nullify the light waves produced by the parent star in favor of infrared emissions from its planets. Officials at the European Space Agency are working on a similar set of instruments known as the Darwin Project.

Spectroscopes on such instruments would be capable of detecting the biosignatures of life. As a planet supporting complex life, the Earth emits a distinctive spectral signature. Free oxygen, methane, nitrous oxide, carbon dioxide, ozone, and water vapor in the atmosphere—in the correct proportions—reveal the Earth to be a living planet. Scientists engaged in astrobiology study the means by which the signatures of Earth-like processes can be differentiated from the chemistry of dead planets that mimic biological processes. Space-based instruments and astrobiology provide the key to locating Earth-like planets around other stars.

How prevalent might such planets be? When Frank Drake proposed his famous formula for calculating the presence of communicative civilizations in the Milky Way, he suggested that the number of candidates might equal a million or more. The Milky Way contains more than 100 billion stars. If planetary systems are common and life arises on bodies possessing liquid water, then the distribution of Earth-like planets might be incredibly wide. Donald Brownlee and Peter Ward, however, have suggested that the development of complex life forms like dinosaurs and humans beings may be extremely rare, requiring oddities like a large moon and the tides it produces on a wet planet.[50]

In the absence of direct investigation, scientists simply do not know which situation prevails. Rocky planets capable of maintaining liquid water around a Sol-like star may be common or rare. If wet, rocky planets are common, simple organisms may create habitable conditions without producing advanced life forms. Despite their lack of complex life, such objects might provide acceptable destinations for humankind, or they may be so widely dispersed as to be inaccessible. If habitable planets are common, some may already be occupied by complex beings.

No single factor is likely to be as profound for the future of humankind, nor as inspirational for the imagination, as knowledge about extra-solar planets. The discovery of Earth-like planets would inspire centuries of investigation and possibly efforts at visitation. Humanity in some form would have the opportunity for dispersion and migration away from its earthly home. Conversely, if Earth-like planets are extremely rare, that too has profound consequences. It means that the Earth is the only home that humans in their bodily form can expect to occupy, with the possible exception of transforming a nearby object like Mars.

Unfortunately, the tendency to squander funds that could be used for seemingly esoteric projects such as the Terrestrial Planet Finder is strong. In 2006, as part of their effort to find funds to replace NASA's aging Space Shuttle and to return to the Moon, government officials indefinitely deferred the Terrestrial Planet Finder.[51] This is the equivalent of a farm family eating its seed corn just before the planting season begins. Throughout history, the various civil space programs of the world have garnered public support by pledging to reveal incredible secrets with accessible technologies that lie only a few perfections away. If civil space leaders ever lose their capacity to push those boundaries of discovery and imagination, they will lose the public interest that supports their endeavors.

With a bit more understanding, the final disposition of the von Braun paradigm will be completely known. Humans may have the opportunity to move their dreams of investigation and migration beyond the solar system and its disappointing spheres to other destinations. Alternatively, the inner solar system may be all that humans have. Either way, humans will need to study many planets if for no other reason than to discover what makes the Earth such a friendly environment for the maintenance of complex life.[52]

Humans stand at a point in their understanding of the galaxy that is similar to that of their understanding of the solar system five hundred years ago. They know a little but not a lot. The little they know is capable of inspiring great feats of discovery and imagination. What they discover will determine how the species eventually proceeds. To repeat, no single priority is more important for the long-term future of space exploration than interstellar studies.

Priority Two

The second priority affecting the new paradigm concerns the search for more effective propulsion systems. Nothing is going to travel very far in space without new methods of propulsion—not humans, robots, androids, cyborgs, DNA, or bacteria. Current spacecraft utilize what might be charitably characterized as the Marco Polo method of travel—slow and plodding. Polo essentially walked to China. In 1271 he and his father and uncle traveled overland to Cathay, China, along the Silk Road. Not until 1275 did they arrive at Kublai Khan's original capital at Shang-tu, then the summer residence. Later Polo moved on to the winter palace at Cambaluc (modern-day Beijing). Polo remained in Khan's court for seventeen years, after which he returned to Venice in a trip that required an equally arduous sea voyage. Polo's adventure was technically feasible given thirteenth-century methods of transportation, but it was certainly not easy.[53] Modern-day spacecraft use methods of transit that—relative to the required distances—are similarly slow. Apollo astronauts embarked on their expeditions to the Moon by first achieving Earth orbit, then firing J-2 engines powered by liquid hydrogen and oxygen, producing sufficient thrust to propel their spacecraft on a three-day journey to the Moon. Robots bound for the planets use a similar method, sometimes with additional gravity assists achieved through planetary flybys. Each of the Mars Exploration Rovers that left Earth in 2003 employed a solid rocket motor burning ammonium perchlorate and aluminum. Ninety seconds of thrust increased the spacecraft velocity to about twenty-five thousand miles per hour, after which the spacecraft cruised to Mars.[54]

Once it has achieved escape velocity, a spacecraft does not maintain that speed for the duration of its flight. The average speed and transit time are affected by a complex set of factors that include initial velocity and the gravitational effects produced by relevant bodies such as the Earth, the sun, and the spacecraft's destination. Spacecraft cruising between the Earth and Mars follow paths that are actually part of very large elliptical orbits around the sun intersecting the points of origin and destination. Like airliners on Earth flying between continents, the most efficient paths are not usually straight lines. (Airliners follow curved routes to avoid the planetary bulge.) Curved routes are energy-efficient given current levels of technology, but by interplanetary standards they remain slow.

In 1999, the Italian physicist and Nobel laureate Carlo Rubbia proposed the use of a nuclear-powered rocket with four to eight times the thrust of chemical fuel for a human expedition to Mars. His proposal avoided the necessity of a long, sweeping transfer orbit—called a Hohmann trajectory—and the requirement that the crew wait on Mars for some 460 days as the planets realign before returning home. Rubbia estimated that the round trip could be completed in about one year, with forty-one days on the planet's surface.[55]

With its Deep Space 1 mission, NASA engineers test-flew an ion propulsion engine in space. This technology uses electric power, generally obtained through solar arrays, to provide a continuous source of acceleration. NASA officials had hoped to test-fly a nuclear-powered engine through their Project Prometheus, but funding cuts delayed this commitment.[56]

The United States spent over $200 billion (in 2007 dollars) developing the Atlas, Titan, *Saturn V*, and reusable Space Shuttle. NASA executives backed out of the X-33/VentureStar effort to replace the Space Shuttle in 2001 when total costs approached $2 billion.[57] As the X-33 experience showed, the history of launch vehicle development in the United States has been one of dwindling investment. Essentially, the U.S. space program—both military and civil—has been living off of its Cold War investment for nearly fifty years. It should come as no surprise, given the reluctance to invest, to learn that NASA's new crew exploration vehicle is intended to rely heavily on already-paid-for shuttle technology.

As the leading spacefaring nation, the United States has deferred the investments necessary to improve rocket technology for far too long. Like the cutbacks in telescope science and astrobiology, the diversion has been prompted by the desire to fund human flight endeavors using old technologies on a limited budget. At some point, someone will need to invest in new methods of rocketry if the full vision of spaceflight is to be realized.

Priority Three

Such an assessment points to the third priority of the new paradigm—the need to reduce all of the expenses associated with space travel. For thirty years, NASA attempted to reduce the cost of spacecraft flying to and from Earth orbit. It adopted three strategies, none of which worked particularly well. NASA engineers built five reusable Space Shuttles on the assumption that savings would accrue if mission chiefs flew the same machines rather than built a new spacecraft for each mission. Flight directors attempted to increase the frequency with which the shuttles flew, utilizing a business model to achieve productivity gains by spreading a larger volume of activity over a relatively fixed work force. The number of missions envisioned never met expectations; neither did the expected savings. Finally, NASA officials attempted to reduce the number of people paid to maintain and fly the shuttle fleet—the total number exceeds twelve thousand—by directing most shuttle responsibilities to a private contractor who imposed new management methods. This strategy did not work particularly well either.

Space expeditions are incredibly expensive. The final flights to the Moon, once the initial investments had been made, cost about $4 billion each in the value of 2007 dollars. A flagship planetary mission, using robotic spacecraft such as employed on the Cassini mission to Saturn, costs about $3 billion. When all of the expenses are factored in, a single flight of the Space Shuttle costs more than $1 billion. About half of that consists of the amortized value of capital investments; the remainder goes for current flight operations. The Mars Exploration Rovers that reached their destination in 2004 cost about $800 million. This is a typical sum for a medium-sized robotic mission. As a point of contrast, a modern sports stadium such as the one approved for the Washington Nationals baseball team in 2006 will cost approximately $700 million.[58]

NASA officials experimented with low-cost missions during the late twentieth century. They dispatched the Mars Pathfinder mission for $265 million and completed the Near Earth Asteroid Rendezvous (NEAR-Shoemaker) for $212 million. These are total program costs and include the expense of designing, building, launching, and operating the spacecraft. Buoyed by enthusiasm for the approach, NASA Administrator Dan Goldin extended the approach to more than a dozen robotic missions, the Spitzer Space Infrared Telescope, and planning efforts for a human mission to Mars. The approach went into disfavor after the failure of four low-cost missions in 1999.[59]

Most of the expense of conducting an extraterrestrial expedition finances the spacecraft. That is where the bulk of the money goes. The Cassini expedition to Saturn is typical. Of the $2.6 billion that the United States contributed toward the Cassini mission (European partners contributed another $660 million, primarily for the Huygens probe), 55 percent, or $1.4 billion, went to pre-launch design and construction. NASA officials devoted 16 percent of the Cassini budget to the *Titan IV* launch vehicle ($420 million), while flight operations ($710 million) and tracking ($54 million) consumed the remaining 29 percent. Although NASA officials worked hard to cut launch expenses during the late twentieth century, far more of their mission outlays went toward the design and construction of spacecraft.[60]

Reductions in the overall cost of designing and building spacecraft would advance the cause of space exploration considerably—no matter which direction exploration goes. That is as true for robotic spacecraft, which have shown the most receptiveness to cost-cutting, as for ones that carry human beings. Ultimately, all current spacecraft are robotic anyway. The ones in which humans travel have levels of automation that are nearly as high as those in which humans stay behind.

The approaches taken toward space expeditions have a considerable effect on overall mission cost. For their planned return to the Moon, NASA officials announced that they would employ lessons learned during the Apollo lunar landing program. "Orion," the crew exploration vehicle, would look like an enlarged Apollo capsule, and the mission would rely upon operational methods associated with Project Apollo. Given the cost of the last few missions to the Moon, this is probably a reasonable approach for a lunar expedition. Yet the purpose of the lunar return effort is to provide "the maximum possible utility for later missions to Mars and other destinations." Apollo-style methods produce total program costs that exceed $500 billion when applied to a Mars landing and return.[61]

Searching for a low-cost method to send humans to Mars, planners at NASA's Johnson Space Center suggested that the International Space Station be used as a model for a Martian expedition. Although cost overruns on the ISS proved embarrassing for the U.S. space agency, NASA officials used the funds to assemble a million-pound spacecraft in Earth orbit. Planners estimated that six spacecraft totaling 850,000 pounds would be necessary for a flight to Mars and back. Assuming that a spacecraft could be produced at space station costs—about $30 billion—the added expense of landers and propulsive devices would place total program costs in the $50 to $100 billion range. Since the exact amount would depend in part on new methods of propulsion, it might make sense to invest toward the upper range of the estimate with the objective of reducing the flight time to Mars and its associated costs.

This is the third great priority in achieving the new spaceflight paradigm. Together with the search for extra-solar planets and improved methods of propulsion, reductions in the cost of spacecraft would permit ventures that simply cannot be completed within the current budgetary framework. NASA officials receive about $16 billion annually from the public treasury. When all of the spending on space is totaled worldwide (civil, military, and commercial)—to say nothing of the investment in related technologies such as artificial intelligence and robotics—the NASA budget becomes a minor part of the overall picture. It is a bump on the log, no more than 10 percent of the whole.[62] Yet NASA officials—and, to a lesser extent, their foreign counterparts—are expected to play a leading role in opening the space frontier.

Priority Four

Civil space agencies—especially NASA—have reached the point where they can no longer do it all. This is a situation identical to the one confronting the promoters of past technologies. The people who sought to build ships, canals, railroads, highways, oil refineries, dams, airports, and airplanes all found methods for enlisting financial support that did not depend wholly upon tax appropriations to a single government agency. In cooperation with public officials, they developed a wide range of incentives that included land grants, tax incentives, loan guarantees, user fees, liability protection, prizes, and favorable government regulations.

As part of their most recent strategic plan, NASA officials announced that they would pursue "appropriate partnerships" that supported the goal of expanded space operations, especially "with the emerging commercial space sector."[63] This may allow NASA officials to purchase the use of launchers from privately owned companies to transport NASA astronauts or allow private firms to operate the bases supplying the basic commodities like water and electric power that astronauts use in space. The partnership model is qualitatively different than the traditional relationship in which government officials direct work to large aerospace firms who as a consequence of their dependence on government contracts can hardly be said to exist in a competitive marketplace. The new model enlists participation from privately owned space industries, entrepreneurs, startups, nonprofits, and foreign governments. If actually implemented, this partnership strategy would be a radical change for government agencies like NASA, which are used to doing it all themselves in conjunction with traditional aerospace firms.

CONCLUSION

Once these four priorities are in place, leaders of the civil space effort will be in a better position to reassess exactly what flies in space and why. This returns us to the original issue posed in our book. Will it be machines, or will the machines carry humans on board? Of the five objectives presented in chapter 3 that motivate nations to fly in space, only one absolutely requires a human presence. To recount, the five reasons are scientific discovery, commercial applications, national security, geopolitical prestige, and survival of the species. The first three can be achieved by robots alone. While humans can often do more work, robots are perfectly adequate for the tasks of scientific research and can visit places that humans are simply not well suited to attend. In spite of the promise of cosmic tourism and microgravity factories in space, all of the commercial applications in space have been accomplished by automated satellites. Military personnel are likewise dependent on remotely controlled satellites for their national security needs, such as defense communication and reconnaissance. The fourth rationale has been used as an argument to maintain human flight programs and is probably sufficient to motivate a lunar research station and to plant a flag on Mars. As recent events have illustrated, however, great national prestige can accrue from robotic capabilities such as those demonstrated by space telescopes and planetary rovers. Humans are certainly useful for these functions, but the only incontestable motivation for the necessity of humans in space is the need to scatter the species.

For many decades government leaders with the means to fulfill that motive avoided any discussion of it, finding it too ethereal for the purpose of obtaining dollars from the public treasury. Only recently have they begun to acknowledge this rationale. Historically, this purpose has not provided sufficient cause to motivate government support for human spaceflight. Advocates of human flight instead attached their ambitions to objectives that did not absolutely require humans in space. Future developments, such as the discovery of habitable planets, may help elevate the survival of humanity to its rightful place as the primary motivator of nonrobotic space travel.

Assuming that survival of the species is an acceptable goal—we think it is—humans are presented with three main options. First, humans can move off the Earth to nearby worlds, beginning with the Moon and Mars. We find this option unlikely. Perhaps humans will go there, perhaps even establish research stations, but conditions in space are such that humans are unlikely to colonize these places in any large numbers anytime in the coming centuries. We doubt that NASA will be able to send astronauts to Mars under the current space exploration regime. The mission is technologically feasible, but NASA is not well positioned to achieve it. The technical obstacles are overwhelming, not at all analogous to sea captains of past centuries spending three years from home who could repair their frequently damaged ships using local materials. Additionally, the United States has not made the necessary financial commitment. Perhaps if NASA had another zero in its budget, it might succeed. If one-third of the defense budget were applied to space, we might get a human expedition to Mars soon, but once the team got there, it would be preoccupied with survival in an extremely hostile realm. Even if humans could visit Mars, they would have an exceptionally difficult time living there on a permanent basis and raising future generations. Any other destinations in the local solar system are pure science fiction.

Second, humans could stay home, dispatching robots and automated spacecraft to do the difficult work of exploration. There are good reasons to do so. Humans stuck on their earthly home need to know as much as they can about how their planet works. Comparative planetary studies offers an excellent approach to a more complete understanding of the models that allow complex life to persist on Earth, and the potential discovery of extraterrestrial life (even in simple form) would provide a much deeper understanding of earthly biology. These are good reasons for an ambitious effort at robotic studies. Robots of ever-increasing capability likely will fill the solar system, most of them on one-way trips.

Yet we must admit that we find this option disappointing. It avoids one of the principal reasons for space travel. Robots are not people. In the end, they become artifacts on other worlds, a technological reminder of a time when humans sent machines to explore. It disappoints us to think that humanity might forever be confined to a single world.

As this book has attempted to present, we think that a third option is worthy of contemplation. In some fashion, the first two options might merge. A part of humanity might move into the galaxy and remain. We are not sure what exact shape such travelers might take, but we believe that they would combine certain characteristics of humans and machines. Our current species, *Homo sapiens,* is ideally suited for the rigors of earthly surface exploration and migration. We are technologically creative and biologically prolific. Asking humans to engage in interplanetary migration, to use an analogy presented earlier in this book, is like asking a fish to live on land by constructing its own bubble of water. Better that the fish should breathe air and grow arms and legs and other more suitable mechanisms.

Homo sapiens is Latin for "wise man." Until recently, *Homo sapiens* have shared the Earth with other hominids. These fellow creatures were not as biologically well suited for multiplication on Earth as modern humans, and they consequently disappeared. They were not our biological ancestors but rather human-like beings that evolved alongside *Homo neanderthalensis* and ourselves. The 2003 finding that *Homo floresiensis,* a species remarkable for its hobbit-sized body, had survived until relatively recent times shocked the public and the scientific community alike. The best evidence suggests that it lived contemporaneously with modern humans on the Indonesian island of Flores as recently as thirteen thousand years ago.[64] Would those who demand space travel as a means for species preservation be satisfied with the migration of another type of being, a related species more adaptable to living conditions on other worlds? Perhaps it will be called *Homo cosmos*.

We do not believe that the development of such beings would occur as a result of the demands of space travel. As presented in the previous chapter, such alterations, to the extent that they occur at all, are likely to arise as a byproduct of efforts to tinker with human biology or artificial intelligence. In the long run, we think such alterations are inevitable, if for no other reason than the knowledge that other hominids existed before us. Given enough time, we find this to be a plausible option.

So exactly how much time do earthlings have to spread humanity in some form into the galaxy? From one point of view, humans have a very long time. To use the words of one rocket pioneer, the necessities of interstellar migration may not manifest themselves until the sun begins to grow cold.[65] According to conventional estimates, the sun will continue to burn for another five to six billion years, much longer than the entire history of complex life on the surface of the Earth and in its seas.

Some scientists think that the available time may be much shorter, marked not by the eventual dissipation of the sun but by its expansion. Peter Ward and Donald Brownlee believe that life on Earth is at its zenith. As the sun grows warmer, the Earth will become unsuitable for complex plant and animal life. This will occur, they predict, in a few hundred million years—about the same length of time that complex life has thrived on the planet. Still, that is a long time to move, although Ward and Brownlee doubt the feasibility of such travel.[66]

Alternatively, humans may have only a few hundred years left. Spaceflight requires advanced technologies and supercharged economies. For the breadth of their history, humans have existed at lesser levels. Preindustrial economies produce great art and stunning cathedrals, but they do not produce space travel. Any number of events could drive humanity back to economic levels where humans would be capable of existing on Earth but not of traveling to the stars. This could occur as a result of an environmental disaster, such as global climate change. A devastating military conflict could drive human civilizations back to the weaker economies that have characterized all but the last century or so.[67] So could widespread disease. Humans, possibly in greatly reduced numbers, would still inhabit the Earth, but they would not fly beyond. Alternatively, people might lose interest in human space travel, deciding instead to concentrate their attention on terrestrial goals. The current fascination with space travel is a cultural phenomenon not universally shared by all societies. Finally, humans could use up the resources necessary to sustain advanced economies. The hydrogen and oxygen that fuel modern rockets are in abundant supply, but the substantial amounts of electric power required to extract them from sources like seawater could grow expensive and rare.

Anticipating potential developments like these, an important bloc of the pro-space community believes that humanity has a finite period of time to colonize other worlds before conditions on Earth no longer sustain human migration.[68] Much of the appeal of the prospective merger of human and robotic capabilities lies in the notion that it could happen soon.

Like the broader realm of invention and discovery, the civil space program works best when it is pursuing the next big thing. It does not lend itself well to repetition. Due to advances in technology and the revelation of reality, the original visions that motivated space exploration are losing their appeal. Perhaps it is time to contemplate what lies beyond those dreams. The window of opportunity for new approaches may be very narrow, providing an opening that advanced civilizations encounter only briefly, at the zenith of their power.

Appendix

INADEQUATE WORDS: A NOTE ON TERMINOLOGY

Anyone seeking to discuss robotic spaceflight quickly enters a terminological morass. The terms traditionally used to describe human and machine flight benefit from exactness but unfortunately lack propriety. When extended into the future, their exactness disappears. Modern substitutes suffer similar deficiencies when confronted with future possibilities. Nothing works well. New terms will undoubtedly emerge, but until they do we are left with the often distressing task of describing unfamiliar events with familiar words, sometimes modified. We offer this note on terminology as a means of dispelling some of the ambiguities and misinterpretations that have inevitably crept in.

To the first people who pursued spaceflight, the distinction between types of activities seemed clear. A spacecraft with humans on board was *manned*. A satellite or spacecraft venturing into the extraterrestrial realm without any humans on board was simply *unmanned*. This pair of antonyms explained subtle distinctions. For example, a spacecraft capable of holding astronauts but fired into space in preparation for their eventual presence reverted to its unmanned form. The distinction was simple, exact, and entirely inappropriate. Not only did it commit an obvious gender impropriety, but such terminology defined automated objects with reference to what they are not—that is, not carrying human beings. While this fit the common but early presumption that serious spaceflight required human participation, it left little room to describe the various means by which spacecraft could be dispatched in alternative ways. Without humans, the spacecraft were simply unmanned, as if possessed by a deficiency.

Early use of the term *robot* was similarly straightforward. A robot was a machine under human control that possessed human capabilities, most often achieving them in ways analogous to human beings. Such robots typically possessed arms and grasping mechanisms, legs or wheels or some other form of locomotion, and a framework holding a number of sensory mechanisms, such as visual and communicative devices. The important early models in *Lost in Space* and *The Day the Earth Stood Still* established the paradigm: robots were machines that resembled people. And robots did what people did—or, in the case of *The Day the Earth Stood Still,* the robot's extraterrestrial commander. At least in their design, they were programmed to obey their creators and do the work that their creators found too tedious or dangerous to perform. They were, in short, machine counterparts of living beings.

At least in the beginning, the common conception of *robot* did not characterize all of the machinery that flew in space without an immediate human presence. Flight designers confined the term *robot* to that subset of devices capable of surface movement and sensory detection (rovers) and to specific appendages on various spacecraft (such as the Viking lander robotic arm). The rest, including nearly all orbiting satellites, fell into the larger and more general category of *unmanned vehicles*.

We have chosen to characterize all spacecraft capable of operating without humans on board as robotic. By this we mean that the spacecraft are capable of working on behalf of human beings, under human control, with many human capabilities—without humans physically present. To avoid repetition, we also characterize such objects as *automated,* since the machines can perform some of their functions without direct guidance from humans located at remote control stations. Our use of *robotic* is a broader characterization than the traditional term. It includes orbiting satellites without obvious arms or legs or mechanisms for surface mobility. Some persons refer to such objects as *proto-robotic,* meaning that they possess some but not all of the qualifying characteristics. Proto-robots may, for example, have imaging devices (eyes) but not grasping ones (hands). We have avoided this in favor of the simpler term *robotic,* as it allows us to characterize spacecraft without people on board under the general category of robotic spaceflight.

We admit that the resulting distinction between robotic and human spaceflight does not eliminate all ambiguities. In particular, the distinction begins to break down as one contemplates the future. That is the central thesis of this book—that given enough time, human and robotic characteristics tend to merge. Frustratingly, we found few commonly used words to describe the resulting products of change. An *android* is a robot that takes the human form; a *cyborg* is a human with machine parts. As part of the common definition, the machine parts on a cyborg must possess the robotic ability to sense and adjust to changes in their overall environment, using the identifiable logic of the feedback loop. Yet these terms have their deficiencies too.

What does one call a machine that takes on the characteristics of humanity, including the capacity for independent thought, but does not take the human form? In fact, such an entity may not be mechanical at all, having achieved an electronic presence or even abandoned its physical form through some as yet unknown process. One hesitates to use the word *robot,* with its underlying presumption of human control and machine parts. *Android* is not appropriate, given that the entity may not resemble a human being. The most commonly proffered term is *postbiological,* which refers to life forms that arise in an era in which living entities, including humans, are able to construct complex beings that are not wholly dependent upon flesh and blood and the biological processes of natural selection for their existence. For at least 500 million years, the Earth has been in its biological phase, producing life forms of ever increasing complexity through the process of biological evolution. Now the Earth may be entering a phase in which some complex life forms appear through alternative means. The term *postbiological* refers to the creations so produced, as well as the era in which they appear.

We find the term useful, but still misleading. Biological and postbiological entities might coexist in a postbiological world, at least for a while, yet the epoch would be defined by the latter. The term carries the unfortunate baggage also possessed by the manned-unmanned distinction, framing its characterizations of new life forms by reference to old ones. This explains what the postbiological world is not, rather than what it is. More important, the term suggests that all postbiological life forms arise from processes beyond biology, such as mechanics or electronics. This might not be true. Some of the transformations characterized as postbiological may arise through advances in biology, such as genetic engineering. In that case, what changes is not the reliance upon biology, but the motivating process of natural selection. We hate the thought that such a development might oblige us to characterize these alterations as *postnatural* or, worse still, as *creationist* in form, with all of the implications those terms carry. For now, we are content to accept the term postbiological, acknowledging its imperfections.

As humanity confronts the postbiological world, it may witness the development of a number of transitory forms. Again, current terms seem inadequate. The term *cybernetic organism* is helpful insofar as it describes humans acquiring advanced mechanical parts. Yet *cyborg* does not cover all possible modifications, including the aforementioned efforts to change the human form through genetic engineering. Other transitory forms may arise as the result of changes in the physical environment in which living beings mature (for example, the effects of being raised on a different planet). Would they be *postterrestrial*? At a different level, *artificial intelligence* is used to describe machines—most commonly computers—that begin to attain thought processes approximating those of human beings. Yet what is artificial about them? The term implies that such entities are unnatural or contrived, a characterization that might not last long with entities so empowered. With superior levels of intelligence, they might view themselves as entirely natural, a possibility Isaac Asimov explored in his earliest robot stories.

In the past few years, a group of people has begun to use the terms *posthuman* and *transhuman* to describe transformations of the human form. Such terms presume alterations significant enough to produce a species with its origins in but distinct from *Homo sapiens*. We use the two terms *posthuman* and *transhuman* interchangeably. If such a super-species ever arises, we suspect that it will want to rename itself.

Humans, robots, proto-robots, cyborgs, androids, postbiological life forms, artificial intelligence, posthuman, and *transhuman* are entering a discussion once confined to the conventional distinction between manned and unmanned spaceflight. Any full discussion of prospective developments in this regard will lead to a search for terms that more aptly describe the amazing possibilities involved.

Notes

INTRODUCTION: A FALSE DICHOTOMY

1. "NASA: Is It Worth It?" Massachusetts School of Law, October 3, 2000; and Roger D. Launius and Howard E. McCurdy, *Imagining Space: Achievements, Predictions, Possibilities, 1950–2050* (San Francisco: Chronicle Books, 2001).

2. See NASA News Release No. 03-227, "Humans, Robots Work Together to Test 'Spacewalk Squad' Concept," July 2, 2003, NASA Historical Reference Collection, NASA History Division, NASA Headquarters, Washington, D.C.

3. Arthur C. Clarke, "Extra Terrestrial Relays: Can Rocket Stations Give World Wide Radio Coverage?" *Wireless World* (October 1945): 306; Wernher von Braun, "Crossing the Last Frontier," *Collier's,* March 22, 1952, 72.

4. "Man Will Conquer Space Soon," *Collier's,* March 22, 1952, cover, 22–23.

5. Isaac Asimov, *I, Robot* (New York: Bantam Books, 1950), 9; Philip K. Dick, *Do Androids Dream of Electric Sheep?* (1968; repr. New York: Ballantine Books, 1996); and Stanley Kubrick, *2001: A Space Odyssey,* MGM, 1968.

6. Jim Oberg, "Space Explorers! The Evolution of Robotic Arms and Mobility in the Space Age," *Robot Magazine* (Winter 2006): 20–24.

7. Ibid., 20; see also Piers Bizony, *The Rivers of Mars: Searching for the Cosmic Origins of Life* (London: Aurum Press, 1997); Donna Shirley, *Managing Martians* (New York: Broadway Books, 1998); Steve Squyres, *Roving Mars: Spirit, Opportunity, and the Exploration of the Red Planet* (New York: Hyperion, 2005); and Peter Putz, "Space Robotics," *Reports on Progress in Physics* 65 (2002): 421–63.

8. Michael Mayor and Didier Queloz, "A Jupiter-Mass Companion to a Solar-Type Star," *Nature* 378 (1995): 355.

9. See Robert Roy Britt, "First Confirmed Picture of a Planet beyond the Solar System," April 1, 2005, available at www.space.com/scienceastronomy/050401_first_extrasolarplanet_pic.html, accessed September 28, 2005.

10. Olaf Stapledon, "Interplanetary Man?" in *An Olaf Stapledon Reader,* ed. Robert Crossley (Syracuse, NY: Syracuse University Press, 1997), 218–41; Olaf Stapledon, *Last and First Men: A Story of the Near and Far Future* (New York: J. Cape and H. Smith, 1931).

11. William Sims Bainbridge, *The Spaceflight Revolution: A Sociological Study* (New York: Wiley-Interscience, 1976).

12. "Man Will Conquer Space Soon," cover, 22–23; Chesley Bonestell and Willy Ley, *The Conquest of Space* (New York: Viking Press, 1949).

CHAPTER 1: THE HUMAN/ROBOT DEBATE

1. This was the central thesis of Howard E. McCurdy, *Space and the American Imagination* (Washington, D.C.: Smithsonian Institution Press, 1997).

2. NASA Facts, "President Delivers Remarks on U.S. Space Policy," January 14, 2004 (text of remarks); see also Space Task Group, "The Post-Apollo Space Program: Directions for the Future," September 1969, NASA Historical Reference Collection, History Division, NASA Headquarters, Washington, D.C.; and Thomas O. Paine, *Pioneering the Space Frontier: The Report of the National Commission on Space* (New York: Bantam Books, 1986).

3. James A. Van Allen, "Is Human Spaceflight Obsolete?" *Issues in Science and Technology* 20, no. 4 (Summer 2004), available at www.issues.org/20.4/p_van_allen.html, accessed September 9, 2004.

4. Laurence Bergreen, *Over the Edge of the World: Magellan's Terrifying Circumnavigation of the Globe* (New York: William Morrow, 2003).

5. James M. Beggs, "The Wilbur and Orville Wright Memorial Lecture," Royal Aeronautical Society, London, England, December 13, 1984, NASA Historical Reference Collection; see also Daniel Goldin, "Celebrating the Spirit of Columbus," *Phi Kappa Phi Journal* (Summer 1992): 8; and Frank Viviano, "China's Great Armada," *National Geographic* (July 2005), 28–53. For the controversial thesis that Chinese explorers reached the New World before Columbus, see Gavin Menzies, *1421: The Year China Discovered America* (New York: William Morrow, 2003).

6. Stephen J. Pyne, "Seeking Newer Worlds: An Historical Context for Space Exploration," in *Critical Issues in the History of Spaceflight,* ed. Steven J. Dick and Roger D. Launius (Washington, D.C.: NASA Special Publication (hereafter SP) 4702, 2006), 7–35.

7. John F. Kennedy, "Address at Rice University on the Nation's Space Effort," September 12, 1962, NASA Historical Reference Collection, also available at www.jfklibrary.org/j091262.htm, accessed July 6, 2005. The phrase is sometimes reproduced as "this new ocean." See Loyd S. Swenson Jr., James M. Grimwood, and Charles C. Alexander, *This New Ocean: A History of Project Mercury* (Washington, D.C.: NASA SP-4201, 1966).

8. See Frank H. Winter, *Prelude to the Space Age: The Rocket Societies, 1924–1940* (Washington, D.C.: Smithsonian Institution Press, 1983).

9. See Frederick I. Ordway III and Randy Liebermann, eds., *Blueprint for Space: Science Fiction to Science Fact* (Washington, D.C.: Smithsonian Institution Press, 1992); and McCurdy, *Space and the American Imagination.*

10. Joseph J. Corn, *The Winged Gospel: America's Romance with Aviation* (New York: Oxford University Press, 1983).

11. See Roger D. Launius and Jessie L. Embry, "The 1910 Los Angeles Airshow: The Beginnings of Air Awareness in the West," *Southern California Quarterly* 77 (Winter 1996): 329–46; and Roger D. Launius, "Planes, Trains, and Automobiles: Choosing Transportation Modes in the 20th-Century American West," *Journal of the West* 42 (Spring 2003): 45–55.

12. "An Airplane in Every Garage?" *Scribner's* (September 1935), 179–82; see also Tom D. Crouch, "An Airplane for Everyman: The Department of Commerce and the Light Airplane Industry, 1933–1937," in *Innovation and the Development of Flight,* ed. Roger D. Launius (College Station: Texas A&M University Press, 1999), 166–87.

13. Air Force Systems Command, "Chronology of Early Air Force Man-in-Space Activity, 1955–1960," AFSC Historical Publications Series 65–21–1, NASA Historical Reference Collection, 22; see also Mae M. Link, *Space Medicine in Project Mercury* (Washington, D.C.: NASA SP-4003, 1965), 25.

14. Michael J. Neufeld, "'Space Superiority': Wernher von Braun's Campaign for a Nuclear-Armed Space Station, 1946–1956," *Space Policy* 22 (February 2006): 52–62.

15. For example, see Robert S. Richardson, "Rocket Blitz from the Moon," *Collier's,* October 23, 1948, 24–25, 44–46. For the alternative and mostly forgotten point of view, see White House, *Introduction to Outer Space* (Washington, D.C.: Government Printing Office, 1958).

16. NASA Office of Program Planning and Evaluation, "Long Range Plan of the National Aeronautics and Space Administration," December 16, 1959, NASA Historical Reference Collection.

17. "Announcement of the First Satellite," *Pravda,* October 5, 1957; also in John M. Logsdon, gen. ed., *Exploring the Unknown: Selected Documents in the History of the U.S. Civil Space Program,* vol. 1 (Washington, D.C.: NASA SP-4218, 1995), 330.

18. Special Committee on Space Technology, "Recommendations to the NASA regarding a National Civil Space Program," October 28, 1958, NASA Historical Reference Collection.

19. Oran W. Nicks, *The Far Travelers: NASA's Exploring Machines* (Washington, D.C.: NASA SP-480, 1985), 88.

20. See Asif A. Siddiqi, *Challenge to Apollo: The Soviet Union and the Space Race, 1945–1974* (Washington, D.C.: NASA SP-4408, 2000).

21. "National Aeronautics and Space Act of 1958," Public Law No. 85–568, 72 Stat., 426. Signed by the President on July 29, 1958, Record Group 255, National Archives and Records Administration, Washington, D.C.

22. Arnold S. Levine, *Managing NASA in the Apollo Era* (Washington, D.C.: NASA SP-4102, 1982), 37.

23. Space Task Group, "Post-Apollo Space Program," ii.

24. Caspar W. Weinberger, Deputy Director, Office of Management and Budget, Memorandum for the President, Subject: Future of NASA, August 12, 1971, in *Exploring the Unknown,* ed. Logsdon, vol. 1, p. 547.

25. Space Task Group, "Post-Apollo Space Program," 5.

26. Ibid.; Office of the White House Press Secretary (San Clemente, California), The White House, Statement by the President, January 5, 1972, NASA Historical Reference Collection.

27. Traci Watson, "NASA Administrator Says Space Shuttle Was a Mistake," *USA Today,* September 27, 2005, 1.

28. James M. Beggs, "Why the United States Needs a Space Station," remarks prepared for delivery at the Detroit Economic Club and Detroit Engineering Society, June 23, 1982, NASA Historical Reference Collection.

29. George H. W. Bush, "Remarks on the Twentieth Anniversary of the Apollo 11 Moon Landing," July 20, 1989, George Bush Presidential Library, College Station, Texas.

30. NASA Facts, "President Bush Delivers Remarks on U.S. Space Policy."

31. See Carl Sagan, *Pale Blue Dot: A Vision of the Human Future in Space* (New York: Random House, 1994); and Robert Zubrin with Richard Wagner, *The Case for Mars: The Plan to Settle the Red Planet and Why We Must* (New York: Free Press, 1996).

32. Sylvia Hui, "Hawking Says Humans Must Go into Space," Associated Press, June 13, 2006; see also Bruno Maddox, "Hawking's Exit Strategy," *Discover* (September 2006), 31.

33. Roger D. Launius, "Compelling Rationales for Spaceflight? History and the Search for Relevance," in *Critical Issues in the History of Spaceflight,* ed. Dick and Launius, 37–70.

34. NASA, "The Vision for Space Exploration," February 2004, 3, available at www.nasa.gov/pdf/55583main_vision_space_exploration2.pdf, accessed May 26, 2005; The President's Science Advisory Committee, "Report of the Ad Hoc Panel on Man-in-Space," December 16, 1960, NASA Historical Reference Collection.

35. See Marilyn J. Landis, *Antarctica: Exploring the Extreme, 400 Years of Adventure* (Chicago: Chicago Review Press, 2001), 150.

36. Isaac Asimov, *Gold: The Final Science Fiction Collection* (New York: HarperCollins, 2003), 196.

37. Thomas Gold, "The Lunar Surface," *Monthly Notices of the Royal Astronomical Society* 115 (1955): 585; Donald A. Beattie, *Taking Science to the Moon: Lunar Experiments and the Apollo Program* (Baltimore: Johns Hopkins University Press, 2001), 17–18, 120–21; and Linda Neuman Ezell, compiler, *NASA Historical Data Book, Vol. II: Programs and Projects, 1958–1968* (Washington, D.C.: NASA SP-4012, 1988), 325–31.

38. Robert W. Smith, *The Space Telescope: A Study of NASA, Science, Technology, and Politics* (New York: Cambridge University Press, 1989), 143–220; Joseph N. Tatarewicz, "The Hubble Space Telescope Servicing Mission," in *From Engineering Science to Big Science: The NACA and NASA Collier Trophy Research Project Winners,* ed. Pamela E. Mack (Washington, D.C.: NASA SP-4219, 1998), 365–96; and Georges Meylan, Juan Madrid, and Duccio Macchetto, "Hubble Science Metrics," *Space Telescope Science Institute Newsletter* 20, no. 2 (2003): 1–8.

39. John Mankins, "Modular Architecture Options for Lunar Exploration and Development," *Space Technology* 21 (2001): 53–64; see also Marc M. Cohen, "Mobile Lunar Base Concepts," Advanced Projects Branch, NASA Ames Research Center, n.d.; and Leonard David, "NASA Goes Lunar: Robot Craft, Human Outpost Plans," March 3, 2004, available at www.space.com/businesstechnology/technology/moonbase_next_040303–1.html, accessed May 24, 2005.

40. Mark Wade, "Phobos Expedition 88," Encyclopedia Astronautica, n.d., available at www.astronautix.com/craft/phoion88.htm, accessed September 1, 2006; and Geoffrey A. Landis, "Robots and Humans: Synergy in Planetary Exploration," *Acta Astronautica* 55 (December 2004): 985–90.

41. National Security Council, NSC 5520, "Draft Statement of Policy on U.S. Scientific Satellite Program," May 20, 1955, Presidential Papers, Dwight D. Eisenhower Library, Abilene, Kansas; also in *Exploring the Unknown,* ed. Logsdon, vol. 1, pp. 308–12.

42. Nicks, *Far Travelers,* 245–46.

43. Lloyd V. Berkner to James E. Webb, March 31, 1961, and Berkner to Alan T. Watter-

man, March 31, 1961, with enclosure, "Man's Role in the National Space Program," both in NASA Historical Reference Collection; Allan A. Needell, *Science, Cold War, and the American State: Lloyd V. Berkner and the Balance of Professional Ideals* (Amsterdam, The Netherlands: Harwood Academic Publishers, 2000), 360–61.

44. National Academy of Science, "Man's Role in the National Space Program," press release NF6(100), August 7, 1961, NASA Historical Reference Collection.

45. Interview with John E. Naugle by Roger D. Launius, October 17, 1995.

46. This story is well told in Joseph N. Tatarewicz, *Space Technology and Planetary Astronomy* (Bloomington: Indiana University Press, 1990); see also Ronald E. Doel, *Solar System Astronomy in America: Communities, Patronage, and Interdisciplinary Science, 1920–1960* (New York: Cambridge University Press, 1996); and NASA Office of Space Science and Applications, "Program Review: Science and Applications Management," June 22, 1967, Space Science and Applications Files, NASA Historical Reference Collection.

47. Roger Launius witnessed this discussion on October 17, 1998, during a meeting about the upcoming flight of John Glenn on STS-95.

48. David S. Akens, *Historical Origins of the George C. Marshall Space Flight Center* (Washington, D.C.: NASA History Office, December 1963), 67–83, quoted in Henry C. Dethloff, *Suddenly Tomorrow Came . . . A History of the Johnson Space Center* (Houston: Lyndon B. Johnson Space Center, NASA SP-4307, 1993), 27.

49. Erasmus H. Kloman, *Unmanned Space Project Management: Surveyor and Lunar Orbiter* (Washington, D.C.: NASA SP-4901, 1972).

50. Nicks, *Far Travelers,* 91.

51. Ibid., 92.

52. Curtis Peebles, "The Manned Orbiting Laboratory," a three-part article in *Spaceflight* vol. 22 (April 1980): 155–60; vol. 22 (June 1980): 248–53; and vol. 24 (June 1982): 274–77.

53. See Robert S. Kraemer, *Beyond the Moon: Golden Age of Planetary Exploration, 1971–1978,* Smithsonian History of Aviation and Spaceflight Series (Washington, D.C.: Smithsonian Books, 2000).

54. James A. Van Allen, *New York Times,* April 1, 1986, A31, and "Space Science, Space Technology, and the Space Station," *Scientific American* (January 1986), 36. Van Allen made his case many times, beginning in 1961 with his opposition to a National Academy of Sciences study that acknowledged the role of humans in space and most recently in Van Allen, "Is Human Spaceflight Obsolete?" See also Needell, *Science, Cold War, and the American State,* 360; Homer E. Newell, *Beyond the Atmosphere: Early Years of Space Science* (Washington, D.C.: NASA SP-4211, 1980), 289; and *Astronautics and Aeronautics, 1970: Chronology of Science, Technology, and Policy* (Washington, D.C.: NASA SP-4015, 1972), 330.

55. Smith, *Space Telescope,* 393–95.

56. Peter J. T. Leonard and Christopher Wanjek, "Compton's Legacy: Highlights from the Gamma Ray Observatory," *Sky and Telescope* 100 (July 2000): 48–54; and Wallace H. Tucker and Karen Tucker, *Revealing the Universe: The Making of the Chandra X-Ray Observatory* (Cambridge, MA: Harvard University Press, 2001).

57. Personal correspondence with Bob Brugger, Johns Hopkins University Press, July 12, 2006.

58. With the advent of the Crew Exploration Vehicle, or Orion spacecraft, far-ranging

observatories such as the James Webb Space Telescope (designed to do its work from a Lagrange point, L-2, beyond the path of the Moon) will come within reach of astronaut repair crews, reigniting proposals for human repair.

59. Maxime A. Faget, Oral History Transcript, interviewed by Jim Slade, Houston, Texas, June 18–19, 1997, Johnson Space Center Oral History Project, 69, 82, 84, 89; Faget, Oral History 2 Transcript, interviewed by Carol Butler, Houston, Texas, August 19, 1998, Johnson Space Center Oral History Project, 21; and Marcus Lindross, "Industrial Space Facility: Rise and Fall of Commercially Developed Space Facility," Encyclopedia Astronautica website, May 31, 2005, available at astronautix.com/craft/indility.htm, accessed June 12, 2007.

60. Jon D. Miller, "Space Policy Leaders and Science Policy Leaders in the United States," 2004, 12–13, copy in possession of authors.

61. Newell, *Beyond the Atmosphere,* 290.

62. James E. Webb to Lee A. DuBridge, August 29, 1961, NASA Historical Reference Collection.

63. NASA, "Role of Man in Space," prepared by the Apollo Program Office, Office of Manned Space Flight, Washington, D.C., n.d., 1, NASA Historical Reference Collection.

64. Richard W. Orloff, *Apollo by the Numbers: A Statistical Reference,* October 2001, available at http://history.nasa.gov/SP-4029/Apollo_00_Welcome.htm, accessed June 2, 2005.

65. Siddiqi, *Challenge to Apollo,* 741; and NASA, Goddard Space Flight Center, "Soviet Lunar Missions," August 17, 2004, available at nssdc.gsfc.nasa.gov/planetary/lunar/lunarussr.html, accessed June 2, 2005.

66. NASA, Jet Propulsion Laboratory, Mars Exploration Rover Mission, Spirit Updates, sol 1145–1151, April 3, 2007, "Spirit Begins to Look for Best Access to 'Home Plate,'" available at http://marsrovers.nasa.gov/mission/status_spiritAll.html; and Opportunity Updates, sol 1126–1130, April 3, 2007, "Looking for an 'In,'" available at http://marsrovers.nasa.gov/mission/status_opportunityAll.html#sol1126, both accessed April 5, 2007.

67. Joel Achenbach, "Engineering: Who's Driving? Things Still Go Better with Humans at the Helm," *National Geographic* (November 2004); and 106 P.L. 398 [H.R. 4205] October 30, 2000, Floyd D. Spence National Defense Authorization Act for FY 2001, sec. 220(a).

68. Robert Zubrin, "Three Cheers for Robots—Until We Get to Mars Ourselves," *Washington Post,* January 11, 2004, B2.

69. On the basis of distance traveled, cost, and time, the human expeditions to the moon remained more cost-effective than the 2004–2007 Mars robotic rovers. Astronauts in the Apollo lunar excursion vehicles covered 91 kilometers over 9 days at a total program cost of approximately $150 billion in inflation-adjusted dollars. As of April 3, 2007, the Mars rovers *Spirit* and *Opportunity* had covered 17.4 kilometers over a period of roughly 39 months at a total program cost of approximately $820 million.

$$\text{Exploration Cost / Effectiveness} = \frac{\text{Distance}}{\text{Time} \times \text{Cost}}$$

70. Howard E. McCurdy, "The Cost of Space Flight," *Space Policy* 10, no. 4 (1994): 277–89. Total space station cost includes $10.2 billion for redesign studies, $14 billion for development, $13 billion for operations through fiscal year 2008, $6 billion for research and technology, and $18 billion (the congressional cap) for transportation.

71. See Watson, "NASA Administrator Says Space Shuttle Was a Mistake."

72. "Bush Goes on the Counterattack against Mars Mission Critics," *Congressional Quarterly Weekly Report,* June 23, 1990, p. 1958; see also Dwayne A. Day, "Whispers in the Echo Chamber," *Space Review,* March 22, 2004, available at www.thespacereview.com/article/119/1, accessed September 2, 2006.

73. See Howard E. McCurdy, *Faster, Better, Cheaper: Low-Cost Innovation in the U.S. Space Program* (Baltimore: Johns Hopkins University Press, 2001).

74. NASA Headquarters, News Conference, Mike Griffin, NASA Administrator; Dean Acosta, Moderator; Exploration Systems Architecture Study, transcript of news conference, September 19, 2005, p. 18.

75. NASA, "Human Exploration of Mars: The Reference Mission of the NASA Mars Exploration Study Team," ed. Stephen J. Hoffman and David L. Kaplan, July 1997.

76. Dennis R. Jenkins, *Space Shuttle: The History of the National Space Transportation System, The First 100 Missions* (Cape Canaveral, FL: D. R. Jenkins, 2001); and Nicks, *Far Travelers.*

77. See Launius and McCurdy, *Imagining Space,* 46–48.

78. See James R. Hansen, *First Man: The Life of Neil A. Armstrong* (New York: Simon and Schuster, 2005).

79. NASA, Mars Program Independent Assessment Team, "Summary Report," March 14, 2000; and NASA, JPL Special Review Board, "Report on the Loss of the Mars Polar Lander and Deep Space 2 Missions," March 22, 2000, both in NASA Historical Reference Collection.

80. John Copeland, *Alien Planet,* Discovery Channel, 2005.

81. See Dwayne A. Day, "Boldly Going: Star Trek and Spaceflight," *Space Review,* November 28, 2005, available at www.thespacereview.com/article/506/1, accessed September 2, 2006.

82. Commission to Assess United States National Security Space Management and Organization, *Report of the Commission,* Donald H. Rumsfeld, chair, January 11, 2001, 22, 32.

83. Joseph Seamans and Jason Spigarn-Koff, "The Great Robot Race," NOVA/WGBH Educational Foundation, 2006.

84. DARPA, Advanced Technology Office, "Tactical Mobile Robotics," n.d., available at http://darpa.mil/ato/programs/tmr.htm, accessed June 3, 2005; Thomas E. Ricks, "U.S. Arms Unmanned Aircraft," *Washington Post,* October 18, 2001; Anne Marie Squeo, "Meet the Newest Recruits: Robots," *Wall Street Journal,* December 13, 2001; and Michael Crichton, *Prey* (New York: HarperCollins, 2002).

85. See Guy Gugliotta, "The Animal Mind inside the Machine: Scientists Start to Fuse Tissue and Technology in Robots," *Washington Post,* April 17, 2001; and Peter Weiss, "Dances with Robots," *Science News,* June 30, 2001, available at http://sciencenews.org/articles/20010630/bob8.asp, accessed June 3, 2005.

86. James Cameron, *The Terminator,* MGM, 1984.

CHAPTER 2: HUMAN SPACEFLIGHT AS UTOPIA

1. Alexis de Tocqueville, *Democracy in America,* ed. J. P. Mayer and Max Lerner (New York: Vintage Books, 1966), 343. Parts of this chapter were previously published in Roger D.

Launius, "Perfect Worlds, Perfect Societies: The Persistent Goal of Utopia in Human Spaceflight," *Journal of the British Interplanetary Society* 56 (September/October 2003): 338–49.

2. See www.nara.gov/exhall/charters/declaration/declaration.html, accessed March 14, 2001.

3. Graham Nash, "Chicago," recorded by Crosby, Stills, Nash, and Young on *4 Way Street*, a live double album originally released by Atlantic Records on April 7, 1971. Also recorded by Graham Nash on *Songs for Beginners*, released by Atlantic Records on May 28, 1971.

4. While we do not wish to overstate this position, it has been affirmed repeatedly; it recently found expression by Nuala O'Faolain, author of *My Dream of You* (New York: Riverhead Books, 2001), in an interview on NPR's "Morning Edition" on March 14, 2001, available at www.npr.org/programs/morning/features/2001/mar, accessed June 12, 2007.

5. Beginning in the Apollo era, the NASA budget equaled approximately 3 percent of the U.S. federal budget. At the conclusion of Apollo, that budget percentage had declined to about 1 percent of the total. But this is a large amount for such a small constituency in the federal government. See *Aeronautics and Space Report of the President, Fiscal Year 1998* (Washington, D.C.: NASA Annual Report, 1999), appendixes E-1a and E-1b.

6. Van Allen, "Is Human Spaceflight Obsolete?"

7. Hans L. D. G. Starlife, "On to Mars," July 27, 2004, Quark Soup, available at http://davidappell.com/archives/00000202.htm, accessed August 3, 2004.

8. See www.informatics.org.museum/tsiol.html.

9. Arkady A. Kosmodemyansky, *Tsiolkovsky: His Life and Work* (Moscow, USSR: Nauka, 1956), 153; see also Konstantin E. Tsiolkovsky, *Aerodynamics* (Washington, D.C.: NASA, TT F-236, 1965); Konstantin E. Tsiolkovsky, *Reactive Flying Machines* (Washington, D.C.: NASA, TT F-237, 1965); Konstantin E. Tsiolkovsky, *Works on Rocket Technology* (Washington, D.C.: NASA, TT F-243, 1965); and Arkady A. Kosmodemyansky, *Konstantin Tsiolkovsky: 1857–1953* (Moscow, USSR: Nauka, 1985).

10. The quotations are from Esther C. Goddard, ed., and G. Edward Pendray, assoc. ed., *The Papers of Robert H. Goddard*, vol. 1 (New York: McGraw-Hill Book Co., 1970), 10, 63–66; see also David A. Clary, *Rocket Man: Robert H. Goddard and the Birth of the Space Age* (New York: Hyperion Books, 2003); and Milton Lehman, *This High Man: The Life of Robert H. Goddard* (New York: Farrar, Straus, 1963).

11. Robert H. Goddard, "The Last Migration," in *The Papers of Robert H. Goddard*, ed. Goddard and Pendray, vol. 3, 1611–12.

12. Sam Moskowitz, "The Growth of Science Fiction from 1900 to the Early 1950s," in *Blueprint for Space*, ed. Ordway and Liebermann, 69–82; Eric Burgess, "Into Space," *Aeronautics* (November 1946), 52–57; and Paul A. Carter, *Politics, Religion, and Rockets: Essays in Twentieth-Century American History* (Tucson: University of Arizona Press, 1991), 143–95.

13. Arthur C. Clarke, *Childhood's End* (New York: Ballantine Books, 1953).

14. *Brown Daily Herald*, March 24, 1995; "Quotations from Ray Bradbury," available at www.brookingsbook.com/bradbury/quotations.htm, accessed March 23, 2001.

15. Speeches at Brown University and to National School Board Association, 1995, "Quotations from Ray Bradbury."

16. Ray Bradbury, *The Martian Chronicles* (Garden City, NY: Doubleday, 1950).

17. There are many studies of this dynamic in human history. Excellent examples in-

clude Norman Cohn, *The Pursuit of the Millennium: Revolutionary Millenarians and Mystical Anarchists of the Middle Ages* (1961; repr. New York: Oxford University Press, 1990); Paul Boyer, *When Time Shall Be No More: Prophecy Belief in Modern American Culture* (1992; repr. Cambridge, MA: Belknap Press of Harvard University, 1994); Norman Cohn, *Cosmos, Chaos, and the World to Come* (1993; repr. New Haven, CT: Yale University Press, 2001); Mark Holloway, *Utopian Communities in America, 1680–1880* (Mineola, NY: Dover Publications, 1966); Donald E. Pitzer, *America's Communal Utopias* (Chapel Hill: University of North Carolina Press, 1997); Rosabeth Moss Kanter, *Commitment and Community: Communes and Utopias in Sociological Perspective* (Cambridge, MA: Harvard University Press, 1972); Gerald Lee Gutek et al., *Visiting Utopian Communities: A Guide to the Shakers, Moravians, and Others* (Columbia: University of South Carolina Press, 1998); Floyd Duncan, *The Utopian Prince: Robert Owen and the Search for the Millennium* (New York: Xlibris Corporation, 2003); and Dale Schrag and James C. Juhnke, eds., *Anabaptist Visions for the New Millennium: A Search for Identity* (Toronto: Pandora Press, 2000).

18. Robert Wise, *The Day the Earth Stood Still,* Twentieth Century Fox, 1951. Quotation from Sergio Leemann, *Robert Wise on His Films: From Editing Room to Director's Chair* (Los Angeles: Silman-James Press, 1995), 107.

19. Wernher von Braun, "The Challenge of the Century," April 3, 1965, Wernher von Braun Biographical File, NASA Historical Reference Collection, NASA History Division, NASA Headquarters, Washington, D.C.

20. Von Braun, "Crossing the Last Frontier," 24–29, 72–73; see also Ernst Stuhlinger and Frederick I. Ordway III, *Wernher von Braun, Crusader for Space: A Biographical Memoir* (Malabar, FL: Robert E. Krieger Company, 1994).

21. Wernher von Braun, "For Space Buffs—National Space Institute, You Can Join," *Popular Science* (May 1976), 73.

22. Roger D. Launius, "The Historical Dimension of Space Exploration: Reflections and Possibilities," *Space Policy* 16 (2000): 23–38, esp. 25.

23. On this subject, see Michael Burleigh and Wolfgang Wippermann, *The Racial State: Germany, 1933–1945* (Cambridge, UK: Cambridge University Press, 1991); George L. Mosse, *Toward the Final Solution: A History of European Racism* (Madison: University of Wisconsin Press, 1985); Paul Weindling, *Health, Race, and German Politics between National Unification and Nazism, 1870–1945* (New York: Cambridge University Press, 1993); Stefan Kühl, *The Nazi Connection: Eugenics, American Racism, and German National Socialism* (New York: Oxford University Press, 1994); Henry Friedlander, *The Origins of Nazi Genocide: From Euthanasia to the Final Solution* (Chapel Hill: University of North Carolina Press, 1995); and James C. Scott, *Seeing Like a State: How Certain Schemes to Improve the Human Condition Have Failed* (New Haven, CT: Yale University Press, 1998), 11–52. See also Nicholas Goodrick-Clarke, *The Occult Roots of Nazism: Secret Aryan Cults and Their Influence on Nazi Ideology* (New York: New York University Press, 1994).

24. See Michael J. Neufeld, *The Rocket and the Reich: Peenemünde and the Coming of the Ballistic Missile Era* (Cambridge, MA: Harvard University Press, 1996).

25. L. Arthur Minnich Jr., "Legislative Meeting, Supplementary Notes," February 4, 1958, President's Personal Files (PPF), Ann Whitman, DDE Diary Series, Box 30, February 1958 Staff Notes, Eisenhower Library, Abilene, Kansas.

26. William J. Laurence, "Two Rocket Experts Argue 'Moon Plan,'" *New York Times,* October 14, 1952; and Robert C. Boardman, "Space Rockets with Floating Base Predicted," *New York Herald Tribune,* October 14, 1952.

27. "Journey into Space," *Time,* December 8, 1952, 62–73, quotation 62.

28. Mike Wright, "The Disney–von Braun Collaboration and Its Influence on Space Exploration," in *Inner Space, Outer Space: Humanities, Technology, and the Postmodern World,* ed. Daniel Schenker, Craig Hanks, and Susan Kray (Huntsville, AL: Southern Humanities Conference, 1993), 151–60; and Randy Liebermann, "The *Collier's* and Disney Series," in *Blueprint for Space,* ed. Ordway and Liebermann, 135–44.

29. *TV Guide,* March 5, 1955, 9.

30. See "Minutes of Meeting of Research Steering Committee on Manned Space Flight," May 25–26, 1959, 2, NASA Historical Reference Collection.

31. "Journey into Space," 72; see also "What Are We Waiting For?" *Collier's,* March 22, 1952, 23.

32. "Pre-Press Briefing," April 16, 1958, PPF, Ann Whitman, DDE Diary Series, Box 32, "April 1958 Staff Notes (2)," Eisenhower Library. This is just about as serious a negative comment regarding an individual as one can find in Eisenhower's papers. See also Col. A. J. Goodpaster, "Memorandum for the Record," January 20, 1956, White House Office of Staff Secretary, Box 18, "Missiles (January 1956–January 1960(1))," Eisenhower Library; and Brig. Gen. A. J. Goodpaster, "Legislative Leadership Meeting, Supplementary Notes," March 18, 1958, PPF, Ann Whitman, DDE Diary Series, Box 32, "April 1958 Staff Notes (2)," Eisenhower Library.

33. Roger D. Launius, "Eisenhower, Sputnik, and the Creation of NASA: Technological Elites and the Public Policy Agenda," *Prologue: Quarterly of the National Archives and Records Administration* 28 (Summer 1996): 127–43.

34. Dwight Eisenhower, letter to Soviet Premier Nikolai Bulganin, January 12, 1958, quoted in Swenson et al., *This New Ocean,* 82.

35. A skeptical take on Eisenhower's leadership in space policy is Glen P. Wilson, "Lyndon Johnson and the Legislative Origins of NASA," *Prologue: Quarterly of the National Archives and Records Administration* 25 (Winter 1993): 362–73, which grants Johnson credit for taking action to recover from the crisis caused by the launching of the Soviet satellite.

36. "National Aeronautics and Space Act of 1958," Public Law No. 85–568, 72 Stat., 426. Signed by the president on July 29, 1958, Record Group 255, National Archives and Records Administration, Washington, D.C.; see also Alison Griffith, *The National Aeronautics and Space Act: A Study of the Development of Public Policy* (Washington, D.C.: Public Affairs Press, 1962), 27–43.

37. Clayton L. Koppes, *JPL and the American Space Program: A History of the Jet Propulsion Laboratory* (New Haven, CT: Yale University Press, 1982), 94; and J. D. Hunley, ed., *The Birth of NASA: The Diary of T. Keith Glennan* (Washington, D.C.: NASA SP-4105, 1993), 181; see also 82, 98, 160.

38. John M. Logsdon, *The Decision to Go to the Moon: Project Apollo and the National Interest* (Cambridge, MA: MIT Press, 1970), 130.

39. *Aeronautics and Space Report of the President, Fiscal Year 1998* (Washington, D.C.: NASA Annual Report, 1999), appendixes E-1a and E-1b.

40. Quoted in *Congressional Quarterly,* July 25, 1969, p. 1311, and *The Futurist* (October 1969), 123.

41. Richard M. Nixon to Lee DuBridge, February 8, 1969; Nixon to Spiro Agnew et al., February 13, 1969, David Papers, OST, SMOF, WHCF, NPM, National Archives.

42. Logsdon, ed., *Exploring the Unknown,* vol. 1, 522.

43. Space Task Group, "Post-Apollo Space Program"; *New York Times,* September 16, 1969, 1; D. E. Crabill, Bureau of the Budget, to Director, Bureau of the Budget, "President's Task Group on Space—Meeting No. 2," March 14, 1969, series 69.1, box 51–78–31, National Archives and Records Administration, Washington, D.C.; Clay T. Whitehead, Staff Assistant, White House, to Peter M. Flanigan, White House, June 25, 1969, series 69.1, box 51–78–31, National Archives; Robert P. Mayo, Director, Bureau of the Budget, to President Richard M. Nixon, "Space Task Group Report," September 25, 1969, series 69.1, box 51–78–31, National Archives; and John Ehrlichman, *Witness to Power: The Nixon Years* (New York: Pocket Books, 1982), 123–24.

44. White House Press Secretary, "The White House, Statement by the President," March 7, 1970, Presidential Files, NASA Historical Reference Collection; see also T. A. Heppenheimer, *The Space Shuttle Decision: NASA's Search for a Reusable Space Vehicle* (Washington, D.C.: NASA SP-4221, 1999); Roger D. Launius, "NASA and the Decision to Build the Space Shuttle, 1969–72," *The Historian* 57 (Autumn 1994): 17–34; John M. Logsdon, "The Decision to Develop the Space Shuttle," *Space Policy* 2 (May 1986): 103–19; and "The Space Shuttle Program: A Policy Failure?" *Science,* May 30, 1986, pp. 1099–1105; George M. Low, NASA Deputy Administrator, "Meeting with the President on January 5, 1972," January 12, 1972, NASA Historical Reference Collection; and Claude E. Barfield, "Technology Report/Intense Debate, Cost Cutting Precede White House Decision to Back Shuttle," *National Journal,* August 12, 1972, pp. 1289–99.

45. On this effort, see Michael A. G. Michaud, *Reaching for the High Frontier: The American Pro-Space Movement, 1972–84* (New York: Praeger, 1986).

46. Two interesting cogitations on the weirdness of some elements in the pro-space movement are Jodi Dean, *Aliens in America: Conspiracy Cultures from Outerspace to Cyberspace* (Ithaca, NY: Cornell University Press, 1998); and Constance Penley, *NASA/Trek: Popular Science and Sex in America* (New York: Verso, 1997).

47. Perhaps nothing expresses this theme more consistently than the LaRouchite periodicals *Fusion* (1977–1987) and *Twenty-first Century Science and Technology* (1988-present). These works were "dedicated to the promotion of unending scientific progress, all directed to serve the proper common aims of mankind."

48. Founding minutes quoted in Edward S. Cornish, "A Quest for the Meaning of Life [review of *The Hunger of Eve*]," *The Futurist* (December 1976), 336–40.

49. Barbara Marx Hubbard, *The Hunger of Eve* (Harrisburg, PA: Stackpole Books, 1976), 144, 150.

50. Hubbard even produced what is known as a "SYNCON wheel." See William Sims Bainbridge, *The Space Flight Revolution* (New York: John Wiley and Sons, 1976), 173.

51. Michaud, *Reaching for the High Frontier,* 43–45.

52. Barbara Marx Hubbard, "The Future—Previews of Coming Attractions," *First Foundation News* 1 (August 1995): 1–4.

53. Gerard K. O'Neill, "The Colonization of Space," *Physics Today* 27 (September 1974): 32–40, and *The High Frontier: Human Colonies in Space* (New York: William Morrow, 1976); and Peter E. Glaser, "Power from the Sun—Its Future," *Science,* November 22, 1968, pp. 857–61, "Solar Power via Satellite," *Astronautics and Aeronautics* (August 1973): 60–68, and "An Orbiting Solar Power Station," *Sky and Telescope* (April 1975), 224–28.

54. Michaud, *Reaching for the High Frontier,* 57–102.

55. Arthur C. Clarke, *Rendezvous with Rama* (New York: Bantam Books, 1973).

56. T. A. Heppenheimer, *Colonies in Space* (Harrisburg, PA: Stackpole Books, 1977), 279–80.

57. This would be completely consistent with their ideology. See Roger D. Launius, "A Western Mormon in Washington, D.C.: James C. Fletcher, NASA, and the Final Frontier," *Pacific Historical Review* 64 (May 1995): 217–41; and "Colonies in Space," *Newsweek,* November 27, 1978, 95–101.

58. Richard D. Johnson and Charles Holbrow, eds., *Space Settlements: A Design Study in Colonization* (Washington, D.C.: NASA SP-413, 1977), a study sponsored by NASA Ames, ASEE, and Stanford University in the summer of 1975 to look at all aspects of sustained life in space. See also John Billingham, William Gilbreath, Gerard K. O'Neill, and Brian O'Leary, eds., *Space Resources and Space Settlements* (Washington, D.C.: NASA SP-428, 1979).

59. With the Apollo program gone and the Space Shuttle not yet flying, the late 1970s might best be viewed as a nadir in human space exploration. See Louis J. Halle, "A Hopeful Future for Mankind," *Foreign Affairs* 59 (Summer 1980): 1129–36.

60. Freeman Dyson, "Obituary, Gerard Kitchen O'Neill," *Physics Today* 46 (February 1993): 97–98.

61. See Launius and McCurdy, *Imagining Space.*

62. National Space Society vision statement, available at www.nss.org, accessed April 11, 2001.

63. Carl Sagan, *The Cosmic Connection: An Extraterrestrial Perspective* (Garden City, NY: Anchor Press, 1973). See also the Planetary Society website at www.planetary.org, accessed April 11, 2001.

64. Space Frontier Foundation website, available at www.space-frontier.org/whoweare.html, accessed March 25, 2007.

65. "Welcome to the Revolution," Message 1 of the Frontier Files, 1995, available at www.space-frontier.org/History/frontierfiles.html, accessed April 11, 2001.

66. Rick N. Tumlinson, "Why Space? Personal Freedom," Message 6 of the Frontier Files, 1995, available at www.space-frontier.org/History/frontierfiles.html, accessed April 11, 2001.

67. Rick N. Tumlinson, "The Foundation Credo—Our View of the Frontier," Part 4 of 4, Frontier Files, 1995, available at www.space-frontier.org/History/frontierfiles.html, accessed April 11, 2001.

68. Robert Zubrin and Richard Wagner, *The Case for Mars: The Plan to Settle the Red Planet and Why* (New York: The Free Press, 1996), 297. See also Robert M. Zubrin, "Colonizing the Outer Solar System," in *Islands in the Sky: Bold New Ideas for Colonizing Space,* ed. Stanley Schmidt and Zubrin (New York: John Wiley and Sons, 1996), 85–94.

69. Robert A. Heinlein, "From the Notebooks of Lazarus Long," *Time Enough for Love* (New York: Putnam, 1973).

70. Robert Zubrin, "The Mars Direct Plan," *Scientific American* (March 2000), 52–55.

71. David S. F. Portree, *Humans to Mars: Fifty Years of Mission Planning, 1950–2000,* Monograph in Aerospace History No. 21 (Washington, D.C.: NASA SP-4521, 2000).

72. Frederick Jackson Turner, "The Significance of the Frontier in American History," *The Frontier in American History* (New York: Holt, Rinehart, and Winston, 1920), 1–38.

73. See Richard Slotkin, *Gunfighter Nation: The Myth of the Frontier in Twentieth-Century America* (New York: Atheneum, 1992).

74. John Glenn Jr., "The Next 25: Agenda for the U.S.," *IEEE Spectrum* (September 1983), 91.

75. James A. Michener, "Looking toward Space," *Omni* (May 1980), 57–58, 121, quotation 58, and "Manifest Destiny," *Omni* (April 1981), 48–50, 102–4.

76. James C. Fletcher, "Our Space Program Is Already Back on Track," *USA Today,* July 28, 1987.

77. National Commission on Space, *Pioneering the Space Frontier* (New York: Bantam Books, 1986), 2.

78. Howard R. Lamar, "Westering in the Twenty-First Century," in *Under an Open Sky: Rethinking America's Western Past,* ed. William Cronon, George Miles, and Jay Gitlin (New York: W.W. Norton and Co., 1992), 257–74.

79. Ellen von Nardoff, "The American Frontier as a Safety Valve: The Life, Death, Reincarnation, and Justification of a Theory," *Agricultural History* 36 (July 1962): 123–42.

80. Patricia Nelson Limerick, "The Final Frontier?" *Wilson Quarterly* 14 (Summer 1990): 83. See also Ray A. Williamson, "Outer Space as Frontier: Lessons for Today," *Western Folklore* 46 (October 1987): 255–67; and Claire R. Farrer, "On Parables, Questions, and Predictions," *Western Folklore* 46 (October 1987): 281–93. For limitations of analogy in historical study, see Bruce Mazlish, ed., *The Railroad and the Space Program: An Exploration in Historical Analogy* (Cambridge, MA: MIT Press, 1965); and Richard E. Neustadt and Ernest R. May, *Thinking in Time: The Uses of History for Decision-Makers* (New York: Free Press, 1986).

81. Bruce Mazlish assembled a team and explored the similarities and differences between the two endeavors. See Mazlish, ed., *The Railroad and the Space Program*; and Robert W. Fogel, *Railroads and American Economic Growth: Essays in Econometric History* (Baltimore: Johns Hopkins Press, 1964). The use of federal patronage and power to do things for the West is an old story. See discussions of it in Howard R. Lamar, *The Far Southwest, 1846–1912: A Territorial History* (New Haven, CT: Yale University Press, 1966); Donald Worster, *Rivers of Empire: Water, Aridity, and the Growth of the American West* (New York: Pantheon Books, 1985); and Gerald D. Nash, *The American West Transformed: The Impact of the Second World War* (Bloomington: Indiana University Press, 1985).

82. See John Mack Faragher, *Rereading Frederick Jackson Turner: The Significance of the Frontier in American History, and Other Essays* (New York: Henry Holt, 1994); Allan G. Bogue, *Frederick Jackson Turner: Strange Roads Going Down* (Norman: University of Oklahoma Press, 1998); and Ray Allen Billington, *America's Frontier Heritage* (Albuquerque: University of New Mexico Press, 1974).

83. John W. Young to Steve Hawley et al., "Why the Human Exploration of the Moon and Mars Must Be Accelerated," March 9, 2001, NASA Historical Reference Collection.

84. Rob South, "Apollo's Goal," on Project Apollo e-mail discussion list, August 4, 2004, available at http://groups.yahoo.com/group/ProjectApollo/, accessed June 9, 2007.

85. Randrepo, "Apollo's Goal," on Project Apollo e-mail discussion list, August 4, 2004, available at http://groups.yahoo.com/group/ProjectApollo/, accessed June 9, 2007.

86. Van Allen, "Is Human Spaceflight Obsolete?"

87. Robert L. Park, "The Dark Side of the Moon," *New York Times,* September 22, 2005. See also Robert L. Park, *Voodoo Science: The Road from Foolishness to Fraud* (New York: Oxford University Press, 2000) and his website at http://bobpark.physics.umd.edu/.

CHAPTER 3: PROMOTING THE HUMAN DIMENSION

1. See Erik Bergaust, *Wernher von Braun* (Washington, D.C.: National Space Institute, 1976); and Stuhlinger and Ordway, *Wernher von Braun.* Despite these two biographies, there is a crying need for a new study of the life and career of this towering figure in the history of spaceflight. A new biography fulfills this need: Michael J. Neufeld's *Von Braun: Dreamer of Space, Engineer of War* (New York: Alfred A. Knopf, 2007). The *Collier's* series of articles were conveniently reprinted in Cornelius Ryan, ed., *Across the Space Frontier* (New York: Viking Press, 1952), and *Conquest of the Moon* (New York: Viking Press, 1953). The three Disney programs have recently been released in DVD format as *Tomorrow Land: Disney in Space and Beyond,* Buena Vista Home Entertainment, 2004.

2. A 1949 Gallup poll found more Americans confident of impending cancer cures and atom-powered railroads and planes than space travel. Only 15 percent of the people polled believed that humans would reach the moon within the next half-century. Americans were equally unconvinced of the wisdom of lavishing government funds on human spaceflight. Fifty-eight percent of the respondents to a 1960 Gallup poll agreed that the United States should not spend billions of dollars to send humans to the Moon. George Gallup, *The Gallup Poll: Public Opinion, 1935–1971* (New York: Random House, 1972), vol. 2, 875; vol. 3, 1720.

3. A fuller discussion of this dimension appears in the next chapter. See also Howard E. McCurdy, *Space and the American Imagination*; Sam Moskowitz, *Science Fiction by Gaslight: A History and Anthology of Science Fiction in Popular Magazines, 1891–1911* (Westport, CT: Hyperion Press, 1968); Paul A. Carter, *The Creation of Tomorrow: Fifty Years of Magazine Science Fiction* (New York: Columbia University Press, 1977); and Ordway and Liebermann, eds., *Blueprint for Space.*

4. On the issue of Big Science, see Peter Galison and Bruce Hevly, eds., *Big Science: The Growth of Large-Scale Research* (Stanford, CA: Stanford University Press, 1992); and Needell, *Science, Cold War, and the American State.*

5. Dael Wolfe, "The Administration of NASA," *Science,* November 15, 1968, p. 753. This is also the thesis of Stephen B. Johnson, *The Secret of Apollo: Systems Management in American and European Space Programs* (Baltimore: Johns Hopkins University Press, 2002).

6. NASA Administrator James E. Webb celebrated the Apollo contribution to "Big Science" in *Space Age Management: The Large-Scale Approach* (New York: McGraw-Hill, 1969). The demise of this approach can be seen in Howard E. McCurdy, *Inside NASA: High Technology and Organizational Change in the U.S. Space Program* (Baltimore: Johns Hopkins University Press, 1993).

7. Quoted in Michael J. Neufeld, "The End of the Army Space Program: Interservice Ri-

valry and the Transfer of the von Braun Group to NASA, 1958–1959," *Journal of Military History* 69 (2005): 753.

8. Dwayne A. Day, "The Von Braun Paradigm," *Space Times: Magazine of the American Astronautical Society* 33 (November-December 1994): 12–15; "Man Will Conquer Space *Soon*," *Collier's*, March 22, 1952, 23–76ff.; and Wernher von Braun with Cornelius Ryan, "Can We Get to Mars?" *Collier's*, April 30, 1954, 22–28.

9. William L. Laurence, "Two Rocket Experts Argue 'Moon Plan.'"

10. James A. Van Allen, *The Origins of Magnetospheric Physics* (Washington, D.C.: Smithsonian Institution Press, 1983).

11. This was a very common idea at the time. As late as 1963 the Jet Propulsion Laboratory published a work that noted: "It was reasoned that if the oceans of Venus still exist, then the Venusian clouds may be composed of water droplets; if Venus were covered by water, it was suggested that it might be inhabited by Venusian equivalents of Earth's Cambrian period of 500 million years ago, and the same steamy atmosphere could be a possibility." See Jet Propulsion Laboratory, *Mariner: Mission to Venus* (New York: McGraw-Hill, 1963), 5. This also became an enormously popular conception in science fiction literature. See the 1949 short story by Arthur C. Clarke, "History Lesson," in *Expedition to Earth* (New York: Ballantine Books, 1953), 73–82, for an explanation of the theory.

12. Swenson et al., *This New Ocean*, 111; John M. Logsdon, *Decision to Go to the Moon: Project Apollo and the National Interest* (Chicago: University of Chicago Press, 1976); and James Webb, "Administrator's Presentation to the President," March 22, 1961, NASA Historical Reference Collection, NASA History Division, Washington, D.C. Before leaving office, Eisenhower did approve the "super booster" program space advocates desired, thereby transferring the von Braun group to NASA. Courtney G. Brooks, James M. Grimwood, and Loyd S. Swenson, *Chariots for Apollo: A History of Manned Lunar Spacecraft* (Washington, D.C.: NASA SP-4205, 1979).

13. Dwight D. Eisenhower, "Are We Headed in the Wrong Direction?" *Saturday Evening Post*, August 11, 1962, 24; Dwight D. Eisenhower, "Spending into Trouble," *Saturday Evening Post*, May 18, 1963, 19; President's Science Advisory Committee, "Report of the Ad Hoc Panel on Man-in-Space," November 14, 1960, 5, NASA Historical Reference Collection; and The White House, *Introduction to Outer Space* (Washington, D.C.: Government Printing Office, 1958), 10.

14. "What Are We Waiting For?" *Collier's*, March 22, 1952, 23; and Statement of Democratic Leader Lyndon B. Johnson to the Meeting of the Democratic Conference on January 7, 1958, 3, statements of LBJ collection, box 23, Lyndon Baines Johnson Library, Austin, Texas. See also McCurdy, *Space and the American Imagination*.

15. President's Science Advisory Committee, *Introduction to Outer Space*, 12. See also Neufeld, "'Space Superiority.'"

16. James Van Allen, "Space Station and Manned Flights Raise NASA Program Balance Issues," *Aviation Week and Space Technology*, January 25, 1988, 153.

17. See John Lear, "Hiroshima, U.S.A.: Can Anything Be Done about It?" *Collier's*, August 5, 1950, 11–15.

18. See, for example, Robert S. Richardson, "Rocket Blitz from the Moon," *Collier's*, October 23, 1948, 24–25, 44–46.

19. Edward U. Condon, *Scientific Study of Unidentified Flying Objects* (New York: E. P. Dutton, 1969); and Donald A. Keyhoe, "The Flying Saucers Are Real," *True* (January 1950), 11–13, 83–87. See also Curtis Peebles, *Watch the Skies! A Chronicle of the Flying Saucer Myth* (Washington, D.C.: Smithsonian Institution Press, 1984).

20. There is a huge literature on the ratification of the Constitution, including discussion of this political debate. See Herbert J. Storing, *What the Anti-Federalists Were For* (Chicago: University of Chicago Press, 1981), and *The Anti-Federalist* (Chicago: University of Chicago Press, 1985); and Jackson Turner Main, *The Antifederalists* (New York: W.W. Norton and Co., 1974). For a classic discussion of the role of issue definition in political affairs, see E. E. Schattschneider, *The Semisovereign People: A Realist's View of Democracy in America* (New York: Holt, Rinehart, and Winston, 1960). See also David A. Rochefort, *The Politics of Problem Definition: Shaping the Policy Agenda* (Lawrence: University Press of Kansas, 1994). The theory of problem definition has been applied to space policy in C. J. Bosso and W. D. Kay, "Advocacy Coalitions and Space Policy," in *Space Politics and Policy: An Evolutionary Perspective,* ed. Eligar Sadeh (Dordrecht, The Netherlands: Kluwer Academic Publishers, 2002), ch. 3.

21. Van Allen, "Space Science," 32.

22. John H. Gibbons, "The New Frontier: Space Science and Technology in the Next Millennium," Wernher von Braun Lecture, March 22, 1995, National Air and Space Museum, Smithsonian Institution, Washington, D.C., available at http://clinton1.nara.gov/White_House/EOP/OSTP/other/space.html, accessed August 1, 2004.

23. The story of one central character in this story is told in Roger D. Launius, "Godfather to the Astronauts: Robert Gilruth and the Birth of Human Spaceflight," in *Realizing the Dream of Flight: Biographical Essays in Honor of the Centennial of Flight, 1903–2003,* ed. Virginia P. Dawson and Mark D. Bowles (Washington, D.C.: NASA SP-4112, 2005), 213–56.

24. Swenson et al., *This New Ocean,* ch. 4.

25. Dwight D. Eisenhower, "Farewell Radio and Television Address to the American People," January 17, 1961, Dwight D. Eisenhower Library, available at www.eisenhower.archives.gov/farewell.htm, accessed July 5, 2005.

26. David Halberstam, *The Best and the Brightest* (New York: Viking, 1973), 57, 153.

27. Ralph E. Lapp, *The New Priesthood: The Scientific Elite and the Uses of Power* (New York: Harper and Row, 1965), 227–28. Similar cautions, but aimed at the use of science and technology to dupe Americans, may be found in Park, *Voodoo Science*; and Amitai Etzioni, *The Limits of Privacy* (New York: Basic Books, 2000).

28. Roger D. Launius, "Eisenhower and Space: Politics and Ideology in the Construction of the U.S. Civil Space Program," in *Forging the Shield: Eisenhower and National Security in the Twenty-First Century,* ed. Dennis E. Showalter (Chicago: Imprint Publications, 2005), 151–82.

29. Robert Dallek, "Johnson, Project Apollo, and the Politics of Space Program Planning," in *Spaceflight and the Myth of Presidential Leadership,* ed. Roger D. Launius and Howard E. McCurdy (Urbana: University of Illinois Press, 1997), 68–91.

30. Linda Neuman Ezell, *NASA Historical Data Book, Vol II: Programs and Projects, 1958–1968* (Washington, D.C.: NASA SP-4012, 1988), 122–23.

31. John Law, "Technology and Heterogeneous Engineering: The Case of Portuguese Expansion," 111–34, and Donald MacKenzie, "Missile Accuracy: A Case Study in the Social

Processes of Technological Change," 195–222, both in *The Social Construction of Technological Systems: New Directions in the Sociology and History of Technology,* ed. Wiebe E. Bijker, Thomas P. Hughes, and Trevor J. Pinch (Cambridge, MA: MIT Press, 1987).

32. There are many examples of this. See Jules Verne, *De la terre a la lune* [From the Earth to the Moon] (Paris: J. Hetzel, 1866); H. G. Wells, *The First Men in the Moon* (London: George Newness, 1901); Robert A. Heinlein, *Rocket Ship Galileo* (New York: Scribner, 1947); and Irving Pichel, *Destination Moon,* Eagle-Lion Films, 1950. General studies of science fiction can be found in Brian Ash, ed., *The Visual Encyclopedia of Science Fiction* (New York: Harmony Books, 1977); James Gunn, *Alternate Worlds: The Illustrated History of Science Fiction* (Englewood Cliffs, NJ: Prentice-Hall, 1975); Ed Naha, *The Science Fictionary* (New York: Wideview Books, 1980); Franz Rottensteiner, *The Science Fiction Book: An Illustrated History* (New York: Seabury Press, 1975); and Jean-Claude Suares, Richard Siegel, and David Owen, *Fantastic Planets* (Danbury, NH: Addison House, 1979).

33. Edward Everett Hazlett to Dwight D. Eisenhower, September 24, 1952, Eisenhower Personal Papers (1916–1953), Eisenhower Presidential Library, Abilene, Kansas, as quoted in Brian Balogh, "Reorganizing the Organizational Synthesis: Federal-Professional Relations in Modern America," *Studies in American Political Development* 5 (Spring 1991): 164.

34. James B. Conant, "The Problems of Evaluation of Scientific Research and Development for Military Planning," speech to the National War College, February 1, 1952, quoted in James G. Hershberg, "'Over My Dead Body': James B. Conant and the Hydrogen Bomb," paper presented to the Conference on Science, Military, and Technology, Harvard/MIT, Cambridge, Massachusetts, June 1987, 50.

35. Space Frontier Foundation website, available at www.space-frontier.org/whoweare.html, accessed March 25, 2007.

36. "Welcome to the Revolution," Message 1 of the Frontier Files, 1995, www.space-frontier.org/History/frontierfiles.html, accessed April 11, 2001; Jeff Foust, "Is the Vision Losing Focus?" *Space Review,* July 26, 2004, available at www.thespacereview.com/article/194/1, accessed August 9, 2004.

37. See McCurdy, *Faster, Better, Cheaper.*

38. Kennedy, "Address at Rice University."

39. Stephen J. Pyne, "Space: A Third Great Age of Discovery," *Space Policy* 4 (August 1988): 187–99.

40. One of the best works on this period is Daniel J. Boorstin, *The Discoverers: A History of Man's Search to Know His World and Himself* (New York: Random House, 1983), 145–201, quotation 145.

41. See William H. Goetzmann, *New Lands, New Men: America and the Second Great Age of Discovery* (New York: Viking Press, 1986).

42. Stephen E. Ambrose, *Undaunted Courage: The Life of Meriwether Lewis* (Garden City, NY: Doubleday, 1996), makes the case that the Lewis and Clark expedition of 1804–6 was successful at many levels but that the most enduring legacy may well have been the scientific information gathered by the team.

43. Stephen J. Pyne, *The Ice: A Journey to Antarctica* (Iowa City: University of Iowa Press, 1986); Nathan Reingold, ed., *The Sciences in the American Context: New Perspectives* (Washington, D.C.: Smithsonian Institution Press, 1979); Norman Cousins et al., *Why Man Explores*

(Washington, D.C.: NASA Educational Publication 125, 1976); Derek Price, *Science since Babylon* (New Haven, CT: Yale University Press, 1975), 10–11; and Roger D. Launius, *Frontiers of Space Exploration* (Westport, CT: Greenwood Press, 1998), 3–5.

44. See Stephen J. Summerhill and John Alexander Williams, *Sinking Columbus: Contested History, Cultural Politics, and Mythmaking during the Quincentenary* (Gainesville: University Press of Florida, 2000); Herman J. Viola and Carolyn Margolis, *Seeds of Change: A Quincentennial Commemoration* (Washington, D.C.: Smithsonian Institution Press, 1991); and M. Jane Young, "'Pity the Indians of Outer Space': Native American Views of the Space Program," *Western Folklore* 46 (October 1987): 269–79.

45. Donald K. Slayton, speech, annual meeting, Society of Experimental Test Pilots, Los Angeles, California, October 9, 1959, NASA Historical Reference Collection.

46. David T. Grober and Edward R. Jones, "Human Engineering Implications of Failures in the Mercury Capsule," August 10, 1959, NASA Historical Reference Collection. Emphasis in original.

47. Edward R. Jones, "Man's Integration into the Mercury Capsule," paper presented at the 14th annual meeting of the American Rocket Society, Washington, D.C., November 16–19, 1959, pp. 1–2, NASA Historical Reference Collection.

48. David A. Mindell, "Human and Machine in the History of Spaceflight," in *Critical Issues in the History of Spaceflight,* ed. Dick and Launius, 141–62.

49. An excellent discussion of all space probes launched to date can be found in Asif A. Siddiqi, *Deep Space Chronicle: Robotic Exploration Missions to the Planets* (Washington, D.C.: NASA SP-4524, 2002).

50. John M. Logsdon, moderator, *The Legislative Origins of the National Aeronautics and Space Act of 1958: Proceedings of an Oral History Workshop* (Washington, D.C.: Monographs in Aerospace History, No. 8, 1998).

51. See W. David Compton, *Where No Man Has Gone Before: A History of Apollo Lunar Exploration Missions* (Washington, D.C.: NASA SP-4214, 1989); David M. Harland, *Exploring the Moon: The Apollo Expeditions* (Chichester, UK: Springer Praxis, 1999); Don E. Wilhelms, *To a Rocky Moon: A Geologist's History of Lunar Exploration* (Tucson: University of Arizona Press, 1993); Paul D. Spudis, *The Once and Future Moon* (Washington, D.C.: Smithsonian Institution Press, 1996); and Donald A. Beattie, *Taking Science to the Moon: Lunar Experiments and the Apollo Program* (Baltimore: Johns Hopkins University Press, 2001).

52. Leonard David, "John Glenn's Shuttle Flight Seen as Boon to Biomedical Research," available at www.space.com/scienceastronomy/planetearth/space_health_000128.html, accessed August 10, 2004.

53. Richard A. Muller, "Space Shuttle Science," *Technology Review Online,* February 10, 2003, available at www-muller.lbl.gov/TRessays/13_Space_Shuttle_Science.htm, accessed August 10, 2004.

54. Alex Roland, "Testimony before the House Committee on Science," October 16, 2003, 1–2, Science Committee Hearing on The Future of Human Space Flight, available at www.house.gov/science/hearings/full03/oct16/roland.pdf, accessed August 9, 2004.

55. J. Lynn Lunsford and Nicholas Kulish, "Shuttle Disaster Revives Debate on Merits of Manned Flight," *Wall Street Journal,* February 4, 2004.

56. This is a statement often repeated in the space science community. For instance, Ed Weiler, then NASA's associate administrator for space science, said in 2002 that "Hubble discoveries have not only rewritten the science textbooks, the stunning images from HST have also become a part of American culture." NASA Press Release 02–224, "Hands-on Book of Hubble images allows the visually impaired to 'Touch the Universe,'" November 19, 2002, NASA Historical Reference Collection, NASA History Division, NASA Headquarters, Washington, D.C. See also the many discussions of discoveries arising from space science missions in J. L. Heilbron, ed., *The Oxford Companion to the History of Modern Science* (New York: Oxford University Press, 2003).

57. There were 747 articles published in *Science* relating to planetary science between October 1996 and August 2004. These include some stunning discoveries: David S. McKay et al., "Search for Past Life on Mars: Possible Relic Biogenic Activity in Martian Meteorite ALH84001," *Science,* August 16, 1996, pp. 924–30; Richard A. Kerr, "Life on Mars: Martian Rocks Tell Divergent Stories," *Science,* November 8, 1996, pp. 918–20; S. Nozette et al., "The Clementine Bistatic Radar Experiment," *Science,* November 29, 1996, pp. 1495–98; Alvin Seiff et al., "Thermal Structure of Jupiter's Upper Atmosphere Derived from the Galileo Probe," *Science,* April 4, 1997, pp. 102–4; James Glanz, "Astronomy: Worlds around Other Stars Shake Planet Birth Theory," *Science,* May 30, 1997, 1336–39; S. J. Weidenschilling et al., "The Possibility of Ice on the Moon," *Science,* October 3, 1997, pp. 144–45; Rover Team, "Characterization of the Martian Surface Deposits by the Mars Pathfinder Rover, Sojourner," *Science,* December 5, 1997, pp. 1765–68; Adam P. Showman and Renu Malhotra, "The Galilean Satellites," *Science,* October 1, 1999, pp. 77–84; Lars E. Borg et al., "The Age of the Carbonates in Martian Meteorite ALH84001," *Science,* October 1, 1999, pp. 90–94; Richard A. Kerr, "Making a Splash with a Hint of Mars Water," *Science,* June 30, 2000, pp. 2295–97; David Stevenson, "Europa's Ocean: The Case Strengthens," *Science,* August 25, 2000, pp. 1305–7; Paul D. Spudis, "What Is the Moon Made of?" *Science,* September 7, 2001, pp. 1779–81; W. V. Boynton et al., "Distribution of Hydrogen in the Near Surface of Mars: Evidence for Subsurface Ice Deposits," *Science,* July 5, 2002, pp. 81–85; Carolyn C. Porco et al., "Cassini Imaging of Jupiter's Atmosphere, Satellites, and Rings," *Science,* March 7, 2003, pp. 1541–47; Timothy N. Titus, Hugh H. Kieffer, and Phillip R. Christensen, "Exposed Water Ice Discovered near the South Pole of Mars," *Science,* February 14, 2003, pp. 1048–51; R. Gellert et al., "Chemistry of Rocks and Soils in Gusev Crater from the Alpha Particle X-ray Spectrometer," *Science,* August 6, 2004, pp. 829–32; R. V. Morris et al., "Mineralogy at Gusev Crater from the Mössbauer Spectrometer on the Spirit Rover," *Science,* August 6, 2004, pp. 833–36; P. R. Christensen et al., "Initial Results from the Mini-TES Experiment in Gusev Crater from the Spirit Rover," *Science,* August 6, 2004, pp. 837–42; H. Y. McSween et al., "Basaltic Rocks Analyzed by the Spirit Rover in Gusev Crater," *Science,* August 6, 2004, pp. 842–45; Steve W. Squyres et al., "The Spirit Rover's Athena Science Investigation at Gusev Crater, Mars," *Science,* August 6, 2004, pp. 794–99; J. F. Bell III et al., "Pancam Multispectral Imaging Results from the Spirit Rover at Gusev Crater," *Science,* August 6, 2004, pp. 800–806; John A. Grant et al., "Surficial Deposits at Gusev Crater along Spirit Rover Traverses," *Science,* August 6, 2004, 807–10; R. Greeley et al., "Wind-Related Processes Detected by the Spirit Rover at Gusev Crater, Mars," *Science,* August 6, 2004, pp. 810–13; R. E. Arvidson et al., "Localization and Physical Properties Experiments Conducted by Spirit at

Gusev Crater," *Science,* August 6, 2004, pp. 821–24; K. E. Herkenhoff et al., "Textures of the Soils and Rocks at Gusev Crater from Spirit's Microscopic Imager," *Science,* August 6, 2004, pp. 824–26; and P. Bertelsen et al., "Magnetic Properties Experiments on the Mars Exploration Rover Spirit at Gusev Crater," *Science,* August 6, 2004, pp. 827–29.

58. These observations are based on the budget data included in the annual *Aeronautics and Space Report of the President, 2003 Activities* (Washington, D.C.: NASA Report, 2004), appendix E, which contains this information for each year since 1959.

59. Wernher von Braun, "Crossing the Last Frontier," *Collier's,* March 22, 1952, 28–29; Wernher von Braun, "Man on the Moon: The Journey," *Collier's,* October 18, 1952, 52; Robert R. Gilruth, "Manned Space Stations," *Spaceflight* (August 1969), 258; and S. Fred Singer, *Manned Laboratories in Space* (New York: Springer-Verlag, 1969).

60. "What Are We Waiting For?" *Collier's,* March 22, 1952; and Oscar Schachter, "Who Owns the Universe?" *Collier's,* March 22, 1952, 36, 70–71.

61. See Roger D. Launius, *Space Stations: Base Camps to the Stars* (Washington, D.C.: Smithsonian Books, 2003), 26–35, 114–21; and James C. Fletcher, NASA Administrator, and William P. Clements, Deputy Secretary of Defense, "NASA/DOD Memorandum of Understanding on Management and Operation of the Space Transportation System," January 14, 1977, NASA Historical Reference Collection.

62. See Roger Handberg, *Seeking New World Vistas: The Militarization of Space* (Westport, CT: Praeger Publishers, 2000), 23–25, 50–52.

63. "Army Missile Transport Program Chronology," Fact Book, Vol. II, Systems Information, AOMC, quoted in John W. Bullard, "History of the Redstone Missile System," Historical Monograph Project Number: AMC 23 M, October 15, 1965, 151; U.S. Army, "Project Horizon Report: A U.S. Army Study for the Establishment of a Lunar Outpost," June 9, 1959, 2; U.S. Army, "Project Horizon, Phase I Report: A U.S. Army Study for the Establishment of a Lunar Outpost," June 8, 1959; Frederick I. Ordway III, Mitchell R. Sharpe, and Ronald C. Wakeford, "Project Horizon: An Early Study of a Lunar Outpost," *Acta Astronautica* 17, no. 10 (1988): 1105–21; and Wernher von Braun, Ernst Stuhlinger, and H. H. Koelle, "ABMA Presentation to the National Aeronautics and Space Administration," Report D-TN-1-59, Army Ballistic Missile Agency, Redstone Arsenal, Alabama, December 15, 1958, all available in NASA Historical Reference Collection.

64. On the United States Air Force's Manned Orbiting Laboratory (MOL) concept, see Howard E. McCurdy, *The Space Station Decision: Incremental Politics and Technological Choice* (Baltimore: Johns Hopkins University Press, 1990), 70, 132–33; Walter A. McDougall, *The Heavens and the Earth: A Political History of the Space Age* (New York: Basic Books, 1985), 340–41; and Roy F. Houchin III, "Interagency Rivalry: NASA, the Air Force, and MOL," *Quest: The Magazine of Spaceflight* 4 (Winter 1995): 40–45. On Dyna-Soar, see Roy F. Houchin III, "Why the Air Force Proposed the Dyna-Soar X-20 Program," *Quest: The Magazine of Spaceflight* 3 (Winter 1994): 5–11; Terry Smith, "The Dyna-Soar X-20: A Historical Overview," *Quest: The Magazine of Spaceflight* 3 (Winter 1994): 13–18; Robert Godwin, comp., *Dyna-Soar: Hypersonic Strategic Weapons System* (Burlington, Ontario: Apogee Books, 2003); and Roy F. Houchin, "Air Force-Office of the Secretary of Defense Rivalry: The Pressure of Political Affairs in the Dyna-Soar (X-20) Program, 1957–1963," *Journal of the British Interplanetary Society* 50 (May 1997): 162–68.

65. Jon D. Miller, "Space Policy Leaders and Science Policy Leaders in the United States," August 9, 2004, 13, copy in possession of authors.

66. Richard DalBello, executive director, Satellite Industry Association, "Satellite Industry Association (SIA)/Futron Satellite Industry Indicators Survey: 2000–2001 Survey Results," Futron Corporation, Bethesda, Maryland; and Launius and McCurdy, *Imagining Space*, 107–10.

67. See Sarah L. Gall and Joseph T. Pramberger, *NASA Spinoffs: 30 Year Commemorative Edition* (Washington, D.C.: NASA, 1992).

68. See Frederick I. Ordway III, Carsbie C. Adams, and Mitchell R. Sharpe, *Dividends from Space* (New York: Thomas Y. Crowell, 1971); and Paul S. Hardersen, *The Case for Space: Who Benefits from Explorations of the Last Frontier?* (Shrewsbury, MA: ATL Press, 1997).

69. See Hugh R. Slotten, "Satellite Communications, Globalization, and the Cold War," *Technology and Culture* 43 (April 2002): 315–60; David J. Whalen, *The Origins of Satellite Communications, 1945–1965* (Washington, D.C.: Smithsonian Institution Press, 2002); Andrew J. Butrica, ed., *Beyond the Ionosphere: Fifty Years of Satellite Communication* (Washington, D.C.: NASA SP-4217, 1997); and Heather E. Hudson, *Communications Satellites: Their Development and Impact* (New York: Free Press, 1990).

70. See W. D. Kay, "Space Policy Redefined: The Reagan Administration and the Commercialization of Space," *Business and Economic History* 27 (Fall 1998): 237–47; and Harrison H. Schmitt, *Return to the Moon* (New York: Copernicus-Praxis, 2005).

71. Patrick Collins, "The Space Tourism Industry in 2030," in *Space 2000: Proceedings of the Seventh International Conference and Exposition on Engineering, Construction, Operations, and Business in Space*, ed. Stewart W. Johnson et al. (Reston, VA: American Society of Civil Engineers, 2000), 594–603.

72. Launius and McCurdy, *Imagining Space*, 114–16.

73. Carl Sagan, *Cosmos* (New York: Random House, 1980), 231–32.

74. David Morrison, "Chicken Little Was Right," presentation to the American Astronautical Society, San Francisco, California, November 17, 1993.

75. John W. Young, "The Big Picture: Ways to Mitigate or Prevent Very Bad Planet Earth Events," *Space Times: Magazine of the American Astronautical Society* 42 (January-February 2003): 22–23.

76. John F. Kennedy, Memorandum for Vice President, April 28, 1961, NASA Historical Reference Collection.

77. John M. Logsdon, "An Apollo Perspective," *Astronautics and Aeronautics*, December 1979, 112–17, quotation 115. See also John M. Logsdon, *The Decision to Go to the Moon: Project Apollo and the National Interest* (Cambridge, MA: MIT Press, 1970).

78. Kennedy, "Address at Rice University."

79. See John Ehrlichman interview by John M. Logsdon, May 6, 1983, NASA Historical Reference Collection. See also George M. Low, NASA deputy administrator, and James C. Fletcher, NASA administrator, "Items of Interest," August 12, 1971; James C. Fletcher, NASA administrator, to Jonathan Rose, special assistant to the president, November 22, 1971, both in NASA Historical Reference Collection.

80. "Revised Talking Points for the Space Station Presentation to the President and the Cabinet Council," November 30, 1983, with attached: "Presentation on Space Station," De-

cember 1, 1983, NASA Historical Reference Collection; see also McCurdy, *Space Station Decision.*

81. Statement by President George W. Bush, Cabinet Room, February 1, 2003, NASA Historical Reference Collection.

82. John M. Logsdon, "A Sustainable Rationale for Human Spaceflight," *Issues in Science and Technology,* Winter 2003, available at www.issues.org/issues/20.2/p_logsdon.html, accessed August 3, 2004.

83. Fritz Lang, *Frau im Mond,* Universum Film (UFA), 1929; Pichel, *Destination Moon;* Kubrick, *2001: A Space Odyssey.*

84. Holman W. Jenkins, "NASA's Coming Crack-Up," *Wall Street Journal,* October 5, 2005.

85. On the hydraulic civilization of the American West, see Donald Worster, *Rivers of Empire: Water, Aridity, and the Growth of the American West* (New York: Pantheon Books, 1985).

86. Matthew Connolly, "We Don't Give Ticker Tape Parades for Robots: Humanity and the Lure of Space Travel," 1997, available at www.rso.cornell.edu/scitech/archive/97sum/man.html, accessed August 12, 2004.

87. See Richard Slotkin, *Gunfighter Nation: The Frontier Myth in Twentieth Century America* (New York: Atheneum, 1992); and Allen Barra, *Inventing Wyatt Earp: His Life and Many Legends* (New York: Carroll and Graf, 1998).

88. See Tom Wolfe, "The Last American Hero," in Wolfe, *The Kandy-Kolored Tangerine-Flake Streamline Baby* (New York: Farrar, Straus, and Giroux, 1965).

89. Chuck Yeager and Leo Janos, *Yeager: An Autobiography* (New York: Bantam Books, 1985).

90. There are some fine histories of this subject—however, most are of a popular nature. Some of the better works are William E. Burrows, *Exploring Space: Voyages in the Solar System and Beyond* (New York: Random House, 1990); Henry S. F. Cooper, *The Evening Star: Venus Observed* (New York: Farrar, Straus, and Giroux, 1993), *Imaging Saturn: The Voyager Flights to Saturn* (New York: Holt, Rinehart, and Winston, 1981), and *The Search for Life on Mars: Evolution of an Idea* (New York: Holt, Rinehart, and Winston, 1980); Steven J. Dick, *The Biological Universe: The Twentieth Century Extraterrestrial Life Debate and the Limits of Science* (New York: Cambridge University Press, 1996); Edward Clinton Ezell and Linda Neuman Ezell, *On Mars: Exploration of the Red Planet, 1958–1978* (Washington, D.C.: NASA SP-4212, 1984); Bevan M. French and Stephen P. Maran, eds., *A Meeting with the Universe: Science Discoveries from the Space Program* (Washington, D.C.: NASA Educational Publication 177, 1981); Paul A. Hanle, Von Del Chamberlain, and Stephen G. Brush, eds., *Space Science Comes of Age: Perspectives in the History of the Space Sciences* (Washington, D.C.: Smithsonian Institution Press, 1981); Jeffrey Kluger, *Journey beyond Selene: Remarkable Expeditions past Our Moon and to the Ends of the Solar System* (New York: Simon and Schuster, 1999); Clayton R. Koppes, *JPL and the American Space Program: A History of the Jet Propulsion Laboratory* (New Haven, CT: Yale University Press, 1982); Bruce C. Murray, *Journey into Space: The First Three Decades of Space Exploration* (New York: W.W. Norton and Co., 1989); Homer E. Newell, *Beyond the Atmosphere: Early Years of Space Science* (Washington, D.C.: NASA SP-4211, 1980); Robert Reeves, *The Superpower Space Race: An Explosive Rivalry through the Solar System* (New York: Plenum Press, 1994); Imke de Pater and Jack J. Lissauer, *Planetary Sciences* (Cam-

bridge, UK: Cambridge University Press, 2001); Ronald E. Doel, *Solar System Astronomy in America: Communities, Patronage, and Interdisciplinary Research, 1920–1960* (New York: Cambridge University Press, 1996); Ronald A. Schorn, *Planetary Astronomy: From Ancient Times to the Third Millennium* (College Station: Texas A&M University Press, 1997); and Robert S. Kraemer, *Beyond the Moon: A Golden Age of Planetary Exploration, 1971–1978* (Washington, D.C.: Smithsonian Institution Press, 2000).

91. Joel Garreau, "Bots on the Ground," *Washington Post,* May 6, 2007, D1.

92. An important object lesson on the decline and collapse of civilizations is shown in the discussion of Easter Island society in Jared Diamond, *Collapse: How Societies Choose to Fail or Succeed* (New York: Viking, 2005), 79–119.

CHAPTER 4: ROBOTIC SPACEFLIGHT IN POPULAR CULTURE

1. Frank H. Winter, *Prelude to the Space Age: The Rocket Societies, 1924–1940* (Washington, D.C.: Smithsonian Institution Press, 1983); and G. Edward Pendray, "The First Quarter Century of the American Rocket Society," *Jet Propulsion* 25 (November 1955): 586–93.

2. Wernher von Braun with Cornelius Ryan, "Baby Space Station," *Collier's,* June 27, 1953, 33, 40.

3. Ibid., 34.

4. See, for example, NASA Press Release, "Humans, Robots Work Together to Test 'Spacewalk Squad,'" release 03–227, July 2, 2003, NASA Historical Reference Collection, NASA History Division, NASA Headquarters, Washington, D.C.; and White House (President George W. Bush), "Bringing the Vision to Reality," n.d., available at www.whitehouse.gov/space/renewed_spirit.html, accessed September 7, 2006.

5. Isaac Asimov, "Strange Playfellow," *Super Science Stories* (September 1940), 67–77, and *I, Robot.*

6. Norbert Wiener, *Cybernetics; or, Control and Communication in the Animal and Machine* (New York: J. Wiley, 1948).

7. Asimov, *Gold,* 195.

8. Writers Guide, *Star Trek* television series, April 17, 1967, lent by Gregory Jein to the National Air and Space Museum, Washington, D.C.

9. Owen Wister, *The Virginian: A Horseman of the Plains* (New York: Macmillan, 1902).

10. John H. Lenihan, *Showdown: Confronting Modern America in the Western Film* (Urbana: University of Illinois Press, 1980), 10–23; and Jon Tuska, *The American West in Film: Critical Approaches to the Western* (1985; repr. Lincoln: University of Nebraska Press, 1988), 3–15.

11. The final phases of this era in the twentieth century are commonly termed the "heroic era of Antarctic exploration." See "Antarctica: History, The 'heroic era' of exploration," Encyclopaedia Britannica Online, accessed July 8, 2004.

12. "We kept our wireless apparatus rigged, but without result. Evidently the distances were too great for our small plant." Christopher Ralling, *Shackleton: His Antarctic Writings* (London: British Broadcasting Corporation, 1983), 180; see also E. W. H. Walton, *Antarctic Science* (Cambridge, UK: Cambridge University Press, 1987), 36.

13. Kurt Neumann, *Rocketship X-M,* Image Entertainment, 1950.

14. The technical appendix to the novel was first published in Germany as Wernher von

Braun, *Das Mars-projekt* (Esslingen: Bechtle Verlag, 1952). The English edition appeared in 1953 and is available as Wernher von Braun, *The Mars Project* (Urbana: University of Illinois Press, 1953).

15. Von Braun, *The Mars Project,* 75–76, provides logistical details. The plan, with the polar modification, appeared in Wernher von Braun with Cornelius Ryan, "Can We Get to Mars?" *Collier's,* April 30, 1954, 22- 28. See also Willy Ley and Wernher von Braun, *The Exploration of Mars* (New York: Viking Press, 1956); and Ron Miller and Frederick C. Durant, *The Art of Chesley Bonestell* (London: Collins and Brown, 2001), 193.

16. Quoted in Daniel Lang, "A Reporter at Large: A Romantic Urge," *The New Yorker,* April 21, 1951, 74.

17. Anna Brendle, "Profile: African-American North Pole Explorer Matthew Henson," *National Geographic News,* January 15, 2003, available at http://news.nationalgeographic.com/news/2003/01/0110_030113_henson.html, accessed July 8, 2004; see also Wally Herbert, "Commander Robert E. Peary: Did He Reach the Pole?" *National Geographic* (September 1988), 386–413.

18. See S. Allen Counter, *North Pole Legacy: Black, White, and Eskimo* (Amherst: University of Massachusetts Press, 1991); and Matthew A. Henson, *A Negro Explorer at the North Pole* (1912; repr. New York: Arno Press, 1969).

19. See Pamela Horn, *The Rise and Fall of the Victorian Servant* (New York: St. Martin's Press, 1975); and Frank E. Huggett, *Life below Stairs: Domestic Servants in England from Victorian Times* (New York: Charles Scribner's Sons, 1977).

20. Landis, *Antarctica,* 174; "Richard E. Byrd, 1888–1957," available at www.southpole.com/p0000107.htm, accessed July 8, 2004.

21. Harriet Beecher Stowe, *Uncle Tom's Cabin* (Boston: J. P. Jewett, 1852). The Spartacus story of the slave uprising against the Roman Empire in 73 B.C.E. was well known to contemporary authors writing about the industrial revolution. See James Leslie Mitchell [Lewis Grassic Gibbon], *Spartacus,* edited by Ian Campbell (1933; repr. Edinburgh: Scottish Academic Press, 1990); and Howard Fast, *Spartacus* (New York: Crown, 1951).

22. Peter Kussi, ed., *Toward the Radical Center: A Karel Čapek Reader* (Highland Park, NJ: Catbird Press, 1990).

23. Elton Mayo, *The Human Problems of an Industrial Civilization* (New York: Macmillan, 1933); Emile Durkheim, *Suicide: A Study in Sociology,* translated by John A. Spaulding and George Simpson. edited with an introduction by George Simpson (New York: Free Press, 1951); Emile Zola, *Germinal,* translated with an introduction and notes by Roger Pearson (New York: Penguin Books, 2004); Upton Sinclair, *The Jungle* (New York: Grosset and Dunlap, 1906); and Richard Llewellyn, *How Green Was My Valley* (New York: Macmillan, 1940).

24. Asimov, *Gold,* 192, 193, 196.

25. Mary Shelley, *Frankenstein: or the Modern Prometheus,* edited and with an introduction and notes by Maurice Hindle (1818; repr. London: Penguin Classics, 1992), 52.

26. Ibid., 47.

27. Asimov, *Gold,* 196.

28. Ibid., 196–97.

29. Asimov, *I, Robot,* 9.

30. Ibid., 26, 28.

31. Ibid., 44–45. A later modification of the laws required very advanced robots to protect and not injure humanity, termed the Zeroth Law.

32. Asimov, *Gold*, 198–99.

33. "Robot B9 from Lost in Space," available at www.jeffbots.com/b9robot.html, accessed July 10, 2004.

34. Wise, *The Day the Earth Stood Still*; and Douglas Trumbull, *Silent Running*, Universal, 1972.

35. George Lucas, *Star Wars*, Twentieth Century Fox, 1977; and "Droid Archive," available at www.starwars.com/databank/droid/, accessed July 10, 2004.

36. Asimov, *I, Robot*, 74, 90. "Reason" first appeared in the April 1941 issue of *Astounding Science Fiction*.

37. "Not Final!" appeared in the October 1941 issue of *Astounding Science Fiction*, while "Victory Unintentional" was published in the August 1942 issue of *Super Science Stories*. See Isaac Asimov, *The Early Asimov; or, Eleven Years of Trying* (Garden City, NY: Doubleday, 1972), 322–38 for "Not Final!" and Asimov, *The Rest of the Robots* (London: Grafton Books, 1964), 37–63 for "Victory Unintentional."

38. Michael Crichton, *Westworld*, Metro-Goldwyn-Mayer, 1973; and Douglas Adams, *The Hitchhiker's Guide to the Galaxy* (New York: Pocket Books, 1979).

39. Dick, *Do Androids Dream of Electric Sheep?*; and Ridley Scott, *Blade Runner*, Warner, 1982.

40. Herbert A. Simon, *New Science of Management Decision Making* (New York: Harper, 1960); see also Charles Perrow, *Normal Accidents: Living with High-Risk Technologies* (New York: Basic Books, 1984).

41. Kubrick, *2001: A Space Odyssey*. See also Arthur C. Clarke, *2001: A Space Odyssey* (New York: New American Library, 1968).

42. John Badham, *WarGames*, Sherwood/MGM/United Artists, 1983.

43. James Cameron, *The Terminator*, Orion Pictures, 1984.

44. Paul Verhoeven, *Robocop*, Orion, 1987; Jonathan Frakes, *Star Trek: First Contact*, Paramount, 1996; and Lucas, *Star Wars*.

45. Isaac Asimov, *Robot Visions* (New York: ROC New American Library, 1990), 7.

46. "Imagining it is what makes it happen." Michael Crichton, *Sphere* (New York: Alfred A. Knopf, 1987), 348. See also McCurdy, *Space and the American Imagination*.

47. Cory Doctorow, "Rise of the Machines," *Wired* (July 2004), 119.

48. Frederick Winslow Taylor, *Principles of Scientific Management* (New York: Harper and Brothers, 1911).

49. Gareth Morgan, *Images of Organization* (Beverly Hills, CA: Sage Publications, 1986).

50. See Susan A. Greenfield, *Journey to the Centers of the Mind: Toward a Science of Consciousness* (New York: W. H. Freeman, 1995), ch. 4.

51. Asimov, *Gold*, 206.

52. Isaac Asimov, *The Complete Robot* (Garden City, NY: Doubleday, 1982), 554. The novella first appeared in 1976.

53. Ibid., 528.

54. Ibid.

55. Robert Scheerer, "The Measure of a Man," production 135, February 13, 1989, Star

Trek Episode Archives, Paramount Pictures, available at www.starfleetlibrary.com/tng/tng2/the_measure_of_a_man.htm, accessed July 13, 2004.

56. Scott v. Sandford, No. 60–393. Supreme Ct. of the U.S., March 5, 1857.

57. On this case, see Don E. Fehrenbacher, *The Dred Scott Case: Its Significance in American Law and Politics* (1978; repr. New York: Oxford University Press, 2001).

58. The perceptions of white nineteenth-century America toward African Americans as something less than human has been analyzed in William Stanton, *The Leopard's Spots: Scientific Attitudes toward Race in America, 1815–1859* (Chicago: University of Chicago Press, 1960); and George M. Frederickson, *The Black Image in the White Mind* (New York: Harper and Rowe, 1971). For an exploration of the issue of sentience related to robots, see Rodney Brooks, *Flesh and Machines: How Robots Will Change Us* (New York: Pantheon Books, 2002).

59. Ridley Scott, *Alien,* Twentieth Century Fox, 1979; and Robert Heinlein, *Starship Troopers* (New York: G. P. Putnam's Sons, 1959).

60. Dick, *Do Androids Dream of Electric Sheep?* 14.

CHAPTER 5: THE NEW SPACE RACE

1. The Apollo Telescope Mount on Skylab required astronaut film recovery and replacement. See John A. Eddy, "Skylab Optics: An Introduction," *Applied Optics* 16, no. 4 (1977): 823; R. Tousey, "Apollo Telescope Mount of Skylab: An Overview," *Applied Optics* 16, no. 4 (1977): 825; W. David Compton and Charles D. Benson, *Living and Working in Space: A History of Skylab* (Washington, D.C.: NASA SP-4208, 1984), 174–79.

2. Space Task Group, "Post-Apollo Space Program," 13, 20, NASA Historical Reference Collection, NASA History Division, NASA Headquarters, Washington, D.C.

3. The Great Observatories and their launch dates are the Hubble Space Telescope (1990), the Compton Gamma Ray Observatory (1991), the Chandra X-Ray Observatory (1999), and the Spitzer Space Telescope (2003).

4. This is the story told in McCurdy, *Faster, Better, Cheaper.*

5. These efforts are documented in John M. Logsdon, "A Failure of National Leadership: Why No Replacement for the Space Shuttle?" in *Critical Issues in the History of Spaceflight,* ed. Dick and Launius, 269–300; and Launius, *Space Stations.*

6. Quoted in McCurdy, *Space Station Decision,* 5, 47.

7. Quoted in ibid., 116.

8. NASA, "Compilation of Papers Presented at the Space Station Technology Symposium," Langley Research Center, February 11–13, 1969, 43–98; W. Ray Hook, "Historical Review," *Journal of Engineering for Industry: Transactions of the ASME* 106 (November 1984): 276–86; and NASA Office of Manned Space Flight, Advanced Manned Missions Program, "Space Station Summary Report," June 1969, NASA Historical Reference Collection. See also McCurdy, *Space Station Decision,* ch. 7.

9. Director, History Division, to Associate Administrator for External Relations, Subject: "Termination of Saturn Vehicle; Skylab as Space Station," March 23, 1988; see also Thomas O. Paine, "NASA Future Plans News Conference," *NASA News,* January 13, 1970, both in NASA Historical Reference Collection.

10. House Science and Technology Committee, Subcommittee on Space Sciences and Applications, *NASA's Space Station Activities,* 98th Cong., 1st sess., 1983, 4.

11. Paul Holloway, "Space Station Technology," paper presented at the Thirty-third International Astronautical Federation Congress, Paris, France, September 16, 1982–October 2, 1982, IAF-82–15, 15.

12. Daniel H. Herman interview conducted by Sylvia D. Fries, March 26, 1985, NASA Historical Reference Collection.

13. James M. Beggs, NASA Administrator, to Craig L. Fuller, Assistant to the President for Cabinet Affairs, April 12, 1984, NASA Historical Reference Collection.

14. NASA, Office of Space Systems Development, "Space Station Freedom Capital Development Plan, Fiscal Year 1993," submitted to the Committee on Science, Space, and Technology, U.S. House of Representatives, and the Committee on Commerce, Science, and Transportation, U.S. Senate, April 1992.

15. See Adam Gruen, "Deep Space Nein? The Troubled History of Space Station Freedom," *Ad Astra,* May / June 1993, 18–23, and "The Port Unknown: A History of the Space Station Freedom Program," 1993, unpublished manuscript available in the NASA Historical Reference Collection.

16. NASA, Office of Space Station, "The Space Station: A Description of the Configuration Established at the Systems Requirements Review (SRR)," June 1986, 20, NASA Historical Reference Collection; and U.S. House, Science and Technology Committee, Space Science and Applications Subcommittee, *NASA's Space Station Activities,* 98th Cong., 1st sess., August 2, 1983, 65. See also Thomas L. Labus, Richard R. Secunde, and Ronald G. Lovely, "Solar Dynamic Power for Space Station Freedom," paper prepared for the International Conference on Space Power, International Astronautical Federation, Cleveland, Ohio, June 5–7, 1989.

17. NASA, News Release no. 62–8, "Mercury-Atlas 6 at a Glance," January 21, 1962, sec. 2, 2, NASA Historical Reference Collection.

18. NASA Facts, "Powering the Future: NASA Glenn Contributions to the International Space Station (ISS) Electrical Power System," FS-2000–11–006-GRC; see also Gruen, "Deep Space Nein?" 22.

19. B. D. V. Marino and H. T. Odum, *Biosphere 2: Research Past and Present,* reprinted from *Ecological Engineering Special Issue* 13, nos. 1–4 (Elsevier Science, 1999). See also John Allen, *Biosphere 2: The Human Experiment* (New York: Viking, 1991) and "Biosphere 2: The Experiment: Invention, Creation, and Mission 1, A Brief Chronology," available at www.biospheres.com / experimentchrono1.html, accessed January 28, 2004.

20. P. O. Wieland, "Living Together in Space: The Design and Operation of the Life Support Systems on the *International Space Station,*" NASA / TM-1998–206956, Marshall Space Flight Center, Huntsville, Alabama.

21. On this subject, see Roger D. Launius and Dennis R. Jenkins, eds., *To Reach the High Frontier: A History of U.S. Launch Vehicles* (Lexington: University Press of Kentucky, 2002).

22. Roger E. Bilstein, *Stages to Saturn: A Technological History of the Apollo / Saturn Launch Vehicles* (Washington, D.C.: NASA SP-4206, 1980), 422.

23. NASA, "Space Shuttle," February 1972, "Space Shuttle Economics, appendix to space shuttle fact sheet," Office of the White House Press Secretary, San Clemente, California;

and White House, "Press Conference of Dr. James Fletcher and George M. Low," San Clemente Inn, San Clemente, California, January 5, 1972, both in NASA Historical Reference Collection.

24. Klaus P. Heiss and Oskar Morgenstern, "Economic Analysis of the Space Shuttle System: Executive Summary," study prepared for NASA under contract NASW-2081, January 31, 1972; and NASA, "Space Shuttle," both in NASA Historical Reference Collection; and Presidential Commission on the Space Shuttle Challenger Accident, *Report of the Presidential Commission* (Washington, D.C.: Government Printing Office, 1986), 164.

25. James C. Fletcher, administrator, NASA, and William P. Clements, deputy secretary of defense, "NASA/DOD Memorandum of Understanding on Management and Operation of the Space Transportation System," January 14, 1977, 8, NASA Historical Reference Collection; and Administration of Ronald Reagan, "United States Space Policy: Fact Sheet Outlining the Policy," *Weekly Compilation of Presidential Documents,* July 4, 1982, 875–76.

26. Alex Roland, "The Shuttle: Triumph or Turkey?" *Discover,* November 1985, 14–24; and Logsdon, "The Space Shuttle Program: A Policy Failure?" 1099–1105.

27. Traci Watson, "NASA Administrator Says Space Shuttle Was a Mistake," *USA Today,* September 27, 2005, A1.

28. Quoted in McCurdy, *Inside NASA,* 87.

29. See, for example, NASA Marshall Space Flight Center, Reusable Launch Vehicles, "10 Years," 2000, in authors' collection; and M. K. Lockwood, "Overview of Conceptual Design of Early VentureStar Configurations," American Institute of Aeronautics and Astronautics, 38th Aerospace Sciences Meeting and Exhibit, Reno, Nevada, January 10–13, 2000. Statement of Daniel S. Goldin, NASA administrator, before the U.S. Senate Subcommittee on Science, Technology, and Space and Committee on Commerce, Science, and Transportation, September 23, 1998, NASA Historical Reference Collection, 3–4.

30. NASA, "Linear Aerospike Engine—Propulsion for the X-33 Vehicle," fact sheet no. FS-2000-09-174-MSFC, August 2000; and Stuart F. Brown, "X-30: Out of This World in a Scramjet," *Popular Science* (November 1991), 70–75, 106–12; see also T. K. Mattingly, "A Simpler Ride into Space," *Scientific American* (October 1997), 121–25.

31. Dennis R. Jenkins, *Space Shuttle: The History of the National Space Transportation System, the First 100 Missions,* 4th ed. (Cape Canaveral, FL: Dennis R. Jenkins, 2001), 77–133.

32. See NASA, "NASA, Lockheed Martin Agree on X-33 Plan," Press Release 00-157, September 29, 2000; and Steven Siceloff, "NASA Kills X-33 Program," *Florida Today,* March 2, 2001.

33. Quoted in Boyce Rensberger, "The Prophet in His Orbit," *Washington Post,* November 7, 1985, C6.

34. NASA, "Space Shuttle"; and Space Task Group, "Post-Apollo Space Program."

35. Thomas Beatty, NASA, "Advanced Propulsion," available at http://lifesci3.arc.nasa.gov/SpaceSettlement/teacher/lessons/contributed/Thomas/Adv.prop/advprop.html, accessed July 7, 2004.

36. See James A. Dewar, *To the End of the Solar System: The Story of the Nuclear Rocket* (Lexington: University Press of Kentucky, 2004).

37. Zubrin used the term *Death Star* frequently in his public appearances. See Ari Arm-

strong, "LP03: Zubrin Reviews Mars Plan," Colorado Freedom Report, May 7, 2003, available at www.freecolorado.com/2003/04/lp03zubrin.html, accessed September 12, 2006. See also NASA, "Report of the 90-Day Study on Human Exploration of the Moon and Mars," November 1989, NASA Historical Reference Collection.

38. Kurt Lancaster, "Pilgrims for Mars," *Christian Science Monitor,* October 21, 1999; and Jim Wilson, "Bringing Life to Mars," *Popular Mechanics* (November 1998, 30–31).

39. S. Olesen, "Electric Propulsion for Project Prometheus," AIAA-2003–5279, 39th AIAA/ASME/SAE/ASEE Joint Propulsion Conference and Exhibit, Huntsville, Alabama, July 20–23, 2003; and Fred Elliot, "An Overview of the High Power Electric Propulsion (HiPEP) Project," paper presented at the 40th AIAA/ASME/SAE/ASEE Joint Propulsion Conference and Exhibit, Fort Lauderdale, Florida, July 11–14, 2004.

40. NASA Jet Propulsion Laboratory, "Spacecraft: Surface Operations: Rover," available at http://marsrovers.jpl.nasa.gov/mission/spacecraft_surface_rover.html, accessed July 28, 2004.

41. NASA Jet Propulsion Laboratory, "Spacecraft: Surface Operations: Rover: The rover's 'body,'" available at http://marsrovers.jpl.nasa.gov/mission/spacecraft_rover_body.html, accessed July 28, 2004.

42. Statement by Steve Squyres, "Sleepy Opportunity," news release 2004–132, Solar System Exploration News Archive, NASA, May 26, 2004.

43. Buddy Nelson, ed., "Hubble Space Telescope Servicing Mission 3B, Media Reference Guide," prepared by Lockheed Martin for the National Aeronautics and Space Administration, n.d., sec. 6, 5.

44. Hubble Project, "Hubble in Cruise Control: The Ins and Outs of a Spacecraft in Orbit," available at http://hubble.gsfc.gov/hubble-operations/cruise-control.html, accessed July 28, 2004; and NASA, "Hubble Telescope Placed into Safe Hold as Gyroscope Fails," press release, November 15, 1999.

45. NASA Jet Propulsion Laboratory, "Farewell to Io," *Galileo News,* January 18, 2002.

46. Tariq Malik, "Thinking on Mars: The Brains of NASA's Red Planet Rovers," January 28, 2004; available at www.space.com/businesstechnology/technology/mer_computer_040128.html, accessed August 3, 2004.

47. NASA, "This Week on Galileo, October 18–24, 1999: Galileo Continues to Return Science Information from Historic Io Flyby," available at http://www2.jpl.nasa.gov/Galileo/today991019.html, accessed July 28, 2004.

48. Howard E. McCurdy, *Low-Cost Innovation in Spaceflight: The Near Earth Asteroid Rendezvous (NEAR) Shoemaker Mission,* Monographs in Aerospace History No. 36 (Washington, D.C.: NASA SP-4536, 2005).

49. Von Braun with Ryan, "Baby Space Station," 40.

50. "Monkeys and Other Animals in Space," Space Today Online, available at www.spacetoday.org/Astronauts/Animals/Dogs.html, accessed December 7, 2005.

51. American Society for the Prevention of Cruelty to Robots, 2003, available at www.ASPCR.com, accessed December 7, 2005.

52. R. Cargill Hall, *Lunar Impact: A History of Project Ranger* (Washington, D.C.: NASA SP-4210, 1977); Paolo Ulivi and David M. Harland, *Lunar Exploration: Human Pioneers and Ro-*

botic Surveyors (Chichester, UK: Springer-Praxis, 2004), 34–39, 47–55; and Keith J. Scala, "Crashing Success: An Overview of Project Ranger," *Quest: The History of Spaceflight Magazine* 1, no. 3 (Fall 1992): 4–11.

53. L. J. Kosofsky, "Ranger 7, Part 1: Mission Description and Performance," Jet Propulsion Laboratory, California Institute of Technology, TR 32–700, Pasadena, California; *December 1964; Ranger VII Photographs of the Moon, Part I: Camera "A" Series* (Washington, D.C.: NASA SP-61, 1964); *Ranger VII Photographs of the Moon, Part I: Camera "B" Series* (Washington, D.C.: NASA SP-62, 1965); and *Ranger VII Photographs of the Moon, Part I: Camera "P" Series* (Washington, D.C.: NASA SP-63, 1965).

54. F. O. Huck, H. F. McCall, W. R. Patterson, and G. R. Taylor, "The Viking Mars Lander Camera," *Space Science Instrumentation* 1 (1975): 189–241; Gerald A. Soffen, "Scientific Results of the Viking Mission," *Science*, December 17, 1976, 1274–76, and "The Viking Project," *Journal of Geophysical Research* 82, no. 28 (September 1977): 3959–70; and Edward Clinton Ezell and Linda Neuman Ezell, *On Mars: Exploration of the Red Planet, 1958–1978* (Washington, D.C.: NASA SP-4212, 1984), 72–76.

55. P. H. Smith et al., "The Imager for Mars Pathfinder Experiment," *Journal of Geophysical Research* 102, issue E2 (1997): 4003–26; and P. H. Smith et al., "Results from the Mars Pathfinder Camera," *Science*, December 5, 1997, pp. 1758–65.

56. J. N. Maki et al., "Mars Exploration Rover Engineering Cameras," *Journal of Geophysical Research* 108, issue E12 (2003): 1–23.

57. The rover pancams captured images 1024 by 1024 pixels in size, providing nearly sixteen times more sensitivity than the Pathfinder camera. NASA, "Spacecraft: Surface Operations: Rover: The Rover's 'Eyes' and Other 'Senses,'" available at http://marsrovers.jpl.nasa.gov/mission/spraceraft_rover_eyes.html, accessed February 4, 2004. See also Planetary Society, "Pancams: Panoramic Cameras aboard the Mars Explorer Rovers," available at www.planetary.org/mars/mer-inst-pancam.html, accessed July 30, 2004.

58. J. Matijevic et al., "The Pathfinder Microrover," *Journal of Geophysical Research* 102, issue E2 (1997): 3989–4001.

59. Goddard and Pendray, eds., *The Papers of Robert H. Goddard*, vol. 1, 394–95; and Maurice K. Hanson, "The Payload on the Lunar Trip," *Journal of the British Interplanetary Society* 1 (January 1939): 16.

60. Edward C. Stone, "News from the Edge of Interstellar Space," *Science*, July 6, 2001, pp. 55–56.

61. R. E. Edelson, B. D. Madsen, E. K. Davis, and G. W. Garrison, "Voyager Telecommunications: The Broadcast from Jupiter," *Science*, June 1, 1979, pp. 913–21.

62. NASA Jet Propulsion Laboratory, "Basics of Space Flight," ch. 10, available at www.jpl.nasa.gov/basics, accessed July 31, 2004.

63. NASA, "Galileo's New Telecommunications Strategy," March 23, 2004, NASA Historical Reference Collection.

64. Isaac Asimov, *Robot Visions* (New York: ROC New American Library, 1990), 7.

65. Bilstein, *Stages to Saturn*, 355; and John D. Hodge and Richard Carlisle interview conducted by Sylvia Fries, March 4, 1985, NASA Historical Reference Collection.

66. Charles Murray and Catherine Bly Cox, *Apollo: The Race to the Moon* (New York: Simon and Schuster, 1989), 351–55. See also Andrew Chaikin, *A Man on the Moon* (New York:

Penguin Books, 1994); and James R. Hansen, *First Man: The Life of Neil A. Armstrong* (New York: Simon and Schuster, 2005).

67. Nicks, *Far Travelers*, 126.

68. NASA, "Mars Exploration Rover Landings: Press Kit," January 2004, NASA Historical Reference Collection.

69. NASA, "Deep Space 1 Launch: Press Kit," October 1998, 27, NASA Historical Reference Collection.

70. See Eldon C. Hall, *Journey to the Moon: The History of the Apollo Guidance Computer* (Reston, VA: American Institute of Aeronautics and Astronautics, 1996).

71. NASA Jet Propulsion Laboratory, "Mars Pathfinder Mission Status," October 22, 1997, NASA Historical Reference Collection.

72. Gary L. Bennett, "Space Nuclear Power: Opening the Final Frontier," 4th International Energy Conversion Engineering Conference and Exhibit (IECEC), San Diego, California, June 26–29, 2006; and R. R. Furlong and E. J. Wahlquist, "U.S. Space Missions Using Radioisotope Power Systems," *Nuclear News* 42 (April 1999): 26–34.

73. N. N. Ponomarev-Stepnoi, V. M. Talyzin, and V. A. Usov, "Russian Space Nuclear Power and Nuclear Thermal Propulsion Systems," *Nuclear News* 43 (December 2000): 33–46.

74. NASA, Jet Propulsion Laboratory, "People Are Robots, Too. Almost," October 28, 2003.

75. Figures for interplanetary flight are derived by dividing launch costs by the weight of the spacecraft dispatched, including spacecraft propellant. See McCurdy, *Faster, Better, Cheaper*, 111–12, 120–21.

76. The White House, Statement by the President, January 5, 1972.

77. Cady Coleman, "Payloads: STS-93 Chandra X-Ray Observatory," July 7, 1999, available at www.shuttlepresskit.com/STS-93/payload45.htm, accessed December 9, 2005; NASA, "Space Shuttle Mission STS-37," press kit, April 1991, available at http://science.ksc.nasa.gov/shuttle/missions/sts-37/sts-37-press-kit.txt, accessed December 9, 2005; and Mark Wade, "Delta 7000," Encyclopedia Astronautica, March 28, 2005, available at www.astronautix.com/lvs/dela7000.htm, accessed December 9, 2005. The Hubble Space Telescope weighed 23,981 pounds at launch. NASA, "Space Shuttle Mission STS-31," press kit, April 1990, available at http://science.ksc.nasa.gov/shuttle/missions/sts-31/sts-31-press-kit.txt, accessed December 9, 2005. See also George Rieke, *The Last of the Great Observatories: Spitzer and the Era of Faster, Better, Cheaper at NASA* (Tempe: University of Arizona Press, 2006).

78. Much of the material in this section is drawn from McCurdy, *Faster, Better, Cheaper* and *Low-Cost Innovation in Spaceflight*.

79. Linda Neuman Ezell, *NASA Historical Data Book: Volume III, Programs and Projects, 1969–1978* (Washington, D.C.: NASA, 1988), 61–63; see also William C. Schneider, Director, Skylab Program, to Public Affairs Officer, Subject: Unit Cost of Skylab Hardware, April 17, 1973, NASA Historical Reference Collection.

80. What was then called Space Station Freedom was to be completed at a cost of between $8.8 and $12 billion by 1994, compared to an inflation-adjusted Skylab cost of about $9 billion. Peggy Finarelli to OMB/Bart Borrasca, Subject: Space Station Funding, September

8, 1983, NASA Historical Reference Collection. See also John J. Madison and Howard E. McCurdy, "Spending without Results: Lessons from the Space Station Program," *Space Policy* 15 (1999): 213–21.

81. Marcia S. Smith, Congressional Research Service, "NASA's Space Station Program: Evolution and Current Status," testimony before the House Science Committee, April 4, 2001; and NASA Advisory Council, "Report of the Cost Assessment and Validation Task Force on the International Space Station," April 21, 1998, both in NASA Historical Reference Collection.

82. McCurdy, *Faster, Better, Cheaper.*

83. Even so, the number of robotic spacecraft fell. See Mark Wade, "The Year in Review: 2004," available at www.astronautix.com / articles / thew2004.htm, accessed September 17, 2006.

CHAPTER 6: INTERSTELLAR FLIGHT AND THE HUMAN FUTURE IN SPACE

1. Robert Goddard, "The Ultimate in Jet Propulsion." in *The Papers of Robert H. Goddard,* ed. Goddard and Pendray, vol. 3, 1612.

2. Sagan, *Pale Blue Dot,* 371, 386.

3. George P. Mueller, "Space: The Future of Mankind," *Spaceflight* 104 (March 27, 1985): 107.

4. George H. W. Bush, "Remarks on the Twentieth Anniversary of the Apollo 11 Moon Landing," July 20, 1989, George Bush Presidential Library and Museum, available at http://bushlibrary.tamu.edu / research / papers / 1089 / 89072000.html, accessed October 26, 2004.

5. E. E. Smith, *The Skylark of Space* (New York: Pyramid Books, 1928); Rick Sternbach and Michael Okuda, *Star Trek: The Next Generation Technical Manual* (New York: Pocket Books, 1991); Isaac Asimov, *Foundation* (New York: Gnome Press, 1951); and StarWars.com, accessed January 5, 2006.

6. Michael Hoskin, ed., *The Cambridge Concise History of Astronomy* (Cambridge, UK: Cambridge University Press, 1999).

7. See NASA, *Why Man Explores,* symposium held at Bechman auditorium, California Institute of Technology, Pasadena, California, July 2, 1976 (Washington, D.C.: Government Printing Office, 1977), 11.

8. Malin Space Science Systems, Mars Global Surveyor, Mars Orbiter Camera, "Earth, Moon, and Jupiter, as Seen From Mars," MGS MOC Release No. NOC2–368, May 22, 2003.

9. Jean Schneider, "Extra-solar Planets Catalog," May 28, 2004 available at http://cfa-www.harvard.edu / planets / catalog.html, accessed January 5, 2006.

10. Epsilon Indi sits at 336 degrees longitude and –48 degrees latitude, Tau Ceti at 173 degrees longitude and –73 degrees latitude.

11. See for example Imaginova, Starry Night astronomy software, 2004.

12. McDonald Observatory, University of Texas at Austin, "Search for Extrasolar Planets Hits Home," August 7, 2000; and Artie P. Hatzes et al., "Evidence for a Long-Period Planet Orbiting Epsilon Eridani," *Astrophysical Journal* 544 (December 1, 2000): L145-L148.

13. "Epsilon Indi," solstation.com, 1998–2004, available at www.solstation.com / stars / eps-indi.htm, accessed April 24, 2007.

14. J. S. Greaves, M. C. Wyatt, W. S. Holland, and W. R. F. Dent, "The Debris Disc around t Ceti: A Massive Analogue to the Kuiper Belt," *Monthly Notices of the Royal Astronomical Society* 351 (2004): L54-L58.

15. "Habitability: Betting on 37 Gem," *Astrobiology,* October 9, 2003; and Margaret C. Turnbull and Jill C. Tarter, "Target Selection for SETI: 1. A Catalog of Nearby Habitable Stellar Systems," *Astrophysical Journal Supplement Series* 145 (March 2003): 181–98.

16. Paul A. Weigert and Matt J. Holman, "The Stability of Planets in the Alpha Centauri System," *Astronomical Journal* 113 (April 1997): 1145–50; see also "Alpha Centauri 3," available at www.solstation.com/stars/alp-cent3.htm, accessed April 24, 2007.

17. Peter D. Ward and Donald Brownlee, *Rare Earth: Why Complex Life Is Uncommon in the Universe* (New York: Copernicus, 2000); see also "Stars and Habitable Planets," available at www.solstation.com/habitable.htm, accessed April 24, 2007.

18. J. F. Davies, "A Brief History of the Voyager Project, Part 2," *Spaceflight* 23 (March 1981): 71–74.

19. NASA, "Voyager Spacecraft Find Clue to Another Solar System Mystery," release 93–099, May 26, 1993; and NASA, "Two Voyager Spacecraft Still Going Strong after 20 Years," release 97–189, September 2, 1997.

20. NASA, "Voyager Spacecraft Find Clue to Another Solar System Mystery," May 26, 1993.

21. J. J. Matese, P. G. Whitman, and D. P. Whitmire, "Cometary Evidence of a Massive Body in the Outer Oort Cloud," *Icarus* 141 (1999): 354–66.

22. C. J. Everett and S. M. Ulam, "On a Method of Propulsion of Projectiles by Means of External Nuclear Explosions," LAMS-1955, Los Alamos Scientific Laboratory, August 1955; see also George Dyson, *Project Orion: The True Story of the Atomic Spaceship* (New York: Henry Holt, 2002).

23. Alan Bond and Anthony R. Martin, "Project Daedalus Reviewed," *Journal of the British Interplanetary Society* 39 (1986): 386.

24. Bond and Martin, "Project Daedalus—Final Report" (1978): S5–S7.

25. Science @NASA, "Reaching for the Stars: Scientists Examine Using Antimatter and Fusion to Propel Future Spacecraft," April 19, 1999, available at http://science.nasa.gov/NEWHOME/headlines/prop12apr99_1.htm, accessed September 29, 2006; and Bill Steigerwald, "New and Improved Antimatter Spaceship for Mars Missions," NASA Vision for Space Exploration, April 14, 2006, available at www.nasa.gov/mission_pages/exploration/mmb/antimatter_spaceship.html, accessed September 29, 2006.

26. George P. Mueller, "Antimatter and Distant Space Flight," *Spaceflight* 25 (May 1983): 203, 206; and Sternbach and Okuda, *Star Trek: The Next Generation Technical Manual.*

27. Mueller, "Antimatter and Distant Space Flight," 204.

28. Robert W. Bussard, "Galactic Matter and Interstellar Flight," *Astronautica Acta* 6 (1960): 179–94.

29. Alan Bond, "An Analysis of the Potential Performance of the Ram Augmented Interstellar Rocket," *Journal of the British Interplanetary Society* 27 (1974): 674–88.

30. Joe Haldeman, "Colonizing Other Worlds," in *Interstellar Travel and Multi-Generational Space Ships,* ed. Yoji Kondo, Frederick Bruhweiler, John Moore, and Charles Sheffield (Burlington, Ontario: Apogee Books, 2003), 65.

31. Robert L. Forward, "Roundtrip Interstellar Travel Using Laser-Pushed Lightsails," *Journal of Spacecraft and Rockets* 21 (March-April 1984): 187–95; and Robert M. Zubrin and Dana G. Andrews, "Magnetic Sails and Interplanetary Travel," *Journal of Spacecraft and Rockets* 28 (March-April 1991): 197–203.

32. K. E. Tsiolkovsky, "The Reaction Machine as Insurance against Possible Disaster," in *Collected Works of K. E. Tsiolkovsky*, vol. 2, ed. A. A. Blagonravov (National Aeronautics and Space Administration, NASA TT F-237, 1965), 164–67.

33. Robert A. Heinlein, *Orphans of the Sky* (New York: G. P. Putnam's Sons, 1964).

34. O'Neill, "Colonization of Space," 32.

35. Ibid., 39.

36. Robert Page Burruss, "Intergalactic Travel," *The Futurist* (September-October 1987): 29–32.

37. Mueller, "Antimatter and Distant Space Flight," 205.

38. Buzz Aldrin and John Barnes, *Encounter with Tiber* (New York: Warner Books, 1996), 221.

39. Kondo et al., eds., *Interstellar Travel*, 6–8, emphasis ("NOW") removed.

40. Robert Forward, "Ad Astra!" in ibid., 36.

41. Mueller, "Antimatter and Distant Space Flight," 205.

42. Project Daedalus Study Group, *Project Daedalus: The Final Report on the BIS Starship Study*, JBIS Interstellar Studies, Supplement (1978).

43. Forward, "Ad Astra!" 47.

44. Robert W. Forward, "Feasibility of Interstellar Travel: A Review," *Journal of the British Interplanetary Society* 39, no. 9 (1986): 379–84.

45. Ibid. See also Paul A. Gilster, "The Interstellar Conundrum: A Survey of Concepts and Proposed Solutions," *New Trends in Astrodynamics and Applications, Annals of the New York Academy of Sciences* 1065 (December 2005): 462–70.

46. House Science and Technology Committee, *The Possibility of Intelligent Life Elsewhere in the Universe*, 94th Cong., 1st sess., 1975.

47. Carl Sagan and Frank Drake, "The Search for Extraterrestrial Intelligence," *Scientific American* (May 1975), 83.

48. James C. Fletcher, "NASA and the 'Now' Syndrome," address to the National Academy of Engineering, Washington, D.C., November 1975, NASA brochure, 7; NASA, "SETI," NASA NP-114, June 1990; see also Philip Morrison, John Billingham, and John Wolfe, eds., *The Search for Extraterrestrial Intelligence (SETI)* (Washington, D.C.: NASA SP-419, 1977).

49. House Science and Technology Committee, *The Possibility of Intelligent Life Elsewhere in the Universe*, 94th Cong., 1st sess., 1975, 24–27; and M. J. Klein et al., "Status of the NASA SETI Sky Survey Microwave Observing Project," *Acta Astronautica* 26 (March-April 1992): 177–84.

50. Frank Drake, quoted in Dava Sobel, "Is Anybody Out There?" *Life*, September 1992, 14. See also William Triplett, "SETI Takes the Hill," *Air and Space/Smithsonian*, November 1992, 80–86; Rob Meckel, "Proxmire 'Fleeces' NASA over Communications," Proxmire biography file, NASA Historical Reference Collection, Washington, D.C.; and Lance Frazer, "Small Change, High Gain," *Ad Astra*, September 1989, 19.

51. SETI Institute, "About Us: Institute History," available at www.seti.org, accessed October 27, 2005.

52. Carl Sagan, *Contact: A Novel* (New York: Simon and Schuster, 1985).

53. Kip S. Thorne, *Black Holes and Time Warps* (New York: W.W. Norton and Co., 1994), 55, 483–84; and Michael S. Morris and Kip S. Thorne, "Wormholes in Spacetime and Their Use for Interstellar Travel: A Tool for Teaching General Relativity," *American Journal of Physics* 56 (May 1988): 395–412. Sagan continued to refer to the tunnel as "the black hole, if that was what it really was." Sagan, *Contact,* 335.

54. Robert Zemeckis, *Contact,* Warner Studios, 1997; see also "Contact: The Movie," available at http://contact-themovie.warnerbros.com, accessed October 27, 2005.

55. Morris and Thorne, "Wormholes in Spacetime," 407.

56. For instance, see Roland Emmerich, *Stargate,* MGM, 1994; David Peckinpah, *Sliders,* Fox/SciFi Channel, 1995–2000; and Ira Steven Behr, Michael Piller, and Rick Berman, *Star Trek: Deep Space Nine,* CBS Studios, 1993–99.

57. NASA, Glenn Research Center, "Ideas Based on What We'd Like to Achieve: Worm Hole Transportation," July 1, 2005, available at www.nasa.gov/centers/glenn/research/warp/ideachev.html, accessed October 28, 2005.

58. Charles Sheffield, "Interstellar Flight in Fact and Fiction," in Kondo et al., eds., *Interstellar Travel,* 25.

59. Olaf Stapledon, "Interplanetary Man?" in *An Olaf Stapledon Reader,* ed. Robert Crossley (Syracuse, NY: Syracuse University Press, 1997), 241.

60. "W. O. Stapledon, 64, Noted Philosopher," obituary, *New York Times,* September 8, 1950, 31.

61. Stapledon, "Interplanetary Man?" 232–33.

62. Ibid., 239.

63. Ibid., 240.

CHAPTER 7: *HOMO SAPIENS*, TRANSHUMANISM, AND THE POSTBIOLOGICAL UNIVERSE

1. Stephen J. Hoffman and David L. Kaplan, eds., *Human Exploration of Mars: The Reference Mission of the NASA Mars Exploration Study Team* (Houston: Lyndon B. Johnson Space Center, July 1997), sec. 1, 9.

2. Albert Einstein, "The Foundation of the General Theory of Relativity," in *The Collected Papers of Albert Einstein,* vol. 6, ed. A. J. Kox, Martin J. Klein, and Robert Schulmann (Princeton, NJ: Princeton University Press, 1997), 146–200; S. Buchman et al., "The Gravity Probe B Relativity Mission," *Advances in Space Research* 25, no. 6 (2000): 1177–80; and "A Matter of Time: Special Issue," *Scientific American* (September 2002).

3. Carl Sagan and Frank Drake, "The Search for Extraterrestrial Intelligence," *Scientific American* (May 1975), 80–89; and Steven J. Dick, *The Biological Universe: The Twentieth-Century Extraterrestrial Life Debate* (Cambridge, UK: Cambridge University Press, 1996), 429–31.

4. This is the question pondered in Ben Zuckerman and Michael H. Hart, eds., *Extraterrestrials: Where Are They?* 2nd ed. (New York: Cambridge University Press, 1995).

5. Notwithstanding the popularity of extraterrestrial visitation in popular culture, no credible evidence supports a conclusion of any extraterrestrial presence whatsoever. See Carl Sagan and Thornton Page, eds., *UFO's: A Scientific Debate* (Ithaca, NY: Cornell University Press, 1972).

6. There is a major segment of the pro-space community that believes humanity has a finite period of time to colonize places beyond Earth before our planet's resources are depleted. This is what happened in many isolated parts of the world as native populations used up the resources of their islands, regions, or continents and had to migrate elsewhere—if they were lucky—or survive as best they could at a much reduced standard of living. This story is well told in Diamond, *Collapse*.

7. H. G. Wells, *The Time Machine* (1895; repr. New York: T. Doherty Associates, 1992); Pierre Boulle, *Planet of the Apes* (New York: Vanguard Press, 1963); and Robert Heinlein, *Orphans of the Sky: A Novel* (New York: Putnam, 1964).

8. Arthur C. Clarke, *2001: A Space Odyssey, 2010: Odyssey Two* (New York: Ballantine Books, 1984), *2061: Odyssey Three* (New York: Ballantine Books, 1988), and *3001: The Final Odyssey* (New York: Del Rey, 1997); and Kubrick, *2001: A Space Odyssey.*

9. Stapledon, *Last and First Men*. Asimov's original trilogy consists of *Foundation, Foundation and Empire* (1952; repr. Garden City, NY: Doubleday, 1970), and *Second Foundation* (New York: Gnome Press, 1953).

10. Frank White, *The Overview Effect: Space Exploration and Human Evolution,* 2nd ed. (Reston, VA: American Institute of Aeronautics and Astronautics, 1998), 73–94; Kevin Fong, "Life in Space: An Introduction to Space Life Sciences and the International Space Station," *Earth, Moon, and Planets* 87 (January 1999): 121–26; Marta Mirazón Lahr and Robert A. Foley, "Towards a Theory of Modern Human Origins: Geography, Demography, and Diversity in Recent Human Evolution," *American Journal of Physical Anthropology* 107, issue S27 (1998): 137–76; and Klaus Legner, *Humans in Space and Space Biology* (Vienna, Austria: United Nations Office for Outer Space Affairs, 2004), 79–133.

11. Steven J. Dick, "Cultural Evolution, the Postbiological Universe, and SETI," *International Journal of Astrobiology* 2, no. 1 (2003): 65. For a popularized version, see Dick, "They Aren't Who You Think," *Mercury* 32 (November/December 2003): 18–26.

12. Dick, "Cultural Evolution," 66.

13. Ward and Brownlee, *Rare Earth*.

14. Dick, "Cultural Evolution," 72.

15. Robert Roy Britt, "Age of Universe Revised, Again," January 3, 2003, available at www.space.com/scienceastronomy/age_universe_030103.html, accessed March 31, 2006.

16. Dick, "They Aren't Who You Think," 22.

17. Ibid., 26.

18. Anyone wishing to study the reorientation of American society in the 1960s should read Milton Viorst, *Fire in the Streets: America in the 1960s* (New York: Simon and Schuster, 1979); Allen J. Matusow, *The Unraveling of America: A History of Liberalism in the 1960s* (New York: Harper and Row, 1984); William L. O'Neill, *Coming Apart* (Chicago: Quadrangle Books, 1971); Godfrey Hodgen, *America in Our Time: From World War II to Nixon, What Happened and Why* (Garden City: Doubleday, 1976); Morris Dickstein, *Gates of Eden: American Culture in the Sixties* (New York: Basic Books, 1977); Irwin Unger, *The Best of Intentions: The Triumph and*

Failure of the Great Society under Kennedy, Johnson, and Nixon (Naugatuck, CT: Brandywine Press, 1995); and Arthur Benavie, *Social Security under the Gun* (New York: Palgrave Macmillan, 2003).

19. Richard R. Nelson, "The Economics of Invention: A Survey of the Literature," *Journal of Business* 32 (April 1959): 101–26; and Halberstam, *Best and the Brightest*, 57, 153.

20. Manfred E. Clynes and Nathan S. Kline, "Cyborgs and Space," *Astronautics*, September 1960, 26–27, 74–75, reprinted in Chris Gables Gray, ed., *The Cyborg Handbook* (New York: Routledge, 1995), 29–34.

21. This infatuation with cyborgs is related in Janice Hocker Rushing and Thomas S. Frentz, *Projecting the Shadow: The Cyborg Hero in American Film* (Chicago: University of Chicago Press, 1995).

22. Clynes and Kline, "Cyborgs and Space," 26.

23. Ibid.

24. Ibid.

25. This argument is laid out in Brooks, *Flesh and Machines*.

26. Robert W. Driscoll et al., "Engineering Man for Space: The Cyborg Study," General Aircraft report to NASA Office of Aerospace Research and Technology, May 15, 1963, 80, NASA Historical Reference Collection, NASA History Office, Washington, D.C.

27. Albert Rosenfeld, "Pitfalls and Perils Out There," *Life*, October 2, 1964, 122–24.

28. Charles D. Laughlin, "The Evolution of Cyborg Consciousness," *Anthropology of Consciousness* 8, no. 4 (1997): 144–59; J. R. McLeod, "The Seamless Web: Media and Power in the Post-Modern Global Village," *Journal of Popular Culture* 25 (Fall 1991): 69–75, esp. 70; John L. Taylor and Scott A. Burgess, "Steve Austin versus the Symbol Grounding Problem," *ACM International Conference Proceeding Series*, vol. 101 (2003): 21–25; and Russell C. Coile Jr., "Impact of the 'New Science' of Genomics," *Journal of Healthcare Management* 46, no. 6 (2001): 365–68.

29. Opening narration, *The Six Million Dollar Man*, 1973–78; the series was based on Martin Caiden, *Cyborg* (New York: Warner, 1972).

30. Chris Hables Gray, "Human-Machine Systems in Space: The Construction of Progress," paper presented at the American Historical Association annual meeting, Washington, D.C., January 4, 1999, copy in possession of authors.

31. Ray Kurzweil and Terry Grossman, *Fantastic Voyage: Live Long Enough to Live Forever* (New York: Rodale Books, 2004).

32. A good overview of the transhumanism movement is Brian Alexander, *Rapture: How Biotech Became the New Religion* (New York: Basic Books, 2003). Joel Garreau, *Radical Evolution: The Promise and Peril of Enhancing Our Minds, Our Bodies—And What It Means to Be Human* (Garden City, NY: Doubleday, 2005), is also very useful.

33. Robert C. W. Ettinger, *The Prospect of Immortality* (Garden City, NY: Doubleday, 1964). See also Robert C. W. Ettinger, *Man into Superman* (New York: St. Martin's Press, 1972).

34. Ray Kurzweil, *The Age of Spiritual Machines: When Computers Exceed Human Intelligence* (New York: Penguin, 1998).

35. "FM-2030," available at http://en.wikipedia.org/wiki/FM-2030, accessed March 22, 2006. See also FM-2030, *Are You a Transhuman? Monitoring and Stimulating Your Personal Rate of Growth in a Rapidly Changing World* (New York: Viking, 1989).

36. Jacques Ellul, *The Technological Society* (New York: Alfred A. Knopf, 1964).

37. John Kenneth Galbraith, *The New Industrial State* (New York: New American Library, 1968), 43.

38. John Kenneth Galbraith, *Economics and the Public Purpose* (New York: New American Library, 1975), 405.

39. Quoted in Ralph E. Lapp, *The New Priesthood: The Scientific Elite and the Uses of Power* (New York: Harper and Row, 1965), 228.

40. "Antitechnology Bias," *Air Force Magazine*, September 1971, 53. This was confirmed in a James C. Fletcher interview by Roger D. Launius, September 19, 1991.

41. Sylvia K. Kraemer, "NASA and the Challenge of Organizing for Exploration," in *Organizing for the Use of Space: Historical Perspectives on a Persistent Issue*, ed. Roger D. Launius, AAS History Series, vol. 18 (San Diego: Univelt, Inc., 1995), 107.

42. Bill Joy, "Why the Future Doesn't Need Us," *Wired* (April 2000), available at www.wired.com/wired/archive/8.04/joy.html, accessed March 22, 2006. The vicissitudes of group success and failure are well laid out in Jared Diamond, *Guns, Germs, and Steel: The Fates of Human Societies* (New York: W.W. Norton and Co., 1999).

43. Joy, "Why the Future Doesn't Need Us."

44. Examples of recent works on this field include Sidney Perkowitz, *Digital People: From Bionic Humans to Androids* (Washington, D.C.: Joseph Henry Press, 2004); Peter Menzel and Faith D'Aluisio, *Robo Sapiens: Evolution of a New Species* (Cambridge, MA: MIT Press, 2000); David Channell, *The Vital Machine: A Study of Technology and Organic Life* (New York: Oxford University Press, 1991); Robbie David-Floyd and Joseph Dumit, eds., *Cyborg Babies: From Techno-Sex to Techno-Tots* (New York: Routledge, 1998); Renee Fox and Judith Swazey, *Spare Parts: Organ Replacement in American Society* (New York: Oxford University Press, 1992); Bruce Mazlish, *The Fourth Discontinuity: The Co-Evolution of Humans and Machines* (New Haven, CT: Yale University Press, 1993); Gregory Stock, *Metaman: The Merging of Humans and Machines into a Global Superorganism* (New York: Simon and Schuster, 1993); N. Katherine Hayles, *How We Became Posthuman: Virtual Bodies in Cybernetics, Literature, and Informatics* (Chicago: University of Chicago Press, 1999); Philip Mirowski, *Machine Dreams: Economics Becomes a Cyborg Science* (New York: Cambridge University Press, 2002); and Thomas E. Georges, *Digital Soul: Intelligent Machines and Human Values* (Boulder, CO: Westview Press, 2003).

45. Andy Clark, *Natural-Born Cyborgs: Minds, Technologies, and the Future of Human Intelligence* (New York: Oxford University Press, 2003), 198.

46. Hayles, *How We Became Posthuman*, 243, 291.

47. James Hughes, *Citizen Cyborg: Why Democratic Societies Must Respond to the Redesigned Human of the Future* (Boulder, CO: Westview Press, 2004).

48. Garreau, *Radical Evolution*.

49. Kurzweil, *Age of Spiritual Machines* and *The Singularity Is Near: When Humans Transcend Biology* (New York: Viking, 2005). A useful critique of posthumanism is Charles T. Rubin, "The Rhetoric of Extinction," *The New Atlantis* 11 (Winter 2006): 64–73. See also Francis Fukuyama, *Our Posthuman Future: Consequences of the Biotechnology Revolution* (New York: Farrar, Straus, and Giroux, 2002).

50. Mary Shelley, *Frankenstein* (1818; repr. New York: W.W. Norton and Co., 1996); Maura Phillips Mackowski, *Testing the Limits: Aviation Medicine and the Origins of Manned Space*

Flight (College Station: Texas A&M University Press, 2005); and José Manuel Rodríguez Delgado, *Physical Control of the Mind: Toward a Psychocivilized Society* (New York: Harper and Row, 1969).

51. Joy, "Why the Future Doesn't Need Us."

52. Adam Keiper, "The Age of Neuroelectronics," *The New Atlantis* 11 (Winter 2006): 4–41.

53. Sagan, *Pale Blue Dot,* 313–34, and *The Dragons of Eden: Speculations on the Evolution of Human Intelligence* (New York: Random House, 1977), 197–248.

54. Stapledon, "Interplanetary Man?"

55. Alexander, *Rapture,* 51.

56. Lawrence M. Krauss, *The Physics of Star Trek* (New York: Harper, 1995); and Sagan, *Contact.*

57. Kondo et al., eds., *Interstellar Travel*; Abigail Alling, Mark Nelson, and Sally Silverstone, *Life under Glass: The Inside Story of Biosphere 2* (Tucson, AZ: Biosphere Press, 1993); B. D. V. Marino and Howard T. Odum, eds., *Biosphere 2: Research Past and Present* (Amsterdam: Elsevier, 1999); and Paul Gilster, *Centauri Dreams: Imagining and Planning Interstellar Exploration* (Chichester, UK: Springer-Praxis, 2004).

58. Jennifer Bails, "Pitt Scientists Resurrect Hope of Cheating Death," *Pittsburgh Tribune-Review,* June 29, 2005; and *Science,* April 22, 2005.

59. Christine Quigley, *Modern Mummies: The Preservation of the Human Body in the Twentieth Century* (Jefferson, NC: McFarland, 1998); and Ronald Bellamy et al., "Suspended Animation for Delayed Resuscitation," *Critical Care Medicine* 24 (February 1996): 24S–47S.

60. The best evidence is that CDs are good for about a hundred years. See Gordon Woolf, "How Long Will a CD-R Last?" *PC Update,* June 2001, 23–26.

61. Robert H. Goddard, "The Ultimate Migration," in *The Papers of Robert H. Goddard,* vol. 3, ed. Goddard and Pendray, 1611–12; see also Stephen Jay Gould, *Wonderful Life: The Burgess Shale and the Nature of History* (New York: W. W. Norton and Co., 1989), a detailed account of the discovery and interpretation of the fossil record of the Burgess Shale site, a limestone quarry in British Columbia.

62. Haldeman, "Colonizing Other Worlds," 67–68.

63. Joseph Campbell with Bill Moyers, *The Power of Myth* (Garden City, NY: Doubleday, 1988).

64. John Noble Wilford, *Mars Beckons: The Mysteries, the Challenges, the Expectations of Our Next Great Adventure in Space* (New York: Alfred A. Knopf, 1990), 210.

65. Michael Collins, *Mission to Mars: An Astronaut's Vision of Our Future in Space* (New York: Grove Weidenfeld, 1990), flyleaf.

66. Dick, "They Aren't Who You Think," 18–26; and Ricky Stilson, "Near Earth Exploration, or Dance with the One that Brought You," *Future Human Evolution,* available at www.human-evolution.org/spaceview.php, accessed March 31, 2006.

CHAPTER 8: AN ALTERNATIVE PARADIGM?

1. Bonestell and Ley, *Conquest of Space*; Arthur C. Clarke, *The Exploration of Space* (London: Temple Press, 1951); *Collier's,* March 22, October 18, October 25, 1952; February 28,

March 7, March 14, June 27, 1953; April 30, 1954; Ward Kimball, *Man in Space*, Walt Disney, 1955, *Man and the Moon* (Walt Disney, 1955; alternate title: *Tomorrow the Moon*), and *Mars and Beyond*, Walt Disney, 1957. See also First Annual Symposium on Space Travel, October 12, 1951; Second Symposium on Space Travel, October 13, 1952; Third Symposium on Space Travel, May 4, 1954, all at the American Museum of Natural History, New York, New York; and Ron Miller and Frederick C. Durant III, *The Art of Chesley Bonestell* (London: Paper Tiger, 2001).

2. Mike Wright, "The Disney–Von Braun Collaboration and Its Influence on Space Exploration," in *Inner Space, Outer Space: Humanities, Technology, and the Postmodern World*, ed. Daniel Schenker, Craig Hanks, and Susan Kray (Huntsville, AL: Southern Humanities Conference, 1993), 151–60; and Roger D. Launius, "Looking Backward/Looking Forward: Spaceflight at the Turn of the New Millennium," *Astropolitics: The International Journal of Space Politics and Policy* 1 (Autumn 2003): 62–73.

3. The Mars-Venus landings were listed by a committee devoted to the task of identifying human flight objectives. NASA, "Minutes of Meeting of Research Steering Committee on Manned Space Flight," NASA Headquarters, Washington, D.C., May 25–26, 1959, 2; NASA Office of Program Planning and Evaluation, "Long Range Plan of the National Aeronautics and Space Administration," December 16, 1959; Albert R. Hibbs, ed., "Exploration of the Moon, the Planets, and Interplanetary Space," Jet Propulsion Laboratory Report 30–1, Pasadena, California, 1959; and NASA Office of Program Planning and Evaluation, "The Ten Year Plan of the National Aeronautics and Space Administration," December 18, 1959, all in NASA Historical Reference Collection.

4. This is exhaustively documented in David S. F. Portree, *Humans to Mars: Fifty Years of Mission Planning, 1950–2000* (Washington, D.C.: NASA SP-4520, 2001).

5. Wernher von Braun with Cornelius Ryan, "Can We Get to Mars?" *Collier's*, April 30, 1954, 22–28.

6. Percival Lowell, *Mars* (Boston: Houghton Mifflin Co., 1895), 201–12; Roger D. Launius, "'Not Too Wild a Dream': NASA and the Quest for Life in the Solar System," *Quest: The History of Spaceflight Quarterly* 6 (Fall 1998): 17–27; and Martin Caiden and Jay Barbree, *Destination Mars: In Art, Myth, and Science* (New York: Penguin Studio, 1997), 83–95.

7. Bonestell and Ley, *Conquest of Space*, 112.

8. See the cover of National Commission on Space, *Pioneering the Space Frontier: An Exciting Vision of Our Next Fifty Years in Space*, (New York: Bantam Books, 1986).

9. Carl Sagan, "Planetary Engineering on Mars," *Icarus* 20 (1973): 513–14.

10. Jet Propulsion Laboratory, *Mariner: Mission to Venus* (New York: McGraw-Hill, 1963), 5. This also became an enormously popular conception in science fiction literature. See the 1949 short story by Arthur C. Clarke, "History Lesson," in *Expedition to Earth* (New York: Ballantine Books, 1953), 73–82, for an explanation of the theory. For a well-grounded scientific discussion of the evolution of understanding about Venus, see Ladislav E. Roth and Stephen D. Wall, *The Face of Venus: The Magellan Radar-Mapping Mission* (Washington, D.C.: NASA SP-520, 1995), 1–9; and Jet Propulsion Laboratory, *Mariner-Venus 1962: Final Project Report* (Washington, D.C.: NASA SP-59, 1965), 6–8.

11. Carl Sagan, "The Planet Venus," *Science*, March 24, 1961, 857–58.

12. V. A. Firsoff, *Strange World of the Moon* (New York: Basic Books, 1959), 172.

13. Patrick Moore, "Life on the Moon?" *Irish Astronomical Journal* 3, no. 5 (1955): 136. See also Ralph B. Bellamy, *A Fundamental Survey of the Moon* (New York: McGraw-Hill, 1965), 115–20.

14. The story of the Mariner mission is recounted in Jet Propulsion Laboratory, *Mariner-Venus 1962*, and *Mariner*, 102–12; and Robert Reeves, *The Superpower Space Race: An Explosive Rivalry through the Solar System* (New York: Plenum Press, 1994), 177–83.

15. "An End to the Myths about Men on Mars," *U.S. News and World Report*, August 9, 1965, 4.

16. Lyndon B. Johnson, "Remarks upon Viewing New Mariner 4 Pictures from Mars," July 29, 1965, *Public Papers of the Presidents of the United States*, 1965 (Washington, D.C.: Government Printing Office, 1965), 806.

17. Billy Cox, "Mars Pioneer Remembered." *Florida Today* (Orlando), December 6, 2000.

18. Bruce Murray, *Journey into Space* (New York: W.W. Norton and Co., 1989), 68–69, 74, 77, quotation 61.

19. For a discussion of the tendency of true believers to reinterpret rather than abandon beliefs in the face of cognitive dissonance, see Leon Festinger, Henry W. Riecken, and Stanley Schachter, *When Prophecy Fails: A Social and Psychological Study* (Minneapolis: University of Minnesota Press, 1956).

20. See William Sheehan, *The Planet Mars: A History of Observation and Discovery* (Tucson: University of Arizona Press, 1996); and Malcolm Walter, *The Search for Life on Mars* (Cambridge, MA: Perseus Books, 1999).

21. Ward and Brownlee, *Rare Earth*.

22. On this issue, see Patricia Nelson Limerick, "The Final Frontier?" *Wilson Quarterly* 14 (Summer 1990): 82–83; Ray A. Williamson, "Outer Space as Frontier: Lessons for Today," *Western Folklore* 46 (October 1987): 255–67; Stephen J. Pyne, "Space: A Third Great Age of Discovery," *Space Policy* 4 (1988): 187–99; and M. Jane Young, "'Pity the Indians of Outer Space': Native American Views of the Space Program," *Western Folklore* 46 (October 1987): 269–79.

23. Statement of Democratic Leader Lyndon B. Johnson to the Meeting of the Democratic Conference on January 7, 1958, statements of LBJ collection, box 23, Lyndon Baines Johnson Library, Austin, Texas.

24. This is documented in many places. For example, see Andrew Smith, *Moondust: In Search of the Men Who Fell to Earth* (New York: Fourth Estate, 2005), 295.

25. See, for example, Tom Wolfe, *The Right Stuff* (New York: Farrar, Straus, Giroux, 1979).

26. O'Neill, "Colonization of Space" and *High Frontier*; and Launius, "Perfect Worlds, Perfect Societies."

27. This has been documented in Roger D. Launius, "Public Opinion Polls and Perceptions of U.S. Human Spaceflight," *Space Policy* 19 (August 2003): 163–75.

28. Although still in vogue by spaceflight advocates, even the aviation analogy is flawed. See Roger E. Bilstein, *The American Aerospace Industry: From Workshop to Global Enterprise* (New York: Twayne Publishers, 1996), 209–23, which notes the general inability of the aviation industry to turn a profit.

29. See McCurdy, *Space Station Decision*, 169–76.

30. While Congress placed a $25 billion cap on the International Space Station, a 2003 es-

timate calculated that the U.S. government would spend $14 billion on component fabrication, $6 billion on research and supporting technology, and $13 billion on operations while assembly was underway. Additionally, Congress set an $18 billion cap on transportation, and NASA spent $10 billion during redesign. The total was $61 billion. See Launius, *Space Stations,* 200–205.

31. Frederick I. Ordway III, "The History, Evolution, and Benefits of the Space Station Concept (in the United States and Western Europe)," *Actes du XIIIe Congrès International d'Histoire des Sciences* 12 (1974): 92–132; Harry E. Ross and Ralph A. Smith, "Orbital Bases," *Journal of the British Interplanetary Society* 8 (January 1949): 1–19; Sylvia D. Fries, "2001 to 1994: Political Environment and the Design of NASA's Space Station System," in *Technology and Choice: Readings from Technology and Culture,* ed. Marcel C. LaFollette and Jeffrey K. Stine (Chicago: University of Chicago Press, 1991), 233–58; and Alex Roland, "The Evolution of Civil Space Station Concepts in the United States," May 1983, NASA Historical Reference Collection.

32. Marcia Smith, "NASA's Space Station Program: Evolution and Current Status," testimony before the House Science Committee, April 4, 2001; and "House Science Committee Hearing Charter: The Space Station Task Force Report," Committee on Science, U.S. House of Representatives, November 7, 2001, both in NASA Historical Reference Collection.

33. "Man Will Conquer Space Soon," *Collier's,* March 22, 1952.

34. NASA, "Space Shuttle," 1972, and "Fact Sheet: The Economics of the Space Shuttle," July 1972; both in NASA Historical Reference Collection. The figures are stated in 1971 dollars, based on a 580-mission model.

35. Total program cost, 2007 dollars: $16.15 billion; inflation index (1971–2007), 5.542 spread over 580 flights equals $154.3 million per flight. NASA, New Start Inflation Index, actual through 2007, Office of the NASA Comptroller, NASA Headquarters, Washington, D.C.

36. Capital investments and upgrades, $36 billion; shuttle operations, $47 billion; facilities, $2 billion; total, $85 billion ($151 billion in 2007 dollars); purchased 110 flights through FY 2002; average total cost per flight, $772 million ($1.38 billion in 2007 dollars).

37. J. K. Davies, "A Brief History of the Voyager Project: The End of the Beginning," *Spaceflight* 23 (February 1981): 35–41; Henry C. Dethloff and Ronald A. Schorn, *Voyager's Grand Tour: To the Outer Planets and Beyond* (Washington, D.C.: Smithsonian Institution Press, 2003); Ben Evans and David M. Harland, *NASA's Voyager Missions: Exploring the Outer Solar System and Beyond* (Chichester, UK: Springer, 2003); and Edward C. Stone, "News from the Edge of Interstellar Space," *Science,* July 6, 2001, 55–56.

38. G. Edward Pendray, "Next Stop the Moon," *Collier's,* September 7, 1946, 12; Robert S. Richardson, "Rocket Blitz from the Moon," *Collier's,* October 23, 1948, 24–25, 44–46; Caleb B. Laning and Robert A. Heinlein, "Flight into the Future," *Collier's,* August 30, 1947, 36; and Curtis Peebles, "The Manned Orbiting Laboratory," a three-part series in *Spaceflight* vol. 22 (April 1980): 155–60, vol. 22 (June 1980): 248–53, and vol. 24 (June 1982): 274–77.

39. See U.S. Department of Commerce, Office of Space Commercialization, "Trends in Space Commerce," prepared by Futron Corporation, Bethesda, Maryland, May 2001.

40. Asimov, *Gold,* 196.

41. Nicks, *Far Travelers,* 245–46.

42. Mindell, "Human and Machine in the History of Spaceflight," 141–68.

43. This is laid out in Brooks, *Flesh and Machines*; Alexander, *Rapture*; Hughes, *Citizen Cyborg*; Perkowitz, *Digital People*; and Menzel and D'Aluisio, *Robo Sapiens*.

44. Haldeman, "Colonizing Other Worlds," 62–68.

45. See Preston Lerner, "Robots Go to War," *Popular Science* (January 2006), 42–49, 96.

46. As an example, the budget for the National Institutes of Health doubled in the five years from FY 1999 through FY 2003, and its total stood at $28.76 billion for fiscal year 2005. See *National Institutes of Health: Summary of the FY 2005 President's Budget* (Washington, D.C.: Government Printing Office, 2004), 2, 7. And this was only one part of the nation's expenditure for health care per year. In 2003, health care spending in the United States reached $1.7 trillion. See "Facts on the Cost of Health Care," National Coalition on Health Care, available at www.nchc.org/facts/cost.shtml, accessed April 3, 2006.

47. "There has been much discussion about attempting to land an astronaut on the planet Mars. How would you feel about such an attempt? Would you favor or oppose the United States setting aside money for such a project?"

	Favor	*Oppose*	*Unsure*
6/2005	40%	58%	2%
7/1999	43%	54%	3%
7/1969	39%	53%	8%

SOURCE: CNN/USA Today/Gallup Poll. June 24–26, 2005. N=1,009 adults nationwide. Margin of error: ±3.

48. See Alan Boss, *Looking for Earths: The Race to Find New Solar Systems* (New York: John Wiley, 1998); Stuart Clark, *Extrasolar Planets: The Search for New Worlds* (New York: John Wiley, 1998); and John S. Lewis, *Worlds without End: The Exploration of Planets Known and Unknown* (Cambridge, MA: Perseus Books, 1998).

49. See J. B. Zirker, *An Acre of Glass: A History and Forecast of the Telescope* (Baltimore: Johns Hopkins University Press, 2005).

50. See Ward and Brownlee, *Rare Earth*; James F. Kasting, "Peter Ward and Donald Brownlee's 'Rare Earth,'" *Perspectives in Biology and Medicine* 44 (Winter 2001): 117–31; and G. D. Brin, "The 'Great Silence': The Controversy concerning Extraterrestrial Intelligence," *Quarterly Journal of the Royal Astronautical Society* 24 (1983): 283–309.

51. According to NASA's 2007 budget documentation, "The Terrestrial Planet Finding project (TPF) has been deferred indefinitely." See Keith Cowing, "Canceling NASA's Terrestrial Planet Finder: The White House's Increasingly Nearsighted 'Vision' for Space Exploration," NASA Watch, February 6, 2006, available at www.spaceref.com/news/viewnews.html?id=1092, accessed April 3, 2006.

52. Science fiction writers like Arthur C. Clarke have suggested a third alternative to the "rare Earth versus common Earths" conundrum. Humans might change so much, even to the extent of leaving their bodily form, as to make their confinement to Earth-like planets unnecessary. In that case, they could live anywhere. Once again, the exact set of necessities and imagined options will be deeply shaped by knowledge about the exact set of possibilities.

53. See John Larner, *Marco Polo and the Discovery of the World* (New Haven, CT: Yale University Press, 1999).

54. NASA, Jet Propulsion Laboratory, "Mars Exploration Rover Mission, Launch Vehicle: Stage III," June 13, 2005, available at http://marsrovers.jpl.nasa.gov/mission/launch_stage3.html, accessed March 9, 2006.

55. Alison Abbot, "Rubbia Proposes a Speedier Voyage to Mars and Back," *Nature* 397 (1999): 374; Carlo Rubbia, "Report of the Working Group on a Preliminary Assessment of a New Fission Fragment Heated Propulsion Concept and its Applicability to Manned Missions to the Planet Mars," Italian Space Agency, Rome, Italy, April 1999; and Tony Reichhardt, "Breaking the Nuclear Taboo," *Nature* 410 (2001): 626.

56. Brian Berger, "Prometheus, ISS Research Cuts Help Pay for Shuttle and Hubble Repair Bills," May 12, 2005, available at www.space.com/news/050512_nasa_prometheus.html, accessed March 3, 2006.

57. Leonard David, "NASA Shuts Down X-33, X-34 Programs," March 1, 2001, available at www.space.com/missionlaunches/missions/x33_cancel_010301.html, accessed March 28, 2003; and Roger D. Launius, "Hypersonic Flight: Evolution from X-15 to Space Shuttle," AIAA-2003–2716, delivered at the "Next Century of Flight" Conference, Dayton, Ohio, July 2003.

58. See McCurdy, "Cost of Space Flight"; and Lori Montgomery and Thomas Heath, "At Long Last, a D.C. Stadium Deal," *Washington Post,* March 8, 2006, A1.

59. See McCurdy, *Faster, Better, Cheaper* and *Low-Cost Innovation in Spaceflight*.

60. NASA, "Cassini-Huygens Saturn Arrival," press kit, June 2004, 6; see also David M. Harland, *Mission to Saturn: Cassini and the Huygens Probe* (Chichester, UK: Springer-Praxis, 2003).

61. NASA, "2006 NASA Strategic Plan," NP-2006–02–423-HQ, 19; and Tariq Malik, "NASA's New Moon Plans: 'Apollo on Steroids,'" September 19, 2005, available at www.space.com/news/050919_nasa_moon.html, accessed April 3, 2006.

62. There was more than $140 billion spent worldwide on spaceflight in 2005; only about $16 billion came from NASA, while some $95 billion was spent on commercial space activities. See Philip McAllister, "Commercial Space Issues," unpublished presentation, AIAA Space 2003, September 24, 2003, copy in possession of authors; FAA Associate Administrator for Commercial Space Transportation, *Commercial Space Transportation: 2003 Year in Review* (Washington, D.C.: Federal Aviation Administration, 2004), 7–9; and John W. Douglas, "41st Annual Year in Review and Forecast Luncheon," Aerospace Industries Association, December 14, 2005, copy in possession of authors.

63. NASA, "2006 NASA Strategic Plan," 17.

64. Ian Tattersall, "Once We Were Not Alone," *Scientific American* (January 2000), 56–62; Kate Wong, "The Littlest Human," *Scientific American* (February 2005), 40–49; P. Brown et al., "A New Small-bodied Hominin from the Late Pleistocene of Flores, Indonesia," *Nature* 431 (2004): 1055–61; M. J. Morwood et al., "Archaeology and Age of a New Hominin from Flores in Eastern Indonesia," *Nature* 431 (2004): 1087–91; and Guy Gugliotta, "More Fossil Evidence from 'Hobbit' Island," *Washington Post,* October 12, 2005, A3.

65. Goddard, "Ultimate in Jet Propulsion," 1612.

66. See Peter D. Ward and Donald Brownlee, *The Life and Death of Planet Earth: How the New Science of Astrobiology Charts the Ultimate Fate of Our World* (New York: Times Books, 2003); and Vince Stricherz, "'The End of the World' Has Already Begun, UW Scientists Say,"

University of Washington News and Events, January 13, 2003, available at www.washington.edu/newsroom/news/2003archive/01–03archive/k011303a.html, accessed March 13, 2006.

67. R. P. Turco, O. B. Toon, T. P. Ackerman, J. B. Pollack, and Carl Sagan, "Nuclear Winter: Global Consequences of Multiple Nuclear Explosions," *Science,* December 23, 1983, 1283–97.

68. Jared Diamond suggests that there is ample evidence in history to believe that a confluence of five major elements leads to a civilization's collapse: environmental damage and resource depletion, climate change, hostile neighbors, loss of trade partners, and a society's responses to its challenges. He applies this model to several past civilizations, including Easter Island (this society collapsed due mostly to environmental damage), the Polynesians of Pitcairn Island (environmental damage and loss of trading partners), the Anasazi of the southwestern United States (environmental damage and climate change), the Maya of Central America (environmental damage, climate change, and hostile neighbors), and the Greenland Norse (all five factors). Diamond sounds a warning that the present course of world civilization is not sustainable and that without corrective actions environmental damage, resource depletion, and climate change are likely to transform human existence in the next century. See Diamond, *Collapse,* 15.

Index